BIOTECHNOLOGY IN AGRICULTURE SERIES

General Editor: Gabrielle J. Persley, Biotechnology Adviser, Environmentally Sustainable Development, The World Bank, Washington DC, USA.

For a number of years, biotechnology has held out the prospect for major advances in agricultural production, but only recently have the results of this new revolution started to reach application in the field. The potential for further rapid developments is, however, immense.

The aim of this book series is to review advances and current knowledge in key areas of biotechnology as applied to crop and animal production, forestry and food science. Some titles focus on individual crop species, others on specific goals such as plant protection or animal health, with yet others addressing particular methodologies such as tissue culture, transformation or immunoassay. In some cases, relevant molecular and cell biology and genetics are also covered. Issues of relevance to both industrialized and developing countries are addressed and social, economic and legal implications are also considered. Most titles are written for research workers in the biological sciences and agriculture, but some are also useful as textbooks for senior-level students in these disciplines.

BIOTECHNOLOGY IN AGRICULTURE SERIES

Titles Available:
1: Beyond Mendel's Garden: Biotechnology in the Service of World Agriculture *G.J. Persley*
2: Agricultural Biotechnology: Opportunities for International Development *Edited by G.J. Persley*
3: The Molecular and Cellular Biology of the Potato *Edited by M.E. Vayda and W.D. Park*
4: Advanced Methods in Plant Breeding and Biotechnology *Edited by D.R. Murray*
5: Barley: Genetics, Biochemistry, Molecular Biology and Biotechnology *Edited by P.R. Shewry*
6: Rice Biotechnology *Edited by G.S. Khush and G.H. Toenniessen*
7: Plant Genetic Manipulation for Crop Protection *Edited by A. Gatehouse, V. Hilder and D. Boulter*
8: Biotechnology of Perennial Fruit Crops *Edited by F.A. Hammerschlag and R.E. Litz*
9: Bioconversion of Forest and Agricultural Plant Residues *Edited by J.N. Saddler*
10: Peas: Genetics, Molecular Biology and Biotechnology *Edited by R. Casey and D.R. Davies*
11: Laboratory Production of Cattle Embryos *I. Gordon*
12: The Molecular and Cellular Biology of the Potato, 2nd edn *Edited by W.R. Belknap, M.E. Vayda and W.D. Park*
13: New Diagnostics in Crop Sciences *Edited by J.H. Skerritt and R. Appels*
14: Soybean: Genetics, Molecular Biology and Biotechnology *Edited by D.P.S. Verma and R.C. Shoemaker*
15: Biotechnology and Integrated Pest Management *Edited by G.J. Persley*
16: Biotechnology of Ornamental Plants *Edited by R.L. Geneve, J.E. Preece and S.A. Merkle*
17: Biotechnology and the Improvement of Forage Legumes *Edited by B.D. McKersie and D.C.W. Brown*
18: Milk Composition, Production and Biotechnology *R.A.S. Welch, D.J.W. Burns, S.R. Davis, A.I. Popay and C.G. Prosser*
19: Biotechnology and Plant Genetic Resources: Conservation and Use *Edited by J.A. Callow, B.V. Ford-Lloyd and H.J. Newbury*
20: Intellectual Property Rights in Agricultural Biotechnology *Edited by F.H. Erbisch and K.M. Maredia*
21: Agricultural Biotechnology in International Development *Edited by C. Ives and B. Bedford*

Agricultural Biotechnology in International Development

Edited by

Catherine L. Ives and Bruce M. Bedford

*Agricultural Biotechnology for Sustainable Productivity
(ABSP) Project, Michigan State University, USA*

CABI *Publishing*

CABI Publishing – a division of CAB INTERNATIONAL

CABI *Publishing*
CAB INTERNATIONAL
Wallingford
Oxon OX10 8DE
UK

CABI *Publishing*
10 E 40th Street
Suite 3203
New York, NY 10016
USA

Tel: +44 (0)1491 832111
Fax: +44 (0)1491 833508
Email: cabi@cabi.org

Tel: +1 212 453 2670
Fax: +1 212 686 7993
Email: cabi-nao@cabi.org

A catalogue record for this book is available from the British Library, London, UK

Library of Congress Cataloging-in-Publication Data
Agricultural biotechnology in international development / edited
by C. Ives and B. Bedford.
    p.   cm. — (Biotechnology in agriculture series; 21)
    "This book is the product of a conference held in California
April 1997, under the auspices of the Agricultural
Biotechnology for Sustainable Productivity (ABSP) project."
    Includes index.
    ISBN 0-85199-278-1 (alk. paper)
    1. Agricultural biotechnology—Developing countries—
Congresses. I. Ives, C. (Catherine) II. Bedford, Bruce.
III. Agricultural Biotechnology for Sustainable Productivity
Project. IV. Series.
S494.5.B563A3728   1998
338.1'6—dc21   98-22771
                                     CIP

ISBN 0 85199 278 1

Typeset in 10/12pt Photina by Columns Design Ltd, Reading
Printed and bound in the UK at the University Press, Cambridge

Contents

Contributors   ix

Preface   xi
Catherine L. Ives

Acknowledgements   xiii

1   The Agricultural Biotechnology for Sustainable Productivity   1
    Project: a New Model in Collaborative Development
    Catherine L. Ives, Bruce M. Bedford and Karim M. Maredia

**Part I: Needs and Potential Uses of Agricultural Biotechnology: 15
Perspectives of Developing Countries**

2   Addressing Agricultural Development in Egypt through Genetic   17
    Engineering
    Magdy Madkour

3   The Release of Transgenic Varieties in Centres of Origin: Effect on 27
    Biotechnology Research and Development Priorities in Developing
    Countries
    Ariel Alvarez-Morales

4   Current Status of Agricultural Biotechnology Research in Indonesia   35
    Achmad M. Fagi and Muhammad Herman

5   Agricultural Needs in Sub-Saharan Africa: the Role of Biotechnology 49
    Cyrus G. Ndiritu and John S. Wafula

Part II: The Application of Biotechnology to Food Security Crops   61

6 Development of Insect-resistant Maize and Its   63
   Potential Benefits to Developing Countries
   Pam Robeff

7 The Application of Biotechnology to Potato   73
   Marc Ghislain, Maddalena Querci, Merideth Bonierbale,
   Ali Golmirzaie and Peter Gregory

8 Development of Virus-resistant Sweetpotato   89
   Maud Hinchee

9 The Application of Biotechnology to Rice   97
   Gurdev S. Khush and Darshan S. Brar

**Part III: The Application of Biotechnology to Non-traditional   123
Crops**

10 Current Advances in the Biotechnology of Banana   125
    Oscar Arias

11 The Application of Biotechnology to Date Palm   133
    Mohamed Aaouine

12 The Use of Coat Protein Technology to Develop Virus-resistant   147
    Cucurbits
    Hector Quemada

13 The Biotechnology of Oil Palm   161
    Suan-Choo Cheah

**Part IV: Issues Surrounding the Development, Transfer,   171
Adaptation and Utilization of Agricultural Biotechnology
for Emerging Nations**

14 Making a Difference: Considering Beneficiaries and Sustainability 173
    while Undertaking Research in Biotechnology
    Joel I. Cohen

15 Rice Biotechnology Capacity Building in Asia   201
    Gary H. Toenniessen

16 International Biosafety Regulations: Benefits and Costs   213
    Robert J. Frederick

17 Cassava Biotechnology Research: Beyond the Toolbox   229
    Ann Marie Thro

18 Fundación Perú: a Path to Capacity Building   247
    Fernando Cillóniz

**Part V: Developing and Accessing Agricultural Biotechnologies: 253
International, US and Developing Country Issues, Perspectives
and Experiences**

19 Transferring Agricultural Biotechnology: US Public/Private Sector   255
    Perspectives
    Frederic H. Erbisch

20 International Intellectual Property and Genetic Resource   273
    Issues Affecting Agricultural Biotechnology
    John H. Barton

21 Developing Capacity and Accessing Biotechnology Research and   285
    Development (R&D) for Sustainable Agriculture and Industrial
    Development in Zimbabwe
    Joseph Muchabaiwa Gopo

22 The Technology Transfer System in Thailand   297
    Lerson Tanasugarn

23 Trade in Conventional and Biotechnology Agricultural Products   311
    Quentin B. Kubicek

**Part VI: Can Developing Countries Turn Biotech into Business?   315
Moving Research Results into Products**

24 Wild Biodiversity: the Last Frontier?   317
    Nicolás Mateo

25 Developing an Agricultural Biotechnology Business: Perspective   335
    from the Front Lines
    Pamela G. Marrone

Index   343

Contributors

Mohamed Aaouinc, Domaine Agricole El Bassatine, BP 299, Meknes, Morocco.

Ariel Alvarez-Morales, CINVESTAV, IPN. Unidad Irapuato, Department of Plant Genetic Engineering, Apdo. Post. 629, Irapuato, Gto., 36500, Mexico (or Sainsbury Laboratory, John Innes Institute, Norwich, UK).

Oscar Arias, Agribiotecnología de Costa Rica, S.A., PO Box 100–4003, Alajuela, Costa Rica.

John H. Barton, Stanford University, Crown Quadrangle, Stanford, CA 94305–8610, USA.

Merideth Bonierbale, International Potato Center (CIP), PO Box 1558, Lima, Peru.

Bruce M. Bedford, ABSP Project, Michigan State University, 324 Agriculture Hall, East Lansing, MI 48824, USA.

Darshan S. Brar, International Rice Research Institute (IRRI), PO Box 933, 1099 Manila, The Philippines.

Suan-Choo Cheah, Palm Oil Research Institute of Malaysia (PORIM), PO Box 10620, 50720 Kuala Lumpur, Malaysia.

Fernando Cillóniz, Fundación Perú, Alcanflores 1245-Miraflores, Lima 18, Peru.

Joel I. Cohen, Intermediary Biotechnology Service, International Service for National Agriculture Research (ISNAR), PO Box 93375, 2509 AJ The Hague, The Netherlands.

Frederic H. Erbisch, Office of Intellectual Property, 238 Hannah Administration Building, Michigan State University, East Lansing, MI 48824–1046, USA.

Achmad M. Fagi, Central Research Institute for Food Crops, Jalan Merdeka 147, Bogor 16111, West Java, Indonesia.

Robert J. Frederick, US EPA/Office of Research and Development, National Center for Environmental Assessment, Washington, DC 20460, USA.

Ali Golmirzaie, International Potato Center (CIP), PO Box 1558, Lima, Peru.

Joseph Muchabaiwa Gopo, Biotechnology Research Institute, SIRDC, PO Box 6640, Harare, Zimbabwe.

Mark Ghislain, International Potato Center (CIP), PO Box 1558, Lima, Peru.

Peter Gregory, International Potato Center (CIP), PO Box 1558, Lima, Peru.

Muhammad Herman, Central Research Institute for Food Crops, Jalan Merdeka 147, Bogor 16111, West Java, Indonesia.

Maud Hinchee, Monsanto Company, 700 Chesterfield Parkway North, St Louis, MO 63198, USA.

Catherine L. Ives, ABSP Project, Michigan State University, 324 Agriculture Hall, East Lansing, MI 48824, USA.

Gurdev S. Khush, International Rice Research Institute (IRRI), PO Box 933, 1099 Manila, The Philippines.

Quentin B. Kubicek, United States Department of Agriculture (USDA), Room 1128-5, Washington, DC 20250, USA.

Magdy Madkour, Agricultural Genetic Engineering Research Institute (AGERI), Agricultural Research Center (ARC), 9 Gamaa Street, Giza 12619, Egypt.

Karim M. Maredia, ABSP Project, Michigan State University, 324 Agriculture Hall, East Lansing, MI 48824, USA.

Pamela G. Marrone, AgraQuest Incorporated, 1105 Kennedy Place, Davis, CA 95616, USA.

Nicolás Mateo, Biodiversity Prospecting Division, National Biodiversity Institute (INBio), Apdo. Postal 22–3100, Santo Domingo, Heredia, Costa Rica.

Cyrus G. Ndiritu, Kenya Agricultural Research Institute (KARI), PO Box 57811, Nairobi, Kenya.

Hector Quemada, Asgrow Seed Company, 2605 East Kilgore Road, Kalamazoo, MI 49002, USA.

Maddalena Querci, International Potato Center (CIP), PO Box 1558, Lima, Peru.

Pam Robeff, Garst Seed Company, 2369 330 St., PO Box 500, Slater, IA 50244, USA.

Lerson Tanasugarn, Office of Intellectual Property Policy Research, Intellectual Property Institute of Chulalongkorn University, Phya Thai Road, Bangkok 10330, Thailand.

Ann Marie Thro, Cassava Biotechnology Network, c/o CIAT, AA 6713 Cali, Colombia.

Gary H. Toenniessen, The Rockefeller Foundation, 420 5th Avenue, New York, NY 10018, USA.

John S. Wafula, Kenya Agricultural Research Institute (KARI), PO Box 57811, Nairobi, Kenya.

Preface

Over the past several decades, the US has maintained a leadership role in biotechnology. The growth of the technology, with its foundations in the US academic community, has been catalysed by a number of legislative initiatives and by aggressive venture capital investment, resulting in the development of a dynamic biotechnology industry. Other industrialized countries have also capitalized on the potential commercial applications of the science and have instituted a number of policy and legislative decisions designed to promote the growth of their respective biotechnology industries.

However, the technological, sociological and economic benefits which may accrue to less developed countries (LDCs) is less assumed. In the absence of efforts to increase the indigenous capacity of LDCs to conduct and manage research in biotechnology, directed efforts of advanced countries toward helping to resolve developing countries' agricultural problems may be short-lived and unsustainable. Technical and institutional constraints to technology development must be addressed. A range of requirements, including increased human resource development, improved facilities and infrastructure development, and development of a favourable policy environment, must be addressed by LDCs in order to benefit from technical advances in agricultural biotechnology.

In addition, new partnerships must be formed to access technology. In a world of increasing global competitiveness and proprietary technologies, the national research systems of developing countries are faced with new 'rules of the road' compared with the open access to improved agricultural varieties available during the Green Revolution of the 1960s. How, then, can private sector companies in the developed world, where much of the technology resides, be engaged to work with developing countries to address agricultural constraints which may not have large enough economic incentives to attract their attention? And how can developing countries participate in this new world?

It is in this spirit that the US Agency for International Development, Michigan State University, the Agricultural Biotechnology for Sustainable Productivity (ABSP) Project, UST and Garst Seed Co. joined to support the 'Agricultural Biotechnology for a Better World' conference. Taking its cue from the ABSP project, the conference integrated a number of essential components including technological developments, regulatory requirements, technology transfer and recognition of commercialization needs. Participants came from both developing and developed countries, public and private sector institutions, and diverse backgrounds to address these issues and explore potential mechanisms for collaborations to address mutual needs and interests.

Catherine L. Ives
Managing Director, ABSP Project

Acknowledgements

This book is based on the proceedings of the 'Biotechnology for a Better World' conference held at the Asilomar Conference Center, Pacific Grove, California April 28–30, 1997. The conference was sponsored and coordinated by the Agricultural Biotechnology for Sustainable Productivity (ABSP) project based at Michigan State University, and was cosponsored by the United States Agency for International Development (USAID), UST, Garst Seed Company and the Institute of International Agriculture at Michigan State University. The ABSP project wishes to express thanks to the members of its management, financial and networking teams and all of the student employees for their great efforts in handling conference logistics.

The chapters in this book are based on presentations given by the conference speakers. The ABSP project wishes to thank the speakers for their valued participation, for their many thought-provoking questions and comments, and for their efforts in developing the chapters. Special thanks are due to Catherine Ives and Bruce Bedford for their editing, to Andrea Johanson for her work on graphics and to ABSP staff members Kristen Sturman, Elva Hernandez and Lisa Cutcher for their tireless efforts in wordprocessing and proofreading through endless revision.

Sponsored in part by the United States Agency for International Development (USAID) and implemented by Michigan State University under cooperative agreement DAN-A-00–97–00126 00.

The Agricultural Biotechnology for Sustainable Productivity Project: a New Model in Collaborative Development

Catherine L. Ives, Bruce M. Bedford
and Karim M. Maredia

ABSP Project, Michigan State University, 324 Agriculture Hall, East Lansing, MI 48824, USA

The Agricultural Biotechnology for Sustainable Productivity (ABSP) project was established in 1991 by USAID, and implemented by a consortium of public and private sector institutions in the US and abroad, to assist developing countries in accessing and generating biotechnology, and in using that technology in an environmentally and legally responsible manner. The project was implemented in two general phases – technology access/generation and technology transfer. ABSP was designed to produce a number of transgenic crops and field test them in the US and collaborating countries. It also aims to develop innovative micropropagation methods for high-value tropical crops and carry out genetic stability tests. Additionally, the training of scientists, administrators and policy makers on the application of biosafety procedures and intellectual property rights in biotechnology is an important priority. Progress to date is summarized in this chapter and the issues that accompanied this integrative agricultural development model are discussed.

Introduction

Today, 800 million people in the developing world – 200 million of them children – are chronically undernourished. Paradoxically, over 60% of the world's poor live in the largely agrarian regions of south Asia and sub-Saharan Africa, where 65% and 79% of the population, respectively, depend on agriculture for their livelihood (Krattinger, 1997).

Food self-reliance continues to be an elusive goal for many developing nations. Obtaining world food security is a complex and difficult task, depending on many interrelated factors encompassing political, social and technical agendas. It is clear that famine relief and food aid, long a staple of many development organization efforts, is not the answer. Developing countries must be encouraged to lay the foundation for a long-term solution which includes support for free market reforms, infrastructure improvements and increased emphasis on agricultural research and education.

Through increased agricultural research and education, agricultural production may be improved. Improvements and new discoveries in biological fertilizers, soil and water conservation, biodiversity conservation, pest control and changes in land ownership and distribution will lead to improved production capacity. Worldwide, agricultural production continues to suffer substantial losses (20–40%) due to pests, weeds and diseases (Walgate, 1990). Biotechnology applications, integrated into conventional systems, hold much promise in augmenting agricultural production and productivity, particularly given the need to protect the environment and biodiversity while increasing production sustainably.

Some of the advantages of agricultural biotechnology include:

- the potential to increase productivity and food availability through better agronomic performance of new varieties, including resistance to pests;
- greater stability in farm production and reduced need for expensive inputs;
- rapid multiplication of disease-free plants;
- ability to obtain natural plant products using tissue culture;
- conversion of crop residues to other value-added foods;
- disease diagnosis in plants and livestock;
- manipulation of reproduction methods, increasing efficiency of breeding;
- low-cost, effective vaccine production in animal husbandry.

There are many potential benefits of biotechnology. *Ex ante* assessments indicate significant economic impacts of the technology, though actual impact remains unclear as many products are just now reaching the marketplace. Adding to this uncertainty is the relatively low priority many countries place on their agricultural research base.

In the US, the agricultural research budget is small when compared with the health research budget. The US Department of Agriculture (USDA) accounts for less than 1% of the federal research budget and has been effectively flatlined for the last two decades. However, the total investment in agricultural research has increased to US$7 billion annually. The private sector now accounts for 60% of this investment, individual states account for 15% and the Federal Government accounts for only 25% of the total investment in agricultural research (C. Wytoki, 1997, personal communication). It seems clear that the private sector will continue to drive the development of new technologies in agricultural research in the US and other developed countries. This increased emphasis on private sector investment has been fuelled by the biotechnology revolution and has been prominent over the past 10–15 years.

In most developing countries, agricultural research is conducted within public sector institutions. The private sector is usually underdeveloped and poorly linked with public sector research institutions, and government policies may not encourage investment in research-intensive industries. The policy environment surrounding agricultural biotechnology which has fuelled public/private sector interactions in the US and other developed countries is often weak or absent in developing countries. The potential benefits of biotechnology hold as much promise for developing countries as for developed countries, especially when the current low-input farming of developing countries is considered. Realizing the positive impact of biotechnology will depend upon the ability of developing countries to access and/or generate technology which is suitable to their needs.

The question for development agencies such as the US Agency for International Development (USAID) and their clients is how researchers in developing countries can access new technology to address local and regional constraints, if this technology is to be found within the private sector (or held as proprietary information in the public sector) in developed countries. How can vastly different cultures, political structures, regulatory regimes and economic philosophies coalesce to address critical problems for farmers and consumers in the developing world?

It was within this new environment that the Agricultural Biotechnology for Sustainable Productivity (ABSP) project was designed by USAID and implemented by a consortium of public and private sector institutions in the US and abroad, with Michigan State University (MSU) as the lead entity. Established in 1991, and currently in the last year of its initial design, the ABSP project takes an integrated approach, combining applied research, product development and policy development in the areas of biosafety and intellectual property rights (IPR), to assist developing countries in accessing and generating biotechnology, and in using that technology in an environmentally and legally responsible manner.

The project has been implemented in two general phases – technology access/generation and technology transfer – although these overlap and individual research projects are currently at different points along the continuum. Technology access and generation activities include the development of technology, human capacity building, policy assistance (biosafety and IPR), and the development of commercial links for access to existing technologies. Technology transfer activities aim to move materials to the developing countries and provide assistance to in-country research programmes in the form of support to research laboratories, legal consultations, biosafety consultations and collaborative field tests.

ABSP was designed to produce a number of transgenic crops and field test them in the US and collaborating countries. The project also aimed to develop innovative micropropagation methods for high value tropical crops and carry out genetic stability tests. Additionally, training scientists, administrators and policymakers in the application of biosafety procedures and intellectual property rights to biotechnology is an important priority. Below is a summary of progress to date and a discussion of issues that have accompanied this integrative agricultural development model.

Research Focus

ABSP has focused its research programme on assisting developing country institutions in adapting applications of biotechnology to address specific constraints on agricultural productivity. Research problems identified for inclusion in ABSP are in two categories: host-plant resistance and micro-propagation. Research was conducted on a limited number of crops (Table 1.1) chosen either for their importance in food security (sweetpotato, potato, maize and banana) or for their potential as a source of economic development (potato, cucurbits, banana, pineapple and tomato).

Host-plant resistance

Host-plant resistance research addresses constraints which cannot be resolved by conventional plant breeding alone. Crops were chosen in two regions of the world (Asia and Africa), based on economic and nutritional significance coupled with severe pest or pathogen constraints on productivity. Collaborations were developed as multiple partnering arrangements involving various agreements between public and private sector partners, as can be seen in Table 1.2.

Potato
Potato research focuses on the development of *Bacillus thuringiensis* (*Bt*) transgenic potato germplasm for resistance to potato tuber moth (PTM), a serious pest of potato worldwide. A collaborating team of plant breeders,

Table 1.1. Crops and constraints of the ABSP Project.

Crop	Insect/virus pest	Technical approach
Potato	Potato tuber moth	*Bt* toxin genes/transgenic plants
Sweetpotato	Feathery mottle virus	Coat protein/transgenic plants
	Sweetpotato weevil	Proteinase inhibitor gene/ transgenic plants
Maize	Corn borers	*Bt* toxin genes/transgenic plants
Cucurbits	Zucchini yellow mosaic virus,	Coat protein/transgenic plants and
(melon, squash)	watermelon mosaic virus, papaya ringspot virus, watermelon strain	traditional breeding of varieties
Tomato	Tomato yellow leaf curl virus	Replicase and other genes/ transgenic plants
Banana		Micropropagation/ bioreactor technology
Pineapple		Micropropagation/ bioreactor technology

Table 1.2. Primary institutional linkages of research groups within the ABSP Project.

US institutions	Sector	Developing country institutions	Sector	Commodities
MSU	Public	CRIFC	Public	Potato
	Public	AGERI	Public	Potato
MSU/Cornell	Public	AGERI	Public	Cucurbits
Scripps	Public	AGERI	Public	Tomato
Garst Seed Company (formerly ICI Seeds Inc.)	Private	CRIFC	Public	Maize
Monsanto	Private	KARI	Public	Sweetpotato
Pioneer Hi-Bred	Private	AGERI	Public	Maize
DNA Plant Technology	Private	Fitotek Unggul	Private	Pineapple
	Private	Agribiotecnología de Costa Rica	Private	Pineapple, banana, coffee

molecular biologists and entomologists in Egypt, Indonesia, Kenya and the US have succeeded in developing improved transformation and regeneration systems for various potato cultivars of importance in developing countries. Additionally, field tests have been conducted in the US and Egypt to evaluate the level of resistance to PTM and the agronomic qualities of transgenic lines. The research team is continuing to develop improved lines through the incorporation of additional resistance genes and improved germplasm to develop a long-term resistance management strategy.

Maize
The maize team focuses on the development of *Bt* transgenic maize germplasm for resistance to corn borers. Molecular biologists, entomologists and breeders at Garst Seed Company (formerly ICI Seeds Inc.) in the US and the Central Research Institute for Food Crops (CRIFC) in Indonesia have successfully trained Indonesian scientists in techniques of maize transformation, cell culture, insect bioassays, molecular characterization (polymerase chain reaction (PCR), enzyme-linked immunosorbent assay (ELISA)), and artificial infestation and field evaluation.

Unfortunately, there has not been a successful transfer of transgenic maize materials to Indonesia for field testing for technical, administrative and legal reasons. While the project initially focused on using tropical germplasm, legal uncertainty surrounding the commercialization of transgenic maize developed using the biolistic gun necessitated the employment of ICI's proprietary transformation technology ('Whiskers' technology) which was successful only in transforming a temperate line of maize. This material will have to be backcrossed into tropical maize for development of material suitable for Indonesia. Additionally, the *Bt* gene which has been incorporated into the maize is proprietary and

Indonesia, while making progress in revising its intellectual property laws, still cannot provide adequate legal protection for this material. As of the writing of this chapter, Indonesia also does not have biosafety guidelines or regulations for field testing genetically engineered plants and many companies are hesitant to test material in countries without adequate biosafety policies.

Recently a collaboration has been established between the Agricultural Genetic Engineering Research Institute (AGERI) in Egypt and Pioneer Hi-Bred to develop *Bt* transgenic maize germplasm with resistance to corn borers endemic to the Middle East. Before this collaboration was established, intellectual property agreements were signed and a technology transfer strategy developed to lessen the constraints described above in the CRIFC/ICI collaboration.

Cucurbits

ABSP cucurbits research seeks to develop potyvirus resistant cucumber, melon and squash via genetic engineering using coat protein genes and conventional breeding. Cucumber mosaic virus (CMV) and the potyviruses zucchini yellow mosaic virus (ZYMV), watermelon mosaic virus (WMV) and the watermelon strain of papaya ringspot virus (PRV-W) can result in yield losses of 50–100% and are found throughout the world. Results from this project are encouraging. Collaborations between MSU and AGERI have resulted in the development of a new leaf-based regeneration system for melon and successful transformation of Egyptian melon genotypes. Multiple-virus resistant cucumber, melon and squash materials resulting from many years of breeding have been exchanged between the breeding programmes at Cornell University and AGERI. Preliminary greenhouse and field trials show increased virus resistance and good agronomic performance from transgenic melons produced by AGERI using a ZYMV coat protein. Additional greenhouse and field trials are necessary to more fully evaluate these materials. The process of identifying a corporate partner to commercialize these materials in the Middle East is underway.

Tomato

Tomato research has focused on the production of tomato yellow leaf curl virus (TYLCV) resistant germplasm using various molecular strategies via an *Agrobacterium*-based transformation system. TYLCV is a devastating disease in the Middle East, Africa and Southeast Asia. In Egypt, yield losses of up to 35% have been reported from this disease for which there are no effective control measures. Unlike constraints to the production of other crops within the ABSP project, molecular methods to control TYLCV had not been developed prior to the start of the project. Progress to date has focused on developing methods to detect and identify geminiviruses producing constructs for transformation of tomato plants and elucidating strategies to control geminivirus infection through genetic engineering. Protocols for transformation and regeneration of Egyptian tomato varieties have been established and transgenic tomatoes produced. Greenhouse testing of these materials is currently under way and additional molecular approaches for the control of geminiviruses are being investigated.

Sweetpotato
The sweetpotato collaboration is composed of researchers from Monsanto Company, the Kenya Agriculture Research Institute (KARI) and CRIFC, and has focused on the development of transgenic sweetpotato germplasm resistant to sweetpotato feathery mottle virus (SPFMV). Transgenic sweetpotato germplasm containing the SPFMV coat protein has been developed and plans to transfer the material for field testing in Kenya are under way. Monsanto has donated their technology for use in sweetpotato in Africa. Additionally, Monsanto and CRIFC are considering the development of sweetpotato resistant to sweetpotato weevil. Progress has been made in regenerating Indonesian varieties of sweetpotato from tissue culture.

Micropropagation technology

Micropropagation research was conducted with the goal of developing a system to rapidly multiply large numbers of plants using bioreactor technology in order to produce high-quality planting stocks of tropical crops (banana, pineapple, coffee, ornamental palm) at a lower economic and environmental cost than conventional methods. These collaborations occurred between a private company in the US (DNA Plant Technology) and private companies in Costa Rica (Agribiotecnología de Costa Rica) and Indonesia (Fitotek Unggul). Pineapple and banana plants derived from the bioreactor technology are currently being evaluated in the field.

Technology Generation and Policy Building

Management and networking

The ABSP philosophy represents an integrated approach to agricultural biotechnology research and development programmes in which research is linked to policy activities and human resource development. It seeks to build and strengthen national agricultural research systems, deliver a specific set of research products and information packages, and develop genuine bridges of collaboration between US and developing country partners. Through short-term training, workshops, internships, distribution of literature and communication, and the ability to meet infrastructure needs within project guidelines (such as computer and e-mail capability), the ABSP management team has met those objectives.

At the outset of the project, ABSP set the following management guidelines:

1. Maintain a highly focused research programme, concentrating on specific crops and technologies of mutual benefit to the US and collaborating developing country partners.

2. Implement a product-oriented research style which links public and private sector institutions in the US and developing countries.

3. Link product-oriented research to policy analysis of intellectual property and biosafety to ensure product commercialization in an environmentally and socially responsible manner.

4. Maintain a geographic focus on specific centres of expertise and develop a critical mass for the multidisciplinary teams responsible for transferring technology.

5. Build a global network that provides global access to information and serves as a forum for the exchange of ideas and information on biotechnology in relation to sustainable agriculture systems.

6. Establish and support formal linkages with organizations such as the Consultative Group on International Agricultural Research (CGIAR), Agricultural Research System, Biotechnology Industry Organization (BIO) and international biotechnology programmes.

The first three objectives have been fulfilled. While the project has focused its efforts on a limited number of countries, there is still a need to assist programmes in developing a critical mass of expertise. Training within project constraints has attempted to build small multidisciplinary teams that will be able to employ the technologies and products developed under the ABSP project. The development of a global network began through the publication of the Bio*Link* newsletter; however, given the rapid changes in technology since the project began, an electronic forum and/or catalogue of potential products may be a useful approach in the future. ABSP sponsored a global conference in 1997, to bring together the public and private sector from both developed and developing countries to address challenges and opportunities in the field of agricultural biotechnology. Formal links with the BIO have been established and informal linkages with international agricultural research centres (IARCs) and other international programmes have been developed.

The flow of information regarding ABSP and other agricultural plant biotechnology research programmes around the world is coordinated by the ABSP network office. The network office coordinates workshops in research and policy areas and has made available to collaborators modular posters which describe ABSP research, management and network activities.

In order to build cooperative teams, the management and networking office has used many different approaches for information exchange and active research collaboration. The project has conducted implementation workshops and/or personal level interactions to plan project activities. Implementation workshops serve a very useful purpose in providing an opportunity for direct input by host country collaborators in the shared development of work plans, identifying personnel and assigning responsibilities, and assessing collaborative research and networking activities already established. As an example, in August 1992, an implementation workshop was held at CRIFC in Bogor, Indonesia, to initiate both the core project and additional projects which received further funding from the USAID mission in Jakarta. In January 1993, an implementation workshop was held at AGERI in Cairo, Egypt, which provided an opportunity for commodity research teams and network and management teams from the US and Egypt to meet and draft work plans for the ABSP/AGERI cooperative project.

Policy building

While research success may be unpredictable, technology generation has been the most straightforward aspect of the ABSP project. The transfer and implementation of technologies and research products has been much less assured since it involves coordination of government policies. Many existing policies in developing countries are not conducive to biotechnology research and development. Any country which seeks to conduct agricultural biotechnology research, develop, test, import, export, use or commercialize transgenic plants should investigate the need for developing biosafety and IPR systems. The innovative management component of ABSP supports client consultations with experts in the policy areas of biosafety and IPR, while the networking office has provided workshops, short courses and internships to assist in the development of sound biosafety and IPR frameworks in the collaborating countries. IPR and biosafety systems will not only allow scientists to develop and test transgenic materials in local environments, but will also help safeguard the environment, attract private sector investment, and facilitate commercialization and international trade.

Biosafety
Through the assistance of ABSP and other projects the collaborating countries have been able to make good progress in the development of national biosafety guidelines. Egypt now has biosafety guidelines that have been approved by the government. Indonesia and Kenya have developed draft biosafety guidelines which are pending governmental approval. In May 1993, the ABSP project organized a 4-week internship programme in the US to help develop these guidelines. Seven scientists and regulatory personnel from Egypt, Indonesia and Kenya were challenged with 'hands-on' applications of biosafety procedures used in safe handling of transgenic plants in laboratories, greenhouses and fields. Interns met with regulatory personnel at state and federal agencies dealing with transgenic field testing permit applications, food safety and other risk assessment issues. They visited and interacted with personnel from the US Department of Agriculture – Animal and Plant Health Inspection Service (USDA–APHIS), Food and Drug Administration (FDA), Environmental Protection Agency (EPA), Michigan Department of Agriculture (MDA), MSU, Asgrow Seed Company, Garst Seed Co. and the National Biological Impact Assessment Programme (NBIAP).

   To further strengthen the foundation for a biosafety regulatory framework in Latin America and the Middle East, ABSP organized two regional biosafety workshops. The workshop for Latin America was organized in Jamaica in May 1993, cosponsored by the ABSP Project, the Bean Cowpea Collaborative Research Support Programme and the Jamaica Agricultural Research Programme. Forty-two representatives from 12 countries in Latin America, the US, Africa, the Middle East and Asia participated in the workshop. In January 1994, the ABSP/AGERI project organized a biosafety workshop in Cairo, Egypt, attended by over 100 biosafety, science and regulatory personnel from Egypt and selected countries in Africa and the Middle East. The workshop addressed policy, risk assessment and field testing issues which surround the management and safe handling of transgenic plants. Funds to support some regional participants were provided by UNIDO/UNEP.

In August 1996, ABSP, in cooperation with the Information Systems for Biotechnology project at Virginia Polytechnic and State University, organized a 2-week internship programme in biosafety. This programme gave participants a thorough grounding in all aspects of biosafety for environmental release of genetically engineered organisms, including the theory and practice of risk assessment and management in agricultural biotechnology applications. Practical experience was gained in biosafety evaluation through case studies.

Through support from USAID/Cairo, the ABSP/AGERI project, in cooperation with the University of Arizona, has constructed a modern biocontainment greenhouse facility in Cairo, Egypt. AGERI scientists are now using this facility for greenhouse testing of transgenic plant materials. Through an appropriate IPR agreement, building plans for the Egypt greenhouse facility have been shared with ABSP collaborators in Indonesia to provide ideas for the construction of a similar facility in Bogor, Indonesia. This facility, financed through The World Bank, is nearing completion.

The lack of institutionalized guidelines and/or field testing regimes, coupled with uncertain policy direction in collaborating country governments, has made the actual transfer of materials difficult.

Intellectual property rights

While many developing countries have patent laws, many of those laws do not adequately address plant breeders' rights or the products of agricultural biotechnology, particularly those developed through genetic engineering. In order to assist in the revision or establishment of such laws, and to provide information and expertise to ABSP focus countries, the ABSP project has provided varied and innovative training.

In April 1993, ABSP and Stanford Law School organized a 4-week internship programme at Stanford University. The internship programme covered both plant breeders' rights and regular patents as they apply to the genetic engineering of plants. The programme included substantial discussion of patent licensing and research agreements, and explored policy issues regarding patent laws for developing countries.

In January 1994, ABSP organized an IPR workshop in Cairo, Egypt. This workshop addressed patent issues related to materials used in and generated by biotechnology projects. The workshop involved scientists, legal professionals and government officials primarily from Egypt. A second IPR workshop was held in Washington, DC in July 1994. Invited participants attended from Brazil, Egypt, Kenya, Indonesia, Costa Rica, Morocco and Thailand as well as institutions and agencies such as The World Bank and USAID.

In February 1996 and August 1997, ABSP and the Institute of International Agriculture at MSU Office of Intellectual Property organized 2-week internship programmes in IPR and technology transfer. The programme provided 'hands-on' experience to participants in the day-to-day handling of intellectual properties in university and private sector settings. The participants gained insight into how technologies generated in the public sector are transferred to the private sector in ways which legally protect innovation and investment.

The lack of relevant national IPR policies has caused difficulty in the actual transfer of proprietary materials accessed or developed via support of the ABSP project.

Private sector linkages

Biotechnology has an orientation in basic research and application in product development that cuts across many scientific disciplines and both the public and private sectors. ABSP is designed to transfer technology leading to product development and has set aside discretionary funds as seed money for the field testing and commercial development of promising research results, including costs for patenting new products. An active involvement of the US private sector was established from the beginning of the project, giving developing country programmes direct access which can serve to drive progress within their own research systems. Collaborating companies in the US include: DNA Plant Technology; Pioneer Hi-Bred; Garst Seed Company (formerly ICI Seeds Inc.); and Monsanto. The collaborating private companies in developing countries include: Fitotek Unggul in Indonesia; and Agribiotecnología de Costa Rica.

Institutions collaborating in ABSP interact to address specific research tasks in relationships which eradicate traditional sector boundaries by establishing public–public, public–private and private–private collaborations between institutions in the US and developing countries. Research is conducted under specific contract agreements designed to accommodate differing needs and approaches in the areas of intellectual property and technology transfer.

To encourage interactions memberships in trade organizations are provided for partner countries. The memberships help in providing new information and opportunities to build new relationships between developed and developing country entities. Membership in BIO provides access to member institutions, a newsletter which highlights research innovations and policy issues, and lowers fees for attending the annual BIO meeting and exposition. ABSP has also supported the participation of individuals from partner countries in annual BIO meetings and has organized sessions highlighting the research contributions, economic opportunities and policy environments in partner countries.

In April 1993, the ABSP project organized a 1-week industrial seminar series (ISS) designed to provide opportunities for participating international leaders to interact with technical and business managers at private US companies which have active agricultural plant biotechnology programmes. Seven senior scientists, administrators, private sector personnel and senior government officials from Costa Rica, Kenya, Egypt and Jamaica participated in the 1993 BIO (then ABC) annual meeting and visited Garst Seed Co., Slater, Iowa, Ecogen in Langhorne, Pennsylvania, and DNA Plant Technology in Cinnaminson, New Jersey. These encounters were instrumental in opening lines of communication between developing country leaders and host companies.

Technology transfer

As research products are developed, the challenge of adapting these technologies and testing the materials under suitable agronomic conditions within host countries becomes paramount. Two major challenges the project now faces are the safe and legal transfer of germplasm, and the integration of new research products into the general agricultural research sector via a mechanism that will deliver the finished products of this research to farmers and consumers in the developing world. Undoubtedly, this has been and continues to be the most challenging aspect of the ABSP project as it involves the successful integration of individuals and institutions with various research, administrative and legislative agendas.

To assist in the transfer of technology, a technology transfer coordinator (TTC) has been appointed in the US, and country coordinators (CCs) have been appointed in participating countries. The TTC and CCs work with principal research investigators and the ABSP management team to implement technology transfer, product development and commercialization activities. The TTC has worked with principal investigators in the US to develop a knowledge-base in terms of technologies, information and training resources available for transfer. The ABSP management team and CCs determine how best to address the needs of the national programmes. This includes selection of appropriate technologies based on the research results from the first phase of the project; it also triggers the need to look for available technologies from sources other than the major participants of the project. Efficient use and management of transferred technologies may require in-country workshops, technical training of laboratory technicians and field testing personnel, or other actions.

Intellectual property and biosafety constraints in technology transfer
Although a significant amount of project resources was directed toward the training of developing country personnel in intellectual property and biosafety,

these two policy areas remain a significant constraint to transferring technologies to partner countries. Individual research institutes cannot move forward in building relationships with partners who hold proprietary technologies unless national policies allow them to do so. Governments generally operate at a glacial pace and, while the technology progresses rapidly, changes in legislation and development of implementation mechanisms proceed slowly. Of our partner countries, only Egypt has revised both its biosafety and IPR laws, which will enable AGERI scientists to access proprietary technologies, protect their own inventions and innovations, and employ them in a safe and responsible manner.

In order to transfer any genetically engineered materials, approval must first be obtained from USAID's Biosafety Committee. This has proved to be a laborious process and has delayed the transfer of materials. However, it has assisted the ABSP project in elucidating what type of training still needs to be conducted in order to assist our partner countries in implementing transgenic field tests which are on a par with those tests conducted in developed countries.

Of more serious concern, the lack of national policies on intellectual property makes transfer of materials difficult or impossible. Without these policies, transfer of technologies, especially to the public sector research institutes, is difficult even for research purposes, and impossible for commercial use. It is certain that, if developing countries are committed to the use of biotechnology to address their agricultural constraints, they must be willing to adopt policies and practices that allow and invite collaboration. It will be nearly impossible for any country to 'go it alone' in the development of agricultural biotechnology products.

Conclusions and Future Efforts

The ABSP project has been successful in most of its goals and objectives. It has produced agricultural products which will address important agricultural constraints in our collaborating countries. It has linked public and private sector institutions together and has linked these research efforts to policy analysis and training in intellectual property and biosafety. It has maintained a geographic focus while building a global network to provide information to a wide cross-section of scientists, administrators and policymakers in both the developed and developing world.

What remains to be done? The project is still striving to develop a critical mass of trained personnel for the multidisciplinary teams responsible for developing and using agricultural biotechnology in developing countries, budget constraints have prevented ABSP from training an optimal number of scientists. Additionally, the goal of establishing links with other organizations in a more formalized manner remains to be accomplished. This is quite difficult, since other organizations have different goals and budget cycles. Finally, and most importantly, the testing and adapting of materials in partner countries, initiated and building momentum in Egypt, needs to be more fully implemented.

From our experience, additional training is certainly needed for developing countries interested in accessing, developing and managing agricultural biotechnology. This training would not only be in the technical sciences, but would include business and management training, the development of entrepreneurial and negotiation skills, continued training on intellectual property as it relates to accessing, using and protecting proprietary technologies, and biosafety as it relates to implementing procedures and conducting meaningful field tests.

References

Krattiger, A.F. (1997) *Insect Resistance in Crops: A Case Study of* Bacillus thuringiensis *(Bt) and its Transfer to Developing Countries.* ISAAA Briefs, No. 2.

Walgate, R. (1990) *Miracle or Menace? Biotechnology and the Third World.* The Panos Institute, London, UK.

Needs and Potential Uses of Agricultural Biotechnology: Perspectives of Developing Countries

Addressing Agricultural Development in Egypt through Genetic Engineering

2

Magdy Madkour

Agricultural Genetic Engineering Research Institute (AGERI), Agricultural Research Center (ARC), 9 Gamaa Street, Giza 12619, Egypt

One of the major targets for biotechnology in Egypt is the production of transgenic plants conferring resistance to: (i) biotic stress resulting from pathogenic viruses, fungi and insect pests; and (ii) abiotic stress such as non-favourable environmental conditions including soil salinity, drought and high temperature. These biotic and abiotic constraints are major agricultural problems leading to deleterious yield losses in a large variety of economically important crops in Egypt. The Agricultural Genetic Engineering Research Institute (AGERI) was established in 1990 to promote the transfer and application of this technology. AGERI aims to adopt the most recent technologies available worldwide and apply them to address existing problems in Egyptian agriculture. This chapter discusses capacity building; human resources developments; research and scientific collaborations; and the goals of AGERI in the agricultural community.

Introduction

The challenge facing the world today is to provide food, fibre and industrial raw materials for an ever growing world population without degenerating the environment or affecting the future productivity of natural resources. This challenge is even more pressing in developing countries, where 90% of the world's population growth will take place within the next two decades. Meeting this challenge will require the continued support of science, research and education.

In Egypt, agriculture spearheads socio-economic development, accounting for almost 28% of national income, employing almost 50% of the workforce,

© CAB INTERNATIONAL 1998. *Agricultural Biotechnology in International Development* (eds C.L. Ives and B.M. Bedford)

and generating more than 20% of the country's total export earnings through agricultural commodities (Fig. 2.1). A limited arable land base (Fig. 2.2) coupled with an ever-growing population with an annual birth rate of 2.7% are the primary reasons for an increasing food production/consumption gap. Egypt's population will grow to about 70 million by the year 2000 and swell to 110 million by the year 2025. In recent years, only 15% of total agricultural production in Egypt has been exported, indicative of the increased domestic demand due to increased population growth. Increasing the agricultural land base from 7.4 million feddans to 14 million feddans (1.3 million ha to 2.6 million ha) would only satisfy 50% of the demands of a current population of 59 million. To bridge the food gap and to fulfil the goal of self-reliance, expanding the land base and optimizing agricultural outputs are urgently needed.

National Perceptions

The Government of Egypt (GOE) is increasingly aware that it must use its own limited resources in a cost effective way. Failure to develop its own appropriate biotechnology applications and inability to acquire technology developed elsewhere could deny Egypt timely access to important new advances that could overcome significant constraints to increased agricultural productivity.

An increase in food production may be achieved by protecting crops from losses due to pests, pathogens and weeds. The total loss of worldwide agricultural production ranges from 20 to 40%, including both preharvest and postharvest losses, despite the widespread use of synthetic pesticides.

It is in the area of crop protection that biotechnology – and especially genetic engineering – could offer great benefits to the environment by replacing the present policy of blanket sprayings of crops with herbicides, fungicides

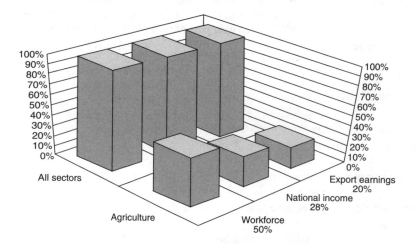

Fig. 2.1. Contribution of the agricultural sector to the national economy.

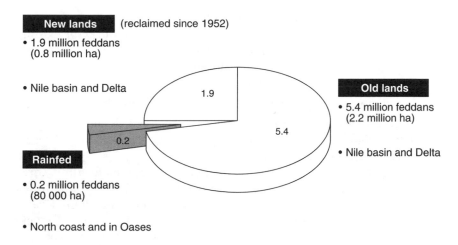

Fig. 2.2. Egypt's agricultural land base 7.5 million feddans (3.1 million ha).

and pesticides with inherent engineered resistance to pests and diseases. Genetic engineering is highly suited to agriculture in the developing world, since it is user-friendly. If it is applied in a sensible manner, there can be no doubt that this technology is 'green'.

One of the major targets for biotechnology in Egypt is the production of transgenic plants conferring resistance to: (i) biotic stress resulting from pathogenic viruses, fungi and insect pests; and (ii) abiotic stress such as non-favourable environmental conditions including soil salinity, drought and high temperature. These biotic and abiotic constraints are major agricultural problems leading to deleterious yield losses in a large variety of economically important crops in Egypt.

The Agricultural Genetic Engineering Research Institute (AGERI), repre-sents a vehicle within the agricultural arena for the transfer and application of this new technology. The original establishment of AGERI in 1990, was the result of a commitment to expertise in agricultural biotechnology. At the time of its genesis, AGERI was named the National Agricultural Genetic Engineering Laboratory (NAGEL). The rapid progress of its activities during the first three years encouraged the Ministry of Agriculture and Land Reclamation to authorize the foundation of AGERI (Fig. 2.3), constituting the second phase of the national goal for excellence in genetic engineering and biotechnology. AGERI aims to adopt the most recent technologies available worldwide and apply them to address existing problems in Egyptian agriculture.

Capacity Building

The physical location of AGERI is within the Agricultural Research Center (ARC) in Giza. This not only facilitates an interface with ARC's ongoing research programmes, but also provides a focal point for biotechnology and genetic engineering for crop applications in Egypt. AGERI has upgraded the existing laboratory and has a total net area of 1116 m^2, consisting of 14 modernly equipped laboratories (Fig. 2.4), the BioComputing and Networks Unit, a central facility, a preparation/washing facility and a supply repository.

In addition, the physical infrastructure at AGERI includes:

- Recently completed 'Conviron' controlled environment chambers (140 m^2) used to acclimatize transgenic plant material.
- A fibreglass greenhouse (307 m^2).
- A containment greenhouse (412 m^2), consisting of eight units, three laboratories and a headhouse and complying with biosafety and United States Department of Agriculture (USDA)/Animal and Plant Health Inspection Service (APHIS) and EPA (Environment Protection Agency) regulations. This greenhouse supports various lines of locally produced transgenic plants with new traits. Experiments which test the level of gene expression in transgenic plants take place in this modern facility.

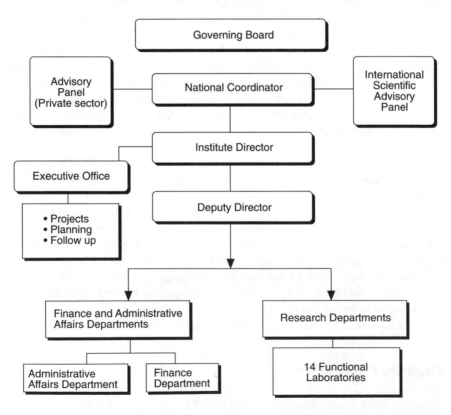

Fig. 2.3. Agricultural Genetic Engineering Research Institute (AGERI).

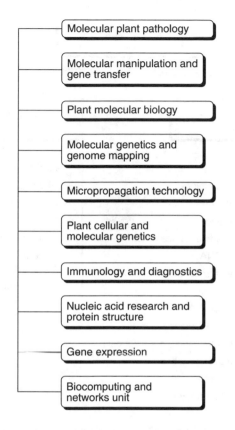

Fig. 2.4. Functional laboratories at AGERI.

To establish the core nucleus of biotechnology in Egypt, AGERI is adopting the most recent technologies available worldwide and applying them to address existing challenges in Egyptian agriculture (Fig. 2.5). This includes employing the BioComputing and Networks Unit, an information centre which supports the research activities at AGERI, to provide worldwide networking capabilities to access databases and biotechnology information centres located abroad, as well as a forum for discussion of technical issues with experts from all over the world through the e-mail facility. The unit provides electronic literature searches on CD-ROM and maintains a software library to meet the biocomputing and publishing needs of the institute.

Human Resources Development

One of many vital contributions made by AGERI is the identification and recruitment of a collective group of 17 high calibre, dedicated senior scientists. Each

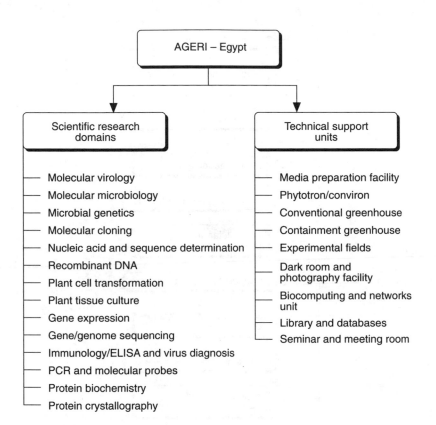

Fig. 2.5. Facilities to support genetic engineering technology transfer.

one is a vital link in achieving programme goals for crop improvement. The senior scientists have institutional affiliations within Egypt as well as their scientific responsibilities within AGERI. They are representatives and practising faculty members from six Egyptian universities, as well as various national agricultural research centres, and work at AGERI on a joint-appointment basis which maximizes their interaction between the academic and research domains. Their high level of international training, in conjunction with their enthusiasm to invest their talents into AGERI's biotechnology programmes, is an encouraging addition to Egypt's agricultural technology development.

Another vital contribution of AGERI has been its role as an interface between the international scientific community and Egypt. Once AGERI became fully commissioned, the research and postdoctoral education components of the project commenced. Various seminars and conferences have been held at AGERI with highly qualified international consultants. Over 60 study tours have taken both senior scientists and junior assistants of AGERI to various international biotechnology centres in Europe, North America and Asia to attend conferences or training courses.

Condensed, short courses and seminars concentrating on the basics of biotechnology have been conducted by members of our local staff. Educational activities have been promoted as a result of this linkage and have encouraged cooperation with international researchers and laboratories. Opportunities have been supplied for the exchange of genetic probes, DNA libraries and vectors. Contacts with research centres worldwide have been encouraged and initiated to facilitate meaningful interactions.

Training courses at AGERI

These include:
- An International Training Course on the Use of Restriction Fragment Length Polymorphisms (RFLPs) and Polymerase Chain Reaction (PCR) for Crop Improvement, November 1991.
- A Regional Training Course on the Application of PCR and Enzyme-linked Immunosorbent Assay (ELISA) in Plant Virus Diagnostics, May 1992.
- A Course in Modern Methods in Microbial Molecular Biology, April 1993.
- A Regional Training Course on Tissue Culture and Micropropagation in Plants with Special Emphasis on Date Palm, May 1994.

Research and Scientific Collaboration

AGERI has been successful in attracting funds to sponsor its research from the following international organizations:

- The United Nations Development Programme (UNDP), as a co-funding agency which supported the initial research at NAGEL, currently AGERI.
- A cooperative research agreement between AGERI and the Agricultural Biotechnology for Sustainable Productivity (ABSP) project based at Michigan State University, which is funded by United States Agency for International Development (USAID)/Cairo, under the Agricultural Technology Utilization and Transfer (ATUT) project. This activity facilitates interaction between AGERI's scientists and researchers from a number of prominent American universities and private industry, i.e. Michigan State University, Cornell University, University of Wyoming, University of Arizona, Pioneer Hi-Bred, and the Scripps Research Institute. Moreover, other USAID funded research has been collaboratively executed with the University of Maryland, and the USDA–ARS, Beltsville.
- Recently, the International Center for Agricultural Research in the Dry Areas (ICARDA), located in Aleppo, Syria, has contracted AGERI to conduct research on their mandated crops.

The projects conducted at AGERI (Table 2.1) are based on the concept of maintaining a programme that is focused on the problems of Egypt. The immediate objectives are to develop and deliver transgenic cultivars of major crops of economic importance in Egypt. The most recent and successful genetic engineering technologies are used to address this need. These projects also represent a spectrum of increasingly complex scientific challenges which require the state-of-the-art technologies of genetic engineering and gene transfer. Gene manipulation techniques such as cloning, sequencing, coding modifications, construction of genomic and cDNA libraries, and plant regeneration in tissue culture, are just a few examples of the cellular and molecular biology methodologies that are utilized for the production of transgenic plants.

The successful implementation of these projects is establishing a national capacity within Egypt for the sustainable production of crops crucial to the economy and a safer, cleaner environment.

Examples of projects at AGERI

- Genetic engineering of virus resistance in a number of crops including: production of potato resistant to important viruses in Egypt (potato potyvirus X (PVX), potato potyvirus Y (PVY), potato leaf roll leutovirus (PLRV)); production of transgenic tomatoes resistant to geminiviruses such as tomato yellow leaf curl virus (TYLCV); introduction of virus resistance in squash and melon against zucchini yellow mosaic virus (ZYMV); and finally the production of transgenic faba bean conferring resistance to bean yellow mosaic virus (BYMV) and faba bean necrotic yellow virus (FBNYV).
- Engineering of insect-resistant plants with *Bacillus thuringiensis* crystal protein genes. *Bt* genes are used in transformed cotton, maize, potato and tomato plants to resist major insect pests.
- Genetic engineering for fungal resistance using the chitinase gene concept for the development of transgenic maize, tomato and faba bean expressing resistance to fungal diseases caused by *Fusarium* sp., *Alternaria* sp. and *Botrytis fabae*.
- Enhancing the nutritional quality of faba bean seed protein by the successful transfer of the methionine gene to faba bean plants.

Table 2.1. Current projects at AGERI on transgenic crops.

	Potato	Tomato	Cotton	Maize	Faba bean	Cucurbits
Virus resistance	✓	✓			✓	✓
Insect resistance	✓	✓	✓	✓		
Stress tolerance		✓	✓		✓	
Genome mapping	✓			✓		
Protein engineering					✓	
Fungal resistance		✓	✓	✓		

- Cloning the genes encoding important economic traits in tomatoes, faba beans and cotton especially those related to stress tolerance i.e. heat shock proteins and genes responsible for osmoregulation.
- Mapping the rapeseed genome in order to develop cultivars adapted to the constraints of the Egyptian environment and thus securing a good source of edible oil.
- Developing efficient diagnostic tools for the identification and characterization of major viruses in Egypt.

These projects are relevant to Egyptian agriculture since they reflect a significant positive impact on agricultural productivity and foreign exchange. To illustrate, Egyptian *Bt* transgenic cotton, resistant to major insect pests, would result in substantial savings of US$50 million spent annually on the purchase of imported pesticides. Mapping of oilseed rape has the potential to substantially reduce the 400,000 tons of edible oil which is imported into Egypt annually. Similarly, transgenic potato varieties resistant to selected viruses and insect pests would prevent the expenditure of approximately US$33 million per annum in the import of seed potatoes.

The Goals of AGERI in the Agricultural Community

- Advance agriculture using biotechnology and genetic engineering capabilities available worldwide to meet contemporary problems of Egyptian agriculture.
- Broaden the research and development capabilities and scope of the Agricultural Research Center in the public and private sectors i.e. initiation of new programme areas and application to a wider array of crop species.
- Expand and diversify the pool of highly qualified trained professionals in the area of biotechnology and genetic engineering.
- Provide opportunities for university trained professionals, e.g. faculty researchers and teachers, the Ministry of Agriculture (professional researchers) and private venture companies to cooperate in agricultural genetic engineering research.
- Promote opportunities for private sector development.
- Achieve the desired level of self-reliance and self-financing within AGERI to mobilize the funds necessary for maintaining laboratories.

Figure 2.6 highlights the role that AGERI is seeking to fulfil in Africa and the Middle East, as an emerging centre of excellence for plant genetic engineering and biotechnology. AGERI will act as an interface between elite centres and laboratories from the international scientific community and research centres, universities and the private sector in Egypt, the Middle East and Africa. The major goal is to assist and provide the mechanism for proper technology transfer to benefit relevant agricultural mandates.

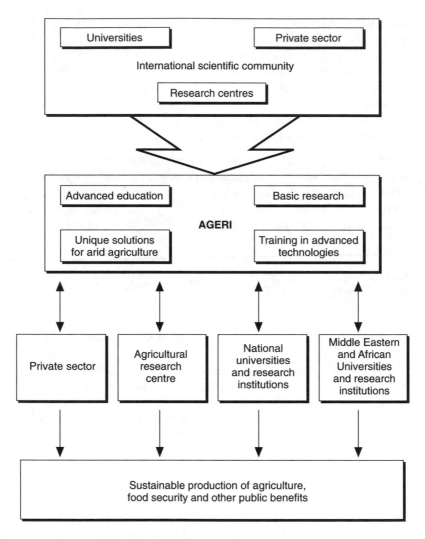

Fig. 2.6. Role of AGERI in the Middle East and Africa.

The Release of Transgenic Varieties in Centres of Origin: Effect on Biotechnology Research and Development Priorities in Developing Countries

Ariel Alvarez-Morales

CINVESTAV, IPN. Unidad Irapuato, Department of Plant Genetic Engineering, Apdo. Post. 629, Irapuato, Gto., 36500 Mexico

Many products derived from biotechnology research and development (R&D) are either in the final stages of product development or already available in the market. Among these are genetically engineered varieties (GEV) of maize, tomato, potato, soybean and cotton, none of which are indigenous to the countries that have developed the new products. In Mexico, the centre of origin of maize, this situation poses a very difficult problem.

In Mexico, research institutions and agricultural universities together with the government agencies involved in regulation of GEV have tried to establish a rational set of guidelines to conduct safe trials with GEV of maize which would provide experience and information in order to, in due course, decide whether or not to deregulate GEV of maize.

GEV of maize have been deregulated in the US and are being rapidly incorporated by the growers. However, since a substantial amount of the maize that Mexico buys every year to satisfy its internal demand comes from the US, the following points become crucial: (i) the maize that is bought from the US is intended for consumption, however, a portion of it finds its way into the seed market; (ii) Mexico may be able to buy non-transgenic maize during the first few years after commercial release in the US; however, very soon the bulk product that is introduced into the country will be a mixture of transgenic and non-transgenic varieties, with the former increasing in proportion every year; (iii) there is growing pressure from Mexican growers to have access to GEV which may lead to the development of a black market for these seeds if access is delayed. Therefore, it seems inevitable that, one way or another, GEV of maize are going to be introduced very soon into Mexican territory and it is necessary to assure the population that it will be safe to the environment, to agriculture and to human and animal health.

In terms of human and animal health the government may not find
it difficult to validate the data which supported the Food and Drug
Administration (FDA) approval in the US. However, in terms of the
environment, a new strategy needs to be implemented based on
immediately allowing field trials of transgenic maize, with different
traits and on different locations, and linking these releases to accurately
monitoring their impact, specifically on the wild relatives and landraces,
and generally on the agronomic and natural environment.

This monitoring should be aiming to detect any significant alteration in the environment, specifically associated to the transgenic
material. The lack of any such alteration over a defined period of time
could indicate whether the monitoring should be continued on a general basis, limited to certain areas or traits, etc., or even completely
stopped. On the other hand, it could result in the early detection of a
potential problem which then could be appropriately addressed, maybe
by conducting a more thorough study, by remediation or even by
indicating the necessity of a permanent or limited ban on the release of
such GEV.

Introduction

Many developing countries see biotechnology as a promising tool, one which
could rapidly enhance their agronomic potential, reduce or eliminate deficits in
their internal supply of staple foods or produce higher quality products and/or
lower production costs of specific crops and thus facilitate entry into global
commodity markets with a higher probability of success.

It is not surprising that all over the world developing countries are
making a serious commitment to the generation of strong research and
development (R&D) capabilities in biotechnology to ensure that they will also
profit from its potential benefits. However, the use and application of
biotechnology has not been straightforward. In order to responsibly engage in
the development of new crop varieties through this technology, a series of laws,
guidelines and specific committees and regulating bodies have been
implemented. This has been done to assess the possible environmental and
legal (patents) consequences of experimenting and using certain transgenic
varieties or methodologies involved in their production, and, prior to the
commercial release of these materials, assessing the possible risks to the
environment, human and animal health as well as the effects on internal and
export markets.

Nevertheless, and in spite of what could have been seen as a major obstacle,
many developing countries on all continents have set up the appropriate laws,
guidelines and official bodies to regulate and oversee the research, development
and use of this new technology, whether derived from their own research
institutes or from international companies wishing to test or commercially
release these materials in their territory.

R&D in Biotechnology

In countries like Mexico, several research institutions began to implement R&D strategies which would allow them to compete more effectively with large international companies. The strategies were to:

- Address local or regional problems affecting staple or other basic crops, such as potato. This would usually be done with resources from the federal or state government, funding agencies and, in some cases, funds from associations of growers.
- Address problems faced by growers involved in large-scale production of export commodities such as hot and sweet peppers, onion, garlic, asparagus and banana. In this case, funding will usually be provided by the growers themselves with little or no involvement of state or federal funds.
- Conduct basic research in areas such as ripening, disease resistance, stress tolerance, pest biocontrol and product development. In most cases, this basic research would be done around a regional crop or product, often aimed at the possibility of later developing a product. An example of this would be the research taking place on *Ustilago maydis* (Huitlacoche) a species which is rapidly gaining acceptance as a specialty food in many places outside Mexico (the only place where this fungus is routinely consumed). The research is aimed at understanding the basic biology, infection mechanism, composition and industrial processing for this product, with some local food companies already expressing interest in the results. Funds for this kind of research will usually be obtained from the federal government or through national or international competitive grants.
- Perform research on biosafety. Mexico is the centre of origin or diversity of many important crops such as maize, cotton, tomato, bean and pepper. It is very important to gain knowledge concerning the possible effects of releasing transgenic varieties for which there are wild species or landraces that might somehow be affected. Of these, maize is undoubtedly the most important crop in Mexico, not only as a staple food but also because it is profoundly embedded in the culture. In this area, both the International Center for Maize and Wheat Improvement (CIMMYT) located in Mexico, and the National Institute for Forestry, Agriculture and Livestock (INIFAP), which is the agricultural research branch of the Ministry of Agriculture, plan to conduct research aimed at identifying or quantifying the possible effects of different transgenes on landraces or teosinte. Again, funds are usually obtained from federal funds or by national or international agencies.

Release of Transgenic Varieties in Mexico for Which There are Wild Relatives

The very rapid pace of development of the new technology, mainly in the US and elsewhere, is having an effect on Mexican agriculture. Mexican growers, aware of the developments in the US, fear that if they do not have access to the new varieties they will not be able to compete and maintain their role as suppliers of the American market; therefore, when they learned that several biotechnology companies wanted to test their products in Mexico and eventually release them to Mexican growers they began to apply pressure on the different government agencies responsible for the decision to approve or disapprove such products.

A specific example of this is the deregulation and approval to commercialize the FLAVR SAVR tomato developed by Calgene. Many Mexican growers believed that because this product had previously been released in the US, the Mexican Ministry of Health and the National Committee for Agricultural Biosafety (NCAB) could add nothing more to this decision; however, the fact that Mexico is a centre of diversity for the genus *Lycopersicon* made a significant difference compared with the assessment carried out by the US. So, despite pressure by growers, the Health Ministry and the NCAB conducted a thorough review of this product. Only after several discussions with local experts and an exhaustive analysis of the available data indicated that the release of this product would not pose a significant risk to the public or the environment, were the proper permits issued.

This episode seemed to set a precedent as to how issues were to be handled concerning the release of transgenic material for which Mexico is a centre of origin or diversity. However, the most important crop of this kind in Mexico, maize, proved to be an extremely difficult case and one that required a complete change of strategy.

Transgenic Maize in Mexico

With the importance attributed to maize, it is not surprising that when the first proposal to release transgenic maize by an international seed company was presented to the Mexican government many discussions and consultations with experts were initiated on the different biological and agronomic aspects of this crop. The traits that had been introduced into the maize were virus resistance along with herbicide resistance as a selectable marker. In this particular case, the permit was denied since many of the experts involved had serious reservations about the possible effects of the herbicide resistance trait being passed on to the wild relatives of maize, which were supposed to be found near the proposed trial site.

Experience from this case clearly shows that not all the experts agreed on many of the relevant issues that would have to be addressed to properly evaluate future proposals, including sites where wild relatives could be found, levels of introgression between the wild teosintes and cultivated varieties, and possible effects of different traits on landraces.

There was no doubt that many more companies were interested in the possibility of testing their products in Mexico, and CIMMYT would certainly try to incorporate the new technology into their research programmes, and thus would be planning to conduct field trials with transgenic varieties derived from their research. In addition, national programmes at INIFAP, several agricultural universities and research institutions were very much concerned with field testing transgenic maize, as they were also involved in maize research.

It was soon realized that there was an urgent need to bring together experts on the different areas of maize technology to discuss the issues involved with the release of transgenic maize in Mexico and try to reach conclusions or guidelines as to the best way to proceed. INIFAP and CIMMYT organized a symposium to which they invited several experts involved in different fields of maize agriculture: biology; ecology; biotechnology; and biosafety. From these discussions important recommendations were made, such as to avoid those areas where important populations of wild relatives were clearly established, and to minimize pollen flow through bagging or detasselling and/or through temporal isolation with respect to the wild and/or landrace maize populations (Serratos *et al.*, 1996). After this symposium, applications to conduct field trials with transgenic maize were analysed by the Biosafety Committee taking into consideration many of the recommendations that evolved.

It was clear that since transgenic maize was being deregulated in the US, there was going to be continued pressure from Mexican growers wanting immediate access to this material as had occurred with the transgenic tomato. A further complication was the fact that Mexican field workers in the US usually take seeds of crops with valuable traits with them on their return to Mexico. Maize is a prominent example of this practice since often the field workers or their close relatives will be growing this crop on a regular basis, even if intended for their own use.

This flux of transgenic maize could pose a substantial complication to the regulation of transgenic varieties because, even though the quantity is small, it represents a loss of control. It was soon realized that the largest concern would come as a consequence of Mexico buying a substantial amount of maize every year to satisfy internal demand. This amount increases every year, as the result of the North American Free Trade Agreement (NAFTA), from 2.5 to 4.5 million tonnes between 1994 and 1996, respectively (US Feed Grain Council, 1997). With the US being the only supplier of maize for the Mexican market, this material will shortly be a mixture of transgenic and non-transgenic maize, with the proportion of the former increasing every year as more growers in the US take full advantage of the new material.

This poses two immediate problems for Mexican authorities: (i) the imported commodity, if a mixture of transgenic and non-transgenic material, must be approved by the authorities as fit for human consumption; and (ii) the importation of any transgenic varieties should be approved by the Ministry of Agriculture as not posing a significant risk to agriculture or the environment.

It is conceivable that the data presented to the FDA by different companies involved in the production of transgenic maize could be readily validated by the Ministry of Health, but the Ministry of Agriculture in Mexico must still examine the possibility of gene flow from transgenic varieties into wild relatives and landraces as a consequence of some of this material finding its way as seed into growing areas lacking control.

It is important to mention that the US Government, when considering deregulating transgenic maize e.g. maize line MON 80100 which has *Bt* insect resistance and is glyphosate tolerant, examined the possibility of this material posing a risk to wild relatives, even though this was not a substantial concern for maize being grown in the US, and found no cause for concern. Their conclusion was 'Gene introgression from corn [maize] line MON 80100 into wild or cultivated sexually-compatible plants is unlikely, and such rare events should not increase the weediness potential of resulting progeny or adversely impact biodiversity' (USDA, 1996). Furthermore, improved varieties have been grown commercially in Mexico for many years and commodity maize or seeds introduced legally or illegally into Mexican territory have also found their way into the peasant's maize plot or 'milpas', and presumably have been grown alongside landraces or teosintes.

So what then are the problems or risks perceived by the people in Mexico? Many people would acknowledge that the possibility of transgenic maize becoming a weed or increasing the weediness of wild relatives is extremely small. However, another concern has been the possible development of insects resistant to the *Bt* toxin. As mentioned in the United States Department of Agriculture (USDA) determination document (USDA, 1996), companies such as Monsanto have a strategy for maximizing the utility of transgenic plants and delaying the development of resistance to the *Bt* protein. This strategy includes labelling, producing brochures that include instructions for the proper use of the transgenic lines, incorporating this material into integrated pest management programmes (IPM), monitoring insect populations for *Bt* susceptibility and providing non-transgenic lines as refugia for sensitive insects. These strategies are certainly likely to prevent or reduce the risk of insect resistance within a very short time; however, such a strategy will work only with the cooperation of the growers.

Unfortunately, as has been discussed earlier, transgenic maize will most likely enter Mexican territory in bulk, as a commodity, and some of it will be used as seed. This may be planted in close proximity to landraces, increasing the possibility of gene flow, and almost certainly will be grown without adequate control and proper use regarding IPM, possibly complicating any strategy aimed at delaying resistance in insect populations exposed to maize carrying the *Bt* gene. Experts at the USDA suggest that while, in the US, gene introgression into other maize cultivars via cross pollination is possible, any resulting seeds would most likely not be used for seed (USDA, 1996); this certainly is not the case in Mexico.

Development of new strategies

Since it is not realistic to think that the flow of transgenic maize can be delayed, what may be more useful and practical is an approach which informs people about this new technology and the consequences of misuse. This should be coupled with a close monitoring of the release of the material and its impact on the environment instead of imposing limits on its use.

INIFAP and the Ministry of Plant Health may already have the infrastructure and expertise to provide information to large- and small-scale growers through their many field stations and different extension programmes. Still, it is most likely that they would require the cooperation of universities/research centres and companies to prepare appropriate information and train personnel.

Monitoring transgenic varieties for their impact on the environment or insect populations should be of interest not only to the Mexican government but to companies involved with the production of the new varieties as well as to the academic community. With regard to the *Bt* gene, for example, knowledge obtained from monitoring could confirm the prediction that gene introgression is minimal and does not pose a threat to any programme aiming to avoid emergence of resistance in target insect populations. If it was found that introgression had a significant effect, early steps could be taken to prevent any irreversible effect and new strategies could be developed. Knowledge gained through this experience could be used to further plan introductions of transgenic material in other centres of origin or diversity.

Effect of large-scale release of maize on biotechnology R&D priorities

The main effect of the rapid introduction and release of transgenic maize in Mexico will be of little direct consequence to ongoing research, except research on biosafety which was mainly concerned with maize. It is possible that the first change in biosafety research will be to shift into developing methodology and strategies to monitor the effect of the introduced maize into the environment rather than trying to determine the extent of gene flow or devise ways to minimize gene transfer as decision elements for risk assessment, since the results obtained may not be relevant in a situation where large quantities of transgenic maize are being introduced into the country.

It is likely that INIFAP, the Ministry of Plant Health and some research institutions which already have expertise in monitoring insect populations and assessing resistance or susceptibility may develop efficient methods to test for the spread of transgenes in areas of suspected gene flow. It is clear that biotechnology companies themselves could play a vital role in the development of such methods since they have at their disposal most, if not all, of the information required.

Besides looking for transgenes in the wild teosintes or landraces, insect monitoring is important. In the case where the transgenes are *Bt* genes, it is important to set up an adequate monitoring system which is able to detect any changes in *Bt* susceptibility in insect populations where maize is being intensely cultivated or where gene flow would have been detected.

Conclusion

Commercial development of transgenic varieties in industrialized countries is advancing at an extraordinary pace. In fact, many developing countries which already have regulatory schemes are currently trying to address the possible problems posed by the release of transgenic varieties in their own particular environment. They may not have a chance to decide on their own whether to accept such products because there may be no alternative.

The best strategy may be to assume that gene flow is going to occur and to develop a monitoring system that would allow for the early recognition of any problems arising from the introduced material or that would confirm the presumed safety of it. In any case, those who oppose the release of genetically engineered materials in centres of origin or diversity on the grounds that it may affect the environment through the wild relatives or landraces, as well as those who favour this introduction on the basis that the possibilities of any deleterious effect are 'minimal', should presumably feel more comfortable knowing that a satisfactory monitoring system is in place. With a satisfactory and efficient monitoring system, governments would be in a position to fulfil their role as the guardian of health for their people and environment. Academics will certainly gain more knowledge from monitoring such releases than by just assuming their safety or their risks. This information should also be useful to the biotechnology industry which has the responsibility of delivering products of high quality and compatibility with a healthy environment. Most important, consumers and people living in these countries should feel their natural resources are being valued.

The real risk in this case is to give the impression that certain political or economic situations can easily override all precautions and concerns about this technology, and that developing countries must passively accept such situations. The challenge is to accept the facts and modify the strategies such that the safety of the technology can still be properly assessed under the set of conditions in each country, thus maintaining the capacity to endorse or denounce it.

Acknowledgement and Note

I would like to thank Dr Philip Dale for his comments on the subject and his revision of the manuscript. The opinions expressed in this chapter are personal and do not necessarily represent the opinion of the National Committee for Agricultural Biosafety of Mexico.

References

Serratos, J.A., Wilcox, M.C. and Castillo, F. (1996) Flujo genético entre maíz criollo, maíz mejorado y teocinte: Implicaciones para el maíz transgénico. CIMMYT, Mexico, D.F. (An English version of this document is available from CIMMYT.)
US Feed Grain Council (1997) (http://www.grains.org/policy/nafta.htm).
USDA (1996) *Response to Monsanto Company Petition for Determination of Nonregulated Status for Insect-protected Corn Line MON 80100*. USDA/APHIS/BBEP/ Document 9509301p.det (http://www.aphis.usda.gov/bbep/bp/).

Current Status of Agricultural Biotechnology Research in Indonesia

Achmad M. Fagi and Muhammad Herman

Central Research Institute for Food Crops, Jalan Merdeka 147, Bogor 16111, West Java, Indonesia

Agriculture is of primary importance in the National Development Programme of Indonesia. Rapid global development in biotechnology prompted the Indoncsian Government to set up a national programme in this field. Biotechnology, as a new frontier in agricultural sciences, has opened new avenues for solutions to agricultural problems. The Central Research Institute for Food Crops (CRIFC), of the Agency for Agricultural Research and Development (AARD), in the Indonesia Ministry of Agriculture (MOA), initiated biotechnology research in 1988 with limited numbers of adequately trained research scientists, discontinuity of funding sources and insufficient support equipment. To achieve the goals and objectives of biotechnology research, CRIFC began developing relationships with nationally and internationally funded research programmes.

National research programmes are funded through the State Earning and Expenditure Budget and include projects such as: Agricultural Research Management; Integrated Superior Research; and Partnership Superior Research. International collaborations are with Michigan State University (MSU), through the Agricultural Biotechnology for Sustainable Productivity (ABSP) Project funded by United States Agency for International Development (USAID), The Rockefeller Foundation (RF), the International Rice Research Institute (IRRI) through the Asian Rice Biotechnology Network (ARBN), Australian Center for International Agricultural Research (ACIAR), Japan International Research Center for Agricultural Sciences (JIRCAS) and Japan International Cooperation Agency (JICA). These programmes and collaborations have been established over the past several years and have proved to be effective mechanisms for increasing the generation of capacity building through human resource development, transfer of technology and availability of facilities and equipment.

Major research findings have been achieved in nationally and internationally funded research programmes: (i) nationally funded programmes – from work funded through the Partnership Superior Research project the formula of *Rhizo*-plus has been patented; (ii) internationally funded programmes – a. (ABSP) transgenic maize resistant to stem borer and potato resistant to potato tuber moth were developed, b. (ACIAR) transgenic groundnut resistant to peanut mottle virus was developed, c. (RF) genetic diversity of rice blast and bacterial leaf blight (BLB) were analysed using molecular markers.

The impacts of improved capacity at the Research Institute for Food Crops Biotechnology (RIFCB), founded in 1995, through international interaction in biotechnology research are reflected in several instances: more concern for intellectual property rights (IPR); initiation of biosafety regulation; construction of biosafety containment facilities; and additional budget allocated by the government. The recently drafted Plant Variety Protection Act (PVPA) will aim to protect genetically engineered plants developed and released in Indonesia by public or private sector institutions.

The current research programme is focused on high productivity, high yield stability and high-quality agricultural commodities. Starting this year, RIFCB has the capability to conduct plant transformation studies and to evaluate transgenic plants. RIFCB, with well-trained scientists and well-equipped laboratories, is now ready to enter the global biotechnology community.

Introduction

The Government of Indonesia (GOI) constructed a long-term national development programme (NDP) which has been implemented in a series of five-year development plans called Repelita, initiated in 1969. Agriculture is of primary importance in the NDP. Rapid developments in biotechnology globally have prompted the GOI to set up a national programme in this field. Biotechnology, as a new frontier in agricultural science, has opened new avenues to solve agricultural problems.

Biotechnology is a relatively new research area in Indonesia; however, it is expected to play an increasingly important role in the near future. Biotechnology will be a principal element in the strategy to increase crop productivity where conventional technology alone may not be adequate to meet present and future challenges. However, research in biotechnology is a very costly investment and requires an outstanding cadre of trained and dedicated scientists. Therefore, priority areas of research in this field should be carefully analysed and identified.

Biotechnology will have a major impact on the future of agriculture. It includes both 'traditional' biotechnology, covering technologies used in fermentation, biological control of pests and conventional animal vaccine production, and 'modern' biotechnology, particularly that based on the use of recombinant DNA technology, monoclonal antibodies (MCA), and new cell and tissue culture techniques including novel bioprocessing techniques. For more efficient and effective use of available resources, the agricultural biotechnology research programme will be focused, for the next five years, on the areas where biotechnology will most likely yield new breakthroughs in agricultural development. To date, the greatest achievements have been in medicine and healthcare, followed closely by agriculture.

Wardani and Budianto (1996) list the priority areas for agricultural biotechnology established by the National Research Council (NRC). They point out that priority setting is crucial to optimize the use of scarce resources from both government and international sources, and they identify the criteria for the establishment of priorities for agricultural biotechnology. Biotechnology must: (i) play a significant role in the innovation and development of science and technology; (ii) have a comparative advantage in solving constraints related to agricultural objectives; (iii) improve the capacity and capability of national institutions in the acquisition, development and utilization of science and technology; and (iv) optimize the use of resources through integrated research among institutes and/or among disciplines at the national, regional or international level. This chapter will briefly describe information related to the current research activities and outputs of biotechnology research on food crops and industrial/estate crops.

Historical Background

In 1988, the Japanese government, through the Japan International Cooperation Agency (JICA), assisted the Central Research Institute for Food Crops (CRIFC) in the initial stage of biotechnology research through the establishment of a laboratory and provision of equipment. CRIFC is under the auspices of the Agency for Agricultural Research and Development (AARD) within the Indonesia Ministry of Agriculture (MOA). A Biotechnology Research Division (BRD) was formed under CRIFC, and has permanent government postdoctorial researchers drawn from different specialties such as soil microbiology, plant physiology, plant pathology, entomology and plant breeding, and from different research institutes under CRIFC in order to form a critical mass of trained scientists. To coordinate the research and development (R&D) of agricultural biotechnology, the State Minister of Research and Technology appointed CRIFC and its associated BRD as the Center for National Biotechnology Network dealing with agricultural applications. Since this time, the BRD of CRIFC, with limited numbers of adequately trained and experienced research scientists, discontinuity of funding sources and/or insufficient equipment, has initiated biotechnology research. In the early years, activities concentrated only in the field of soil microbiology. Since 1990, R&D activities related to tissue culture have been conducted, especially for rice. Molecular biology techniques are being developed and laboratory activities were started in 1992. In 1993, the Agriculture Research Management Project 1 (ARMP 1) funded the construction of a molecular biology laboratory, providing facilities and equipment. In 1995, the BRD of CRIFC was merged with the Bogor Research Institute for Food Crops to become the Research Institute for Food Crops Biotechnology (RIFCB), with a mission to conduct research on food crops biotechnology. The chronological events are described in Fig. 4.1. Research activities are now being diversified for other crops such as industrial and estate crops.

To achieve the goals and objectives of the biotechnology research programme, CRIFC began developing relationships and cooperative arrangements with various national institutions including other research institutes, the Inter-University Centers (IUC) of the Bogor Agricultural Institute (IPB), Gadjah Mada University (UGM), Bandung Technology Institute (ITB) and the Agency for Technology Assessment (ATA). CRIFC developed both nationally and internationally funded research programmes. National research programmes are funded through the State Earning and Expenditure Budget and include various projects such as: ARMP; Integrated Superior Research (1993–1997); and Partnership Superior Research (1995–1997). International collaborations, including partnerships with Michigan State University (MSU) through the Agricultural Biotechnology for Sustainable Productivity (ABSP) Project (funded by United States Agency for International Development (USAID) from 1993 to 1998), the Rockefeller Foundation (RF) (from 1990 to 1996), the International Rice Research Institute (IRRI) through the Asian Rice Biotechnology Network (ARBN) (funded by Asian Development Bank (ADB) from 1993 to 1996), the Australian Center for International Agricultural Research (ACIAR) (from 1990 to 1997), and the Japan International Research Center for Agricultural Sciences (JIRCAS) (from 1992 to 1996) has been established over the past several years. Such collaborations have proved to be an effective mechanism for increasing the generation of human and institutional capacity through human resource development, transfer of technology and improved availability of facilities and equipment.

Research projects

National

INTEGRATED SUPERIOR RESEARCH (ISR). Research topics: (i) improvement of rice varieties for drought and aluminium (Al) toxicity tolerance and resistance to blast disease; (ii) improvement of rice varieties in tidal swamp, acid and peat soils; (iii) improvement of lowland rice varieties for resistance to biotype 3 of brown plant hopper and strain IV of bacterial leaf blight (BLB); (iv) application of system analysis and modelling for the development of marginal soils on the southern area of west and central Java; (v) improvement of soybean varieties for tolerance to acid soils; and (vi) use of molecular markers for rice tolerance to iron (Fe) toxicity.

PARTNERSHIP SUPERIOR RESEARCH (PSR). Research topics: development of *Rhizo*-plus (a microbial fertilizer which contains *Bradyrhizobium japonicum* and phosphate solubilizing bacteria) for increasing production and fertilizer

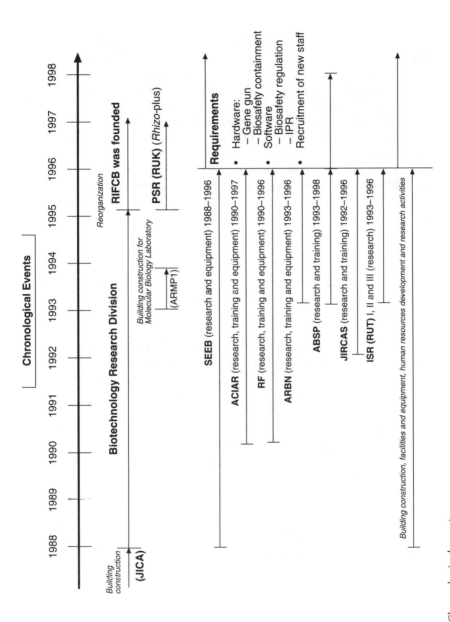

Fig. 4.1. Chronological events.

efficiency to support a sustainable soybean production system. This project is a collaboration between public and private sector institutions, RIFCB and Hobson Interbuana Indonesia Co.

International

ABSP. There were three research projects conducted in the USA: (i) genetic engineering of maize resistant to stem borer using *Bacillus thuringiensis* toxin genes, conducted at ICI Seed Company (now Garst Seed Co.), Slater, Iowa; (ii) genetic engineering of potato resistant to potato tuber moth using *Bt* genes, conducted at MSU; and (iii) genetic engineering of sweetpotato resistant to weevil using proteinase inhibitor genes, conducted at MSU and Monsanto Company. ABSP also funded a small grants programme for four research projects conducted in Indonesia. These projects were: (i) a regeneration study of Indonesian sweetpotatoes (*Ipomoea batatas* L.); (ii) the use of a immunoassay probe for detection and monitoring of *Phytophthora* spp., the causal agent of pod rot disease of *Theobroma cacao*; (iii) development of a micropropagation method for ginger using a bioreactor system; and (iv) the development of citrus vein phloem degeneration (CVPD)-free citrus seedlings from protoplast fusion and embryogenic callus culture.

RF AND ARBN. Research projects under RF and ARBN include: (i) determination of the genetic variability of the rice blast pathogen (*Pyricularia oryzae*) using molecular markers and the testing for pathogen virulence using a differential host; (ii) determination of the genetic variability of the BLB pathogen (*Xanthomonas oryzae* pv. *oryzae*) using molecular markers and testing for pathogen virulence using a differential host (local variety and isogenic lines); (iii) gene tagging for BLB resistance on Cisadane rice using molecular markers, crossing F_1–F_7 populations; (iv) improvement of resistance for BLB on the rice IR64 variety, development of a resistance test for BLB on F_7 plants by crossing and backcrossing populations, and resistant test for BLB on back cross 7 (BC_7) population; (v) development of molecular markers to screen for drought tolerance and yield potential of javanica and upland rice; and (vi) development of an efficient rice regeneration system using anther culture.

ACIAR. Research projects funded by ACIAR include: (i) a study on the biological nitrogen fixation of *Bradyrhizobium* Indonesian strain on soybean; (ii) the genetic engineering of groundnut resistant to peanut stripe virus; and (iii) a study on strain variability of bacterial wilt (*Pseudomonas solanacearum*) using molecular markers.

Human resource development

Biotechnology is a relatively new research area in Indonesia, but it is expected to play a more important role as more well-qualified research staff and additional resources become available. Due to the increased emphasis on biotechnology research, several senior scientists need to gain more experience by working together with international scientists in developed countries. Six senior scientists developed improved skills by working as postdoctoral scientists in genetic engineering, molecular biology and tissue culture in advanced laboratories such as MSU and ICI Seed Company (now Garst Seed Co.) through the ABSP project for periods of 3 months to 2 years. Additional training was conducted at IRRI, Cornell University and The Scripps Research Institute for 3 months to 1.5 years, funded by the RF. JIRCAS also sponsored two senior scientists for 1–2 years postdoctoral study in microbiology, tissue culture and molecular biology at its research centre in Okinawa, Japan. ABSP, funded by USAID, has also supported several senior scientists in an intensive programme at several advanced laboratories and universities in the US on biosafety and intellectual property rights (IPR) for 2–10 weeks. In addition to research projects, both the RF and ARBN sponsored several junior and senior scientist for short- and long-term training in The Philippines and/or the US. Ten junior and senior scientists were sent to the University of Queensland, Australia for 3–6 months training in soil microbiology, genetic engineering and molecular pathology through ACIAR's programme.

Major Accomplishments

Food crops

Major research accomplishments have been achieved through both nationally and internationally funded research programmes. Through the nationally funded programmes, the formula for *Rhizo*-plus was developed and patented. This work was supported by the Partnership Superior Research (PSR) project. Through internationally funded programmes, transgenic maize resistant to stem borer and potato varieties resistant to potato tuber moth were developed at ICI Seed Company (now Garst Seed Co.) and MSU, respectively, through the ABSP project. Through support from the RF, an anther culture technique to shorten breeding time was developed, molecular markers for quantitative traits such as yield and drought tolerance were developed, and molecular markers to study the population structure of blast and BLB were developed. Through the ACIAR programme, transgenic groundnuts with resistance to peanut mottle virus were developed.

Industrial crops

Biotechnology research on industrial crops was initiated in 1986, but is still restricted to tissue culture due to the limitations of research facilities and lack of experienced staff. Plant propagation through tissue culture has been successfully applied to non-woody plants, including: *Angelica acutiloba*, *Mentha* spp., *Pelargonium* spp. and *Chrysanthemum cinerariifolium* (introduced species); ginger, cardamom, vanilla, pepper, abaca, patchouli and rami (widely cultivated by smallholders); and *Alexya stellata*, *Rouwolfia serventina*, *Curcuma petiolata*, *Pimpinella pruacan*, *Rutta angustifolia* and *Gynura procumbens* (endangered species of medicinal crops). Seedlings from tissue culture of ginger, cardamom, abaca, vanilla and pepper have been tested for their growth performance and productivity under field conditions. For perennial crops, tissue culture techniques have been investigated in the propagation of clove, cananga, cashew, cinnamon, kapok, gnettum and tamarind.

Estate crops

The estate sub-sector has a very important role in earning national revenue and supporting agroindustries. Production costs have increased due to the increased use of marginal lands by plantation growers, increased cost of waste management, termination of subsidies for chemical fertilizers and increased cost of energy. Marketing of estate products is becoming difficult due to stiffer international competition in the global market system. Quality requirements of the buyer have evolved and include low free fatty acids in palm oil and improved plasticity and viscosity properties of rubber. Biotechnology is an essential tool in addressing these challenges. During the last 3 years, research activities have yielded significant results which could be adopted by the estate industry. Achievements include the development of: (i) technologies for *in vitro* cloning of oil palm, rubber, coffee, cocoa and tea; (ii) techniques for early diagnosis of tapping panel dryness (TPD) incidence in rubber; (iii) techniques for early detection of *Ganoderma* root infection in oil palm; (iv) biopulping technology for oil palm empty bunches; (v) vesicular arbuscular (VA) mycorrhiza production for growth enhancement of micropropagated oil palm in acid soils; (vi) methods for reducing cocoa acidity through improved aeration during fermentation; and (vii) flavour improvements in cocoa beans through combined pod storage and fermentation.

Impacts

The impacts of improved capacity in biotechnology research at RIFCB through international interactions are reflected in several achievements such as improved awareness of IPR, development of biosafety regulations, construction of biosafety containment facilities, additional funding allocations by the government and extension of international collaborations with RF and MSU. Unlike conventional agricultural research, where the public sector was the major stakeholder, the private sector is the major investor in biotechnology due to the high cost of product development. Consideration must be given to IPR and biosafety, and appropriate legal and regulatory documents are being prepared.

Indonesia presently has a patent law, but no plant breeders' rights (PBR) or plant variety protection (PVP). Drafting teams for the Plant Variety Protection Act (PVPA) and biosafety regulation have been separately appointed and formed by the Director General of AARD. The draft of the PVPA consist of 78 articles, divided into 12 chapters: general terms; scope of PVP; application for PVP; inspection/examination for PVP; transfer of PVP; cancellation of PVP; fees and charges; management of PVP; claim of rights; criminal proceedings; investigation; and a closing chapter. The final academic draft of the PVPA has been submitted to the government. It is anticipated that through the PVPA, any genetically engineered plant developed by the private sector and released in Indonesia will be protected.

A biosafety workshop was conducted at the Safari Garden Hotel, Cisarua, West Java, Indonesia on May 28–29, 1996, where the AARD biosafety writing team decided to combine two versions of the AARD's biosafety draft. The guidelines for *Planned Introductions into the Environment of Organisms Genetically Modified by Recombinant DNA Techniques* and *Guidelines for Biosafety and Bioethics for Biotechnology Products of Animal Husbandry* from the Directorate General of Livestock were combined. The combined biosafety draft was renamed the *Guidelines for Biosafety of Agriculture Biotechnology Products Modified through Genetic Engineering*. This draft was reviewed by Dr Faisal Kasryno, the Director General of AARD, and the Bureau of Law at the Ministry of Agriculture. The biosafety guidelines were released by Ministerial decree on September 2, 1997 (No. 856/Kpts/Hk.330/9/1997).

ARMP II funded the construction of biosafety containment facilities. The site of the containment facility is close to the Molecular Biology Division building at RIFCB. The facility is being constructed in two phases. The first phase of construction, including a head house (preparation building) and three greenhouses, was completed in early March 1997. The second phase of construction will include an additional three greenhouses and commenced in the middle of 1997.

Research Programme

According to Wardani and Budianto (1996), Indonesia is the fourth largest country in the world based on its population, but it is still a developing country economically and in relation to science and technology capacity. They identified three major constraints to research productivity: lack of human resources, especially research scientists; lack of research facilities, particularly well-equipped laboratories; and underdeveloped research management skills to provide an environment that is conducive to innovative research in support of national development goals. An additional deficiency, which could be added to this list, is the poor development of information technologies that would provide access to essential information resources (genomic databases, etc.) and allow for communication and collaboration with advanced research programmes throughout the world. Kasryno (1996) outlines general policies for biotechnology research in Indonesia. One of the five areas of strategically important agricultural research in Indonesia is the improvement of the genetic potential of plants and animals to generate efficient and environmentally friendly technologies for sustainable agricultural development. In line with general agricultural research policies, the AARD will implement the following policies in agricultural research and development: (i) improvement of science and technology in agriculture; (ii) improvement of research–extension linkages; (iii) consistent application of the participatory approach; (iv) rationalization and expansion of the mandated research tasks; and (v) establishment of Assessment Institutes for Agricultural Technology at the regional level. Major programme areas include (i) agricultural resources; (ii) improvement of genetic potential; (iii) development and management of agribusiness systems; (iv) socio-economics and policy; (v) communication and distribution of research results; and (vi) institutional and human resource development. The improvement of genetic material will involve the application of both conventional and modern approaches, with special emphasis on the application of biotechno-logical approaches to germplasm conservation and breeding. The research programmes will be focused on achieving high productivity, high yield stability and improved product quality of agricultural commodities.

Food crops

A great number of problems confront efforts to increase the production of various commodities. These constraints can be physical, biological or socio-economic in nature. Although agricultural production in some parts of Indonesia has benefited from the use of improved technologies, in certain growing areas low yields still persist for several reasons, including wide environmental diversity. This situation lessens the value of general recommendations for varieties and management practices and requires a location-specific approach. Breeding new varieties is extremely complex because there is a need for a range of varieties adapted to widely varying agricultural production systems. All of these varieties should also be resistant to pests and diseases and tolerant of environmental stresses.

The intensity of agricultural production, particularly in rice, leads to the build-up of pests and diseases, which becomes a bottleneck to higher productivity and causes yield instability. These pests and diseases include viruses, bacteria, fungi, insect pests such as brown plant hopper and rice stem borer, and rodents. Breeders face difficulties in obtaining sources of resistance to insect pests, diseases and environmental stresses through conventional breeding. It is anticipated that techniques in molecular biology will be used to isolate genes, such as viral coat-protein genes, which have proven effective against various pests and diseases. Maintaining food self-sufficiency and improving the status of human nutrition are important challenges that need special efforts to be successful. Enhanced agricultural development is facing several barriers in Indonesia, such as a high population growth rate (1.9%), reduction of fertile agricultural land, outbreaks of insect pests and diseases, declining productivity and natural disasters such as floods and drought. The research strategy to increase and sustain food crop productivity will have to: raise yield potential; incorporate durable resistance to disease and insects and tolerance to abiotic stresses to provide greater yield stability; and increase productivity in the less favourable production environment. In order to use biotechnology effectively and efficiently, the following research programmes will be implemented to address major constraints:

- Explore, collect, characterize and document germplasm; identify economic traits and manage databases; and improve germplasm for breeding purposes.
- Identify molecular markers and map genes controlling important traits for marker-aided selection; improve tissue culture methods for efficient regeneration and breeding purposes; and develop efficient transformation methods.
- Identify effective and efficient native microorganisms for use as biofertilizers; develop biofertilizers and associated production technology; and use recombinant DNA technology to increase the effectiveness and efficiency of microbes for bioconversion, bioprocessing and biofertilizer.
- Identify secondary metabolites; develop *in vitro* rapid tests for abiotic and biotic stress tolerance; scale-up micropropagation; and develop cellular biology methods to increase genetic variation.
- Develop biopesticides and associated production technology, and develop tools/kits for early detection of pathogenic plant viruses and bacteria.

Industrial crops

Research in industrial crops is directed toward increasing farming efficiency through improved productivity. Improved planting materials are essential to this approach but many constraints exist. In perennial crops such as coconut, clove and cashew, conventional varietal improvement is slow. Crops such as pepper, vanilla, ginger and some medicinal crops are generally propagated vegetatively, resulting in low genetic variability. Wild relatives of the crops are poorly explored. The development of biotechnology has provided possibilities for overcoming certain time constraints in crop improvement, leading to increased planting material for production of industrial crops. Breeding times could be shortened and high quality planting material could be produced in a relativity short time. Commodities and their associated research priorities (delineated below) show a focus on varietal improvement through cell and tissue culture, micropropagation and *in vitro* germplasm conservation.

- Pepper (*Piper nigrum*): molecular biological approach toward the engineering of pepper resistant to *Phytophthora capcici*, and the use of antagonistic micro-organisms for biological control of nematodes in pepper.
- Vanilla (*Vanilla planifolia*): somaclonal variation to increase genetic variability in vanilla; culture and fusion of vanilla protoplasts to produce hybrid plantlets; embryo rescue of cultivated vanilla crossed with wild vanilla to produce hybrid plantlets; and molecular approaches toward the engineering of vanilla resistant to *Phytophthora oxysporum*.
- Ginger *(Zingiber officinale):* somaclonal variation to increase genetic variability of ginger, and development of seed ginger production system through tissue culture.
- Cashew (*Anacardium occidentale*): plant propagation through tissue culture to develop methods of micropropagation and micrografting.

Estate crops

In the long run, the research programme for estate crops aims to develop technologies and products which support the sustainable development of the plantation industry. Through improvements in productivity, product quality and management efficiency, competitiveness of the estate agroindustry will be enhanced. Priority research comprises activities in plant, microbial and process biotechnology, encompassing oil palm, rubber, coffee, cocoa, tea and coconut as highlighted below.

- Plant Biotechnology: micropropagation system for rubber, oil palm, coffee, cocoa and tea; molecular marker-based mapping of genes associated with valuable traits; genetic engineering of cocoa for resistance to pod borer; and genetic engineering for drought resistance.
- Microbial Biotechnology: cattle feed from rubber processing waste; paper from oil palm empty bunches; development of biopesticides for insects and fungi; nematode-resistant coffee; and diagnostic kits for root diseases.
- Process Biotechnology: upscaling of all of the technologies being developed in the laboratory for mass application, and mass micropropagation of coffee and tea using bioreactors.

Summary Points

- Indonesia's research programme is focused on increasing productivity, improving yield stability and improving product quality of agricultural commodities.
- Beginning in 1997, RIFCB has developed the capacity to conduct plant transformation and to evaluate transgenic plants.
- RIFCB has developed trained scientists and well-equipped laboratories ready to participate in the global biotechnology community.

- RIFCB's biotechnology research programme in food crops seeks to establish and strengthen close collaborations with other institutions for research programmes, networking, training and other novel arrangements for the development and application of biotechnology in Indonesia.
- Both within and without Indonesia, scientific exchanges and participation in conferences and scientific meetings, particularly related to food crop improvement, are encouraged and solicited.

Acknowledgements

We are very grateful to USAID, JICA, RF, ADB, ACIAR and JIRCAS for providing support in strengthening and promoting research activities in agricultural biotechnology to attain sustainable food production systems in Indonesia.

References and Further Reading

Cohen, J.I. (1994) *Biotechnology priorities, planning, and policies. A Framework for Decision Making. A Biotechnology Research Management Study.* ISNAR Research Report No. 6. International Service for National Agricultural Research, The Hague, The Netherlands.

James, C. and Krattiger, A.F. (1996) *Global Review of the Field Testing and Commercialization of Transgenic Plants: 1986 to 1995.* The First Decade of Crop Biotechnology, ISAAA, Ithaca, New York.

Kasryno, F. (1996) Current status of agricultural biotechnology research in Indonesia. In: Darussamin, A., Kompiang, I.P., Moeljopawiro, S., Prasadja, I. and Herman, M. (eds) *Proceedings of the Second Conference on Agricultural Biotechnology.* Vol 1, AARD, Ministry of Agriculture, Jakarta, Republic of Indonesia, pp. 1–12.

Komen, J. and Persley, G. (1993) *Agricultural Biotechnology in Developing Countries: a Cross-country Review.* ISNAR Research Report 2, International Service for National Agricultural Research, The Hague, The Netherlands.

Manwan, I., Brotonegoro, S. and Moeljopawiro, S. (1996) Needs for biosafety guidelines for enhancing agricultural biotechnology development. In: Darussamin, A., Kompiang, I.P., Moeljopawiro, S., Prasadja, I. and Herman, M. (eds) *Proceedings of the Second Conference on Agricultural Biotechnology.* Vol 1, AARD, Ministry of Agriculture, Republic of Indonesia, pp. 131–150.

Moeljopawiro, S. and Manwan, I. (1995) Agricultural biotechnology in Indonesia: New approach, innovation and challenges. In: Altman, D.W. and Watanabe, K.N. (eds) *Plant Biotechnology Transfer to Developing Countries.* R.G. Landes Co., Austin, Texas, pp. 97–116.

Soendoro, T. (1996) Human resources development and training in agricultural biotechnology research activities. In: Darussamin, A., Kompiang, I.P., Moeljopawiro, S., Prasadja, I. and Herman, M. (eds) *Proceedings of the Second Conference on Agricultural Biotechnology.* Vol 1, AARD, Ministry of Agriculture, Republic of Indonesia, pp. 77–80.

Wardani, A. and Budianto, J. (1996) Priority setting and program development in
    biotechnology: a national perspective for Indonesia. In: Darussamin, A., Kompiang,
    I.P., Moeljopawiro, S., Prasadja, I. and Herman, M. (eds) *Proceedings of the Second
    Conference on Agricultural Biotechnology*. Vol 1, AARD, Ministry of Agriculture,
    Republic of Indonesia, pp. 109–116.

Agricultural Needs in Sub-Saharan Africa: the Role of Biotechnology

Cyrus G. Ndiritu and John S. Wafula

Kenya Agricultural Research Institute (KARI), PO Box 57811, Nairobi, Kenya

The countries of sub-Saharan Africa (SSA) depend largely on agriculture for their economic prosperity and the welfare of the people. The current agricultural production efforts and strategies, however, are unable to cope with the demands for food of the rising population in the region. Research and development (R&D) incorporating innovative technologies is urgently needed if the decline in agricultural production in SSA is to be reversed.

Most countries in the region consider biotechnology promising in contributing to increased production of food and other agricultural commodities without relying on the use of rare but very costly agricultural inputs. However, the application of biotechnology in SSA has been very slow to develop. Many countries have ventured into tissue culture application but very few have established programmes in molecular-based and genetic engineering technologies.

Challenges to the effective and useful development and application of biotechnology in SSA are many and include: lack of policies enabling biotechnology development; lack of expertise in biotechnology-related disciplines; an environment not conducive to the advancement of biotechnology; and a lack of information and biosafety regulatory structures. The development of links for technology transfer encompassing capacity building, establishing mechanisms for technology identification and transfer, and developing a safe environment for biotechnology research and application are crucial for the countries of SSA if they are to reap the benefits of modern biotechnology.

Challenges Facing Economic Development in Sub-Saharan Africa

The four major challenges to development facing sub-Saharan Africa (SSA) today are widespread poverty, worsening economic development, population explosion and declining agricultural growth.

Poverty and food security

It is estimated that some 24% of the total population in the developing world will be living below the officially defined poverty line by the year 2050. Today, over 180 million people from SSA live below the poverty line and the number is expected to exceed 300 million people by the year 2020. The World Bank estimates that one of every two Africans, roughly 250 million people, and 1.3 billion people worldwide subsist on a per capita income of less than US$1 per day. Food insecurity in SSA affects health and nutrition, resulting in poor maternal health, high malnutrition in children and poor productivity. These conditions exacerbate the hunger problem.

The problems of poverty, malnutrition, undernutrition and hunger worsen every year in many parts of the developing world. Today, about 200 million people in Africa are malnourished; a number that may increase to over 350 million in the next 25 years. Overcoming poverty in Africa clearly calls for the development of strategies that promote increasing food production and allow for its even distribution. Such measures should, at the same time, ensure the sustainability of the basic natural resources upon which agricultural production depends.

Economic crisis in SSA

Poverty is linked to poor national economic performance and unequal distribution of income. This often leads to disempowerment of people who then become less active in development issues. In SSA, the average rate of economic growth stands at a paltry 3.3% while the gross domestic product (GDP) has declined drastically in recent years relative to that realized in the 1960s. Studies indicate that 31 of the 44 SSA countries fall under the heading of 'least developed countries' (LDCs), and 32 of 48 low-income countries (with GDP per capita less than $500) are found in SSA (FAO, 1994). External debt in the region rose from $48 billion to $423 billion during the 1980s, and now between 55% and 60% of its rural population is considered to live in absolute poverty. The prospects for the next century are not bright.

It is now clear that the root cause of slow development, poverty and hunger which are widespread in SSA is the decline in agriculture arising principally from high population levels, worsening land degradation and inadequate application of new and emerging agricultural technologies.

Challenges to agricultural development

Agriculture remains the major driving force in the development of the economies of SSA. Its role in these countries is to contribute to the overall national goals of food security and poverty alleviation. Its other stated goals are to facilitate provision of employment, increase foreign exchange earnings through export promotion and import substitution, and conserve natural resources. Table 5.1 shows the contribution of agriculture to the national economies of some countries in SSA.

While world food production has continued to increase over the years (FAO, 1986), a number of recent studies have shown that the per capita food production in SSA has declined by 20% over the period 1961–1986, and this trend continues. This adverse and deteriorating performance of agricultural growth arises from a number of reasons, notably the high population growth rate, increasing land degradation, narrow technological base and inadequate policy support.

Agriculture and population explosion

The world population is expected to increase from 5.5 billion people in 1994, to 8.5 billion by the year 2025. In Africa, the population was 630 million people in 1990, and is expected to increase to 1.4 billion people by the year 2020, i.e. it will more than double in the short span of 30 years. The threat of continuing population expansion in SSA is significant in that the region's food production will have to increase by at least 4% per year to meet demand. Unfortunately, agricultural performance is declining and is unable to cope with the sharp increases in food demand. In Kenya, for example, the total population is expected to rise to 36.9 million people by the year 2010, yet the agricultural growth rates have slowed down considerably in the recent past (3.6% in the 1970s, 3.4% in the 1980s and −1.4% by the early 1990s). The pressure put on productive land by an exploding population has led to degradation of land

Table 5.1. Contribution of agriculture to gross domestic product (GDP) of some sub-Saharan African countries. Source: FAO (1994).

Country	% Employment in agriculture	% Contribution to export	% Contribution to GDP
Burkina Faso	80	60	30
Côte d'Ivoire	70	80	n/a
Ethiopia	85	80	40
Kenya	75	60	30
Mali	80	n/a	46
Nigeria	70	n/a	30
Tanzania	90	80	55
Zimbabwe	26	40	13

n/a: not applicable.

resources, reduction in soil fertility and a decline in the productive capacity of the land. Table 5.2 illustrates the current status of population and land availability for some countries of SSA.

Agriculture and improved technological base
The medium- and long-term solution to halt or even reverse the poor economic performance lies partly in the transformation of SSA agriculture through the application of technologies which aim to produce more and at the same time conserve natural resources.

There are two principal ways of increasing agricultural output: (i) expansion of land under cultivation; or (ii) intensification on land already under cultivation. The first scenario is not feasible due to the limited available land. That leaves long-term increases per unit area as the only solution to increasing food

Table 5.2. Population and land resource of sub-Saharan Africa by sub-region. Source: FAO (1994).

Sub-region	Total population (million people)	Total land area (million ha)	Total agricultural land (million ha)
Sudan Sahelian Burkina Faso, Chad, Mali, Cape Verde, Djibouti, Gambia, Niger, Senegal, Somalia, Sudan	81.20	744.40	282.40
Central and Western Benin, Cameroon, Congo, CAR*, Côte d'Ivoire, Eq. Guinea, Gabon, Ghana, Guinea Bissau, Liberia, Nigeria, São Tomé, Sierra Leone, Togo, Zaire	216.00	595.50	169.40
Eastern Mountains Burundi, Comoros, Ethiopia, Kenya, Madagascar, Mauritius, Rwanda, Seychelles, Uganda	127.60	257.10	96.80
Southern Angola, Botswana, Lesotho, Malawi, Mozambique, Namibia, Swaziland, Tanzania, Zambia, Zimbabwe	90.42	571.00	247.20
Total	515.22	2168.00	795.80

*CAR = Central African Republic.

production and thus reducing hunger and poverty. This can only be achieved through more 'science-based' as opposed to 'traditional' agriculture.

The contribution of research to the development of agriculture is mainly through technological improvement and its application for intensified productivity. Traditional methods of research continue to contribute to the development of high-yielding crop varieties and livestock breeds in SSA but the processes involved are extremely tenuous and expensive, taking well over 15 years to successfully breed and release a variety and undertake on-farm verifications and multiplication trials. The accelerating demand for food exerts immense pressure for increased production at affordable costs. Africa has to exploit more vigorously new frontiers in technological know-how to complement the more traditional approaches if she is to achieve rapidly increasing agricultural productivity.

There is now a body of evidence showing that the new technological frontiers present unique and highly beneficial means of significantly and positively impacting agricultural productivity worldwide – possibly to a greater extent in the developing countries of SSA. The two major technologies now available which can be employed in different ecosystems to increase production at low costs while protecting the environment include physico-biology and biotechnology.

Physico-biology encompasses space science and radiobiology. Space science can facilitate technology development through the use of geographical information systems (GIS) which can provide weather and climate predictions, early warning systems, population projections and land resource surveys including crop suitability applications for specific areas. Radiobiology may help provide varieties which require little water through the application of natural and/or induced mutations followed by selection of resistant survivors that can be propagated for multiplication. This science may also lead to the development of pest- and disease-resistant food crops.

Biotechnology (which may be more powerful than physico-biology), especially genetic engineering, creates new frontiers of knowledge in plant and animal science. This knowledge is providing tools to improve animal and plant productivity. Use of microorganisms, and cell and tissue culture in biochemical processes and in microbiology has brought desired effects both in food production and productivity as well as in industrial applications.

The Role of Biotechnology in Agriculture in SSA

Hvoslet-Eide and Rognli (1996) have elaborated on the significance of biotechnology for developing countries using the Asian agricultural development perspective as an example. With more than 50% of the world's population and less than 25% of the world's arable land, the only way Asia, particularly countries such as China and India, has been able to feed its enormous population has been through continuous improvements in crop yields (Sasson and

DaSilva, 1994). China has conducted the most extensive field trials of genetically modified organisms of any developing country to date (Altman, 1993).

The challenges to agriculture arising from a limited arable land-base, severe disease and pest challenges, frequent droughts and land degradation are high in SSA. Most countries in the region must aim to increase their agricultural productivity through higher yields per unit area of land. Biotechnology clearly presents opportunities to produce more food and other agricultural commodities using less land and water resources without the adverse ecological effects associated with high-input agriculture. Biotechnology offers possibilities for reducing over-reliance on inorganic fertilizers and chemical pesticides by incorporating new traits through genetic engineering; for example, nitrogen fixing capability in the plant itself or alteration of genes that confer disease resistance thus potentially protecting crops more effectively than pesticide spraying.

Although many agricultural systems throughout SSA are striving to adopt new and better technologies for more efficient food production for self-sufficiency and export, the uptake and utilization of biotechnology to increase agricultural outputs has been strikingly slow. Different countries are today at different stages of developing and applying biotechnology. Only very few have made tangible efforts in moving into the early stages of its application.

There are many factors which have contributed to the slow integration of biotechnology in the agricultural systems of the SSA countries. Apart from working within constrained funding circumstances, agricultural research in SSA is mainly in the public domain and has to be accomplished within an under-developed infrastructure. Modern biotechnology is a knowledge-intensive, high-cost undertaking. There is a need, therefore, not only to establish policies that promote the development and use of biotechnology but also to define priorities for biotechnology within the national agricultural research objectives. This will ensure that biotechnology research and development (R&D) provides impacts at the production level and that it is supported within the overall national investment for agricultural research.

It is with these considerations that most countries in SSA have tended to venture into biotechnology application mainly at the level of tissue culture/micro-propagation, particularly for commercial seed supply. Thus Burundi, Cameroon, Ethiopia, Kenya, Nigeria, Zimbabwe and Uganda have initiated *in vitro* commercial micropropagation of rice, maize, banana, sorghum, yams, cassava, potato, pyrethrum, cocoa, strawberry, sugarcane, ornamentals, sweetpotato and various tree species in their national agricultural research institutions directed at the multiplication and distribution of clean seed materials (Wafula, 1995).

The application of biotechnology techniques such as molecular markers for selection of desirable characters and identification of pests, pathogen races and biotypes is greatly underdeveloped in SSA. Such applications are best nested within clearly focused and functional breeding programmes which at the moment are found in only a few SSA countries like South Africa, Kenya and Zimbabwe where they are focused on drought, salinity and pest resistance mechanisms in cereals.

The development of the capacity to integrate genetic engineering technology, encompassing transformation and regeneration of crops, in agricultural research and development in SSA is almost non-existent. Few countries in the whole of Africa have initiated research in genetic engineering, the most advanced cases being the Egyptian and South African transgenic crop development in potato, tomato, cotton, faba bean and maize (Wafula and Ndiritu, 1996). These projects target pests, herbicides, and fungal and viral disease resistance (Table 5.3). Countries such as Kenya, Zimbabwe and Uganda seem poised to follow as they have initiated investments in specific gene technology projects, and in policy development, capacity building and technology transfer programmes in biotechnology as well.

Challenge to the Development and Use of Biotechnology in Sub-Saharan Africa

The policy environment

The goal of agricultural research in Africa is to address the immediate needs of technology development in order to reverse declining food security. Africa is very poor and challenges to the development and effective use of biotechnology are engrained not only in financial limitations but also in policy, national capacities, information access and the regulatory environment.

The majority of countries in SSA do not have a clearly articulated policy and strategy for developing and integrating biotechnology into their national agricultural research systems, but rather operate fragmented and uncoordinated research activities which only lead to unnecessary duplication and waste. Only a few countries like Zimbabwe, Kenya, Uganda and Nigeria have formulated, or are at an advanced stage of developing, policies for defining priorities for biotechnology within the context of their national agricultural research objectives and investment portfolio.

Table 5.3. Genetically engineered crops and microorganisms in Africa. Source: Wafula and Ndiritu (1996).

Country	Transgenics	Characteristic
Egypt	Cotton	Stress and insect resistance
	Maize	Insect and fungal resistance
	Faba beans	Stress and virus resistance
Kenya	Sweetpotato	Virus resistance
	Capriprox-RVF	Recombinant vaccine
South Africa	Potato, tomato and tobacco	Virus and insect resistance
Zimbabwe	Cotton and tobacco	Insect resistance*

*Only field testing of imports.

Without visible government commitment there is little hope for biotechnology development in SSA. An environment must be created in which R&D can flourish by focusing resources, through policy and priority definition, and by providing incentives for the development of a successful biotechnology industry. A biotechnology policy and strategy is needed to give direction to and determine priorities for promotion of biotechnology R&D within national institutions and programmes. In this manner, resources can be mobilized and allocated to areas of greater potential impact. Policy direction in the context of priority setting for biotechnology presents a means of focusing on the end-users and their recognized agricultural needs. Formulating biotechnology research activities or programmes should follow priority setting.

Information and awareness

Technology has driven changes in agriculture towards more effective and efficient production practices. Agricultural systems throughout SSA are striving to adopt new and better technologies that enable them to become more efficient in producing adequate food for self-sufficiency and surpluses for the marketplace. That biotechnology has not taken hold and provided immediate results for agricultural improvement in SSA emphasizes the gap that exists in knowledge of new technologies between developed and developing countries. Because of unreliable and inadequate communication, countries of SSA have received limited up-to-date information on the importance of biotechnology, its current and evolving tools, intellectual property rights and the biotechnology regulatory environment. It is important that awareness is created and aimed at familiarizing the public, particularly policy makers, researchers, extension workers, farmers and the general public, with the usefulness of biotechnology, advances in biotechnology and its related disciplines, including the pertinent ethical issues.

Human resource and infrastructure capacity

Considering the urgent need for new sustainable technologies in SSA and the scarce resources available, the development of agricultural biotechnology should be harnessed and directed towards productivity, sustainability and equity of agricultural products. The human resource development dimension poses one of the most important challenges to the application of biotechnology and, for SSA, this is a seriously limiting resource.

The total number of agricultural scientists in SSA in 1994 was 7594, of whom 1739 (23%) and 2530 (33%) had PhD and masters degrees respectively (Table 5.4; FAO 1994). Most of the countries in the region have yet to attain the level of staffing necessary to conduct effective agricultural research. At present, these countries have grossly inadequate or no expertise in the disciplines of modern biotechnology.

Table 5.4. Total number of National Agricultural Research Systems (NARS) scientists in SSA and percentage at PhD and MSc level. Source: FAO, 1994.

African region	Total scientists			% of total		
	PhD	MSc	Total	PhD	MSc	PhD+MSc
Sahelian	348	284	1444	24	20	44
Central/West	1155	1216	3642	32	33	65
Eastern	136	629	1538	9	41	50
Southern	100	401	970	10	41	52
Total	1739	2530	7594	23	33	56

Research objectives in biotechnology cannot be achieved without the development of scientific knowledge in advanced biotechnology techniques. Manpower development programmes are required to ensure a strong base in advanced molecular genetics, breeding, pathology, entomology and their related fields. University courses, refresher and postdoctoral courses need to be enriched to impart knowledge on up-to-date biotechnology techniques to scientists while providing the flexibility to adapt to new challenges.

Biotechnology R&D has a high investment cost and demands considerable inputs in physical facilities, operations and capacity building. A laboratory capacity audit in SSA would show large deficiencies in the basic laboratory equipment and operational funds required to support and run a medium-level biotechnology laboratory. It is important to take stock of available equipment in existing laboratories and to compare this to the level needed. Strategies for acquisition of the right equipment should be put in place and funds sought to ensure that gaps in operational expertise are rapidly addressed through staff training and development.

Biotechnology Impediments and Partnerships

There is an often-expressed argument that biotechnology research in the African setting is not a need and that the prevalent food shortage is merely a distribution problem solvable through international and national trade policies. This indeed makes the world a very backward place as regards distribution because thousands of tonnes of food are destroyed each day in rich countries while millions of people go hungry throughout the world. If Africa is to feed itself, it must marshall the relevant apparatus and meet its own food requirements in a sustainable manner. The continent must be enabled to increase its agricultural productivity to feed its own population and to put in place long-lasting sustainable development. Africa is painfully aware of the difficulties and prohibitive costs invested in food importation. This situation raises questions as to whether SSA should develop its own biotechnology capacities to enhance its agricultural production or whether it should stand by and only access the technologies through importation. In other words, how much should the countries of SSA invest in modern biotechnology developments vs. technology importation or application of existing techniques?

Investment in biotechnology in Africa must be considered in the context of the unique priorities and needs of her countries, the nature of the biotechnology application in question, and the availability and accessibility of the required technology. If development of a product or variety is a priority for a country and the time and cost of such development through conventional approaches are prohibitive, biotechnology approaches should be considered. On the other hand, if such development entails application of transformation, regeneration and molecular maps, and the cost becomes too high, outweighing the benefits, consideration could be given to accessing such technologies through transfer, if they have already been developed elsewhere and are accessible.

It is important that the agricultural research institutions in SSA organize their biotechnology research so that there are effective links with traditional research programmes and end-users, and so appropriate links are established with publicly funded international programmes and the private sector. Links with private sector companies provide opportunities for accessing specific technologies. However, their products are largely proprietary in nature and costly for many resource-poor countries in SSA. Access to such technologies has largely required broker agreements between companies that hold patents for the technology in question and the developing country's research institutions. Such services have been provided by the International Service for the Acquisition of Agri-biotech Applications (ISAAA) and other organizations like the United States Agency for International Development (USAID)-funded Agricultural Biotechnology for Sustainable Productivity (ABSP) project based at Michigan State University (MSU) in the US. These kinds of partnerships should be pursued actively by all parties.

Biotechnology Enabling Environment

There is a strong link between biotechnology developments and safe research environments. It has not been possible to separate biosafety requirements from biotechnology, the two need to be supported together. At present the regulatory framework and national capacities for biosafety are either not available or are very weak in SSA countries. Only very few countries have draft, let alone operational, biosafety guidelines (Wafula, 1993). Egypt, Kenya, Uganda, South Africa and Zimbabwe are pertinent examples (Table 5.5).

Challenges to establishing national biosafety capacities in SSA countries include a lack of high quality trained expertise in molecular biology, ecology and other disciplines that are crucial in national development. These are the people who understand the potential risks of biotechnology activities and, therefore, take initiative to develop, plan and implement oversight mechanisms. Their

Table 5.5. Supervision mechanisms of biotechnology in African countries as of October 1995. Source: Wafula (1994).

Country	Present regulatory structures
Egypt	National Biosafety Guidelines
Kenya	Kenya Standing Committee on Import and Export
	Institutional Biosafety Committee in KARI*
	National Biosafety Regulations and Guidelines
South Africa	Plant and Animal Quarantine Systems
	National Genetic Examination Committee
Tanzania	Phytosanitary and Animal Disease Regulations
Uganda	Phytosanitary and Animal Disease Regulations
	Draft Biosafety Guidelines
Zimbabwe	Phytosanitary and Animal Disease Regulations
	National Biosafety Guidelines

*KARI = Kenya Agricultural Research Institute.

absence or deficiency constitutes a major obstacle in the development of biosafety supervision.

In building national capacity in risk assessment and management, some countries have incorporated their biosafety needs within the framework of national biotechnology research policies. This has enabled governments to seek support for capacity building from relevant international organizations and special programmes. For example, some countries such as Kenya and Egypt have joined with the ABSP Project in training personnel through biosafety internships and risk assessment and management workshops. Similarly, ISAAA has held a number of regional and national biosafety workshops in which scientists from SSA have participated. The Special Programme on Biotechnology of The Netherlands (DGIS) and the Biotechnology Advisory Commission (BAC) of the Stockholm Environment Institute (SEI) have coordinated regional biosafety workshops, supported the development of a harmonized regional biosafety focal point in Southern and Eastern Africa and supported the formulation of national biosafety guidelines in some of the countries in SSA.

Conclusions

Clearly the countries of SSA are faced with enormous challenges to their economic development arising principally from a declining agricultural sector. The main constraints to the development of agriculture in SSA are a severe degradation in the land resources, frequent occurrence of adverse climatic conditions, numerous crop and livestock pests and diseases, and a lack of enabling technologies to support yield increase per unit area of land. These countries wish, and should be encouraged and supported, to advance innovative technologies for their own purposes, particularly biotechnology which they believe will assist them in their quest for development. The donor community, collaborators and SSA nations must understand the challenges, needs, nature and purpose of the development process in each country, and acknowledge that biotechnology provides but one of the many elements that can influence change at the farm level.

References

Altman, D.W. (1993) Plant biotechnology transfer to developing countries. *Current Opinions in Biotechnology* 4, 177–179.

FAO (1986) African agriculture: the next 25 years. In: *Main Report. Atlas of African Agriculture; Annex I, Socio-economic and Political Dimensions; Annex II, The Land Resource Base and Rising Productivity*. FAO, Rome.

FAO (1994) Funding agricultural research in sub-Saharan Africa. In: *Proceedings of an FAO/SPAAR/KARI Expert Consultation, Nairobi*. FAO, Rome.

Hvoslet-Eide, A.K. and Rognli, O.A. (1996) Environmental issues for plant biotechnology transfer: a Norwegian perspective. In: Altman, D.W. and Watanabe, K.N. (eds) *Plant Biotechnology Transfer to Developing Countries*. R.G. Landes Company, Austin, Texas, pp. 38–50.

Roman, K.V. (1996) Facilitating plant biotechnology transfer to developing countries. In: Altman, D.W. and Watanabe, K.N. (eds) *Plant Biotechnology Transfer to Developing Countries*. R.G. Landes Company, Austin, Texas, pp. 268–277.

Sasson, A. and DaSilva, E.J. (1994) Achievement, expectations and challenges. *The UNESCO*, 11–15.

Wafula, J.S. (1993) Introduction to national and regional needs, constraints and priorities related to safety in biotechnology. In: *Proceedings of the African Regional Conference for International Cooperation on Safety in Biotechnology*. Harare, Zimbabwe, pp. 107–114.

Wafula, J.S. (1995) Opportunities for regional planning for biotechnology under ASARECA. In: Komen, J., Cohen, J.I. and Ofir, Z. (eds) *Turning Priorities into Feasible Programs. Proceedings of the Seminar on Planning, Priorities and Policies for Agricultural Biotechnology for East and Southern Africa. South Africa, 23–28 April 1995*. Intermediary Biotechnology Service/Foundation for Research Development, The Hague/Pretoria, pp. 136–142.

Wafula, J.S. and Ndiritu, C.G. (1996) Capacity building needs for assessment and management of risks posed by living modified organisms, perspectives of a developing country. In: Mulongoy, K.J., van der Meer, P. and Zannoni, L. (eds) *Transboundary Movement of Living Modified Organisms Resulting from Modern Biotechnology*. International Academy of the Environment, Geneva, Switzerland, pp. 63–70.

The Application of Biotechnology to Food Security Crops

II

Development of Insect-resistant Maize and Its Potential Benefits to Developing Countries

6

Pam Robeff

Garst Seed Company, 2369 330 St., PO Box 500, Slater, IA 50244, USA.

Bacillus thuringiensis (*Bt*) has been used extensively for the control of insect pests both as a formulated biocontrol agent and, more recently, via plant transformation with specific endotoxin genes. In the US, *Ostrinia nubilalis* (European corn borer, ECB) is a major pest of *Zea mays* (maize), and its control with *Bt* proteins has been well documented (Shnepf and Whiteley, 1985; Ge *et al.*, 1989). A close relative, *Ostrinia furnacalis* (Asian stem borer, ASB) is a pest in Indonesia, where its life cycle allows it to threaten maize virtually all year round.

This paper provides an overview of a collaboration between the Central Research Institute for Food Crops (CRIFC), Bogor, Indonesia, and Garst Seed Company, Slater, Iowa, USA (previously ICI/Zeneca Seeds) to develop insect-resistant maize using the *cryV Bt* gene (Tailor *et al.*, 1992). The CryV protein was known to be active against ECB and later shown to be similarly active against ASB. The technical aspects involved in the production of insect-resistant germplasm include the selection of suitable germplasm for transformation, selection of resistance genes and construct components, plant transformation, analysis of transformants and finally introgression into a breeding programme for inbred and hybrid maize production. An overview of these technical areas is presented and the potential benefits to developing countries discussed.

Introduction

The collaboration between Garst and CRIFC started in 1993, and ended in September 1997. The three main objectives of the agreement were to produce insect-resistant maize via transformation, to transfer the technologies involved by training Indonesian scientists and to explore possible commercialization of products derived from the programme. The project is complete in terms of training and production of transgenic plants. Issues relating to commercialization and freedom to operate are being addressed within the context of an ever-changing intellectual property arena. Some of these issues, although considered at the onset of the programme, became and continue to be of paramount importance to the progression of this work and will be discussed later.

To date, many transgenic plants have been produced which vary a great deal in their potential for inclusion in breeding programmes. Laboratory insecticidal activity, plasmid insertion pattern, protein production and field efficacy must all be considered before a transformation event is progressed. Generally, single or low transgene copy number is the aim of transgenic plant programmes. In an attempt to cover and summarize the full programme it is only possible to present a broad overview of the research.

Production of Insect-resistant Maize via Transformation

Technical components of the programme are broadly described and include five main areas:

- selection of germplasm for transformation
- selection of resistance genes and construct components
- transformation
- analysis of transformants
- introgression into elite germplasm.

Selection of tropical germplasm

Initial research involved the selection of tropical germplasm suitable for transformation. Bahagiawati Amirhusin, our first visiting scientist, conducted a tissue culture study using 12 inbred lines from both Zeneca's portfolio and other tropically adapted material. Primary objectives were to evaluate embryonic initiation, embryogenic-proliferative capacity and regenerative ability. Some of these tropical lines were unable to complete their life cycle in the United States Midwest region and were eliminated from the study. Lines selected for further study, based on the above criteria, were PN2119, YB2113 and LM2112. Those eliminated were HN2193, PN2116, WM2154, EV2150, LM2111, UM2160, UM2141, MN2151 and UM2198. Zeneca proprietary line PN2119 was used for further experiments based on its potential for commercialization. Unfortunately, quarantine restrictions prevented the use of Indonesian germplasm at that time and it was agreed that the *Bacillus thuringiensis* (*Bt*) transgene would be introgressed into local (Indonesian) material through breeding programmes at a later stage.

Selection of *Bt* genes and construct components

The main considerations involved in the selection of genes for transformation constructs were activity against the target species *Ostrinia furnacalis*, access to *Bt* genes and codon optimization of *Bt* genes. Activity studies using *E. coli*-derived protein extracts in diet incorporation assays were conducted in the US against the proxy species *O. nubilalis*, and in Indonesia against the target species *O. furnacalis*. *CryV* and *CryIA(c)* gene products were tested and found to be effective against both species (Table 6.1).

The intent was to use the *cryV* insecticidal product because it is proprietary to Zeneca and would potentially reduce downstream intellectual property issues. The *cryV* gene was optimized under contract to increase the overall guanine:cytosine (GC) content (Table 6.2). While the codon optimization was in progress, a construct containing *cryIA(c)* donated by Dr Pam Green (Michigan State University) was used for initial research and proof of concept studies.

A review of promoters was conducted based on tissue-specific and constitutive expression. Among promoters tested for their suitability were *cab* (chlorophyll a and b), *MR7* (maize root), *CaMV 35S* (derived from the cauliflower mosaic virus) and maize polyubiquitin. At the time of evaluation, no patents had issued in the US for the *CaMV 35S* and maize polyubiquitin promoters; however, in October 1994, Monsanto Company was granted a patent for the *CaMV 35S* promoter and use of all constructs containing it was discontinued. The research construct used after that time was designated pAID5PUB (*cryV* gene driven by the maize polyubiquitin promoter, which was used under licence from Mycogen Inc.). In accordance with company policy, further changes were made to the final constructs eliminating antibiotic genes

Table 6.1. Bioassay of CryV protein (*E. coli* crude extract) vs. *O. nubilalis* and *O. furnacalis* showing mortality 6 days after treatment.

	O. nubilalis	*O. furnacalis*
pIC224 (*cryV*)	100	97
pIC18 (*cryIA(c)*)	100	74
pIC224E5 (disabled *cryV*)	0	0

Table 6.2. Codon optimization of the *cryV* gene.

	Wild type gene	Optimized gene
Amino acid homology	–	100%
DNA homology	–	69%
GC content	37%	62%
Poly-adenylation sites	24	2
ATTTA	7	6

from the plasmid and bacterial selection was performed using a nutritional selection method. These adjustments were made because of biosafety concerns held primarily in Europe. The final constructs used were designated pIGPDCV and pIGPDC5 (polyubiquitin-*cryV* gene cassette in two orientations with respect to the plant selectable marker).

Transformation

Stable transformation of A188 × B73 cell suspensions and PN2119 zygotic embryos via bombardment
All early transformation work was conducted using the Biolistic™ gun (via research-only licence) and transgenic plants generated were from both A188 × B73 and PN2119 germplasm. A total of 60 gun-derived *cryIA(c)* plants and 40 *cryV* plants in A188 × B73 background, and 163 gun-derived *cryIA(c)* plants and three *cryV* plants in PN2119 germplasm, were produced.

However, as the project progressed, it was decided to change over to the Garst proprietary Whiskers transformation system in order to overcome intellectual property issues relating to the use of the Biolistic gun, which could only be used to generate research (non-commercial) material. The system, although less efficient than the gun, had the distinct advantage of circumventing the Dupont gun patent. Whisker clones could only be created using A188 × B73 cell suspensions; therefore, conventional or marker-assisted breeding will be used to introgress the insect resistance trait into the appropriate germplasm.

Stable transformation of A188 × B73 suspension cells via Whiskers technology
Transformation with Whiskers using the constructs pIGPDCV and pIGPDC5 ended in August 1996. This completed the front-end effort to generate enough expressing transformation events from which to select plants for inclusion in a breeding programme. Associated downstream tissue culture and regeneration continued through 1997. A summary of transformation events generated is given in Table 6.3.

Table 6.3. Summary of A188 × B73 transformation (Whiskers transformation to April 1997).

Plasmid	PCR positive *cryV* clones	Leaf assay positive*
pIGPDCV	370	147
pIGPDC5	225	90
Total	595	237

* 0–5% leaf feeding damage by *O. nubilalis* 3 days after infestation in plate bioassay.

The number of plants regenerated per transformation event (clone) varied from five to 15 depending upon clone health and initial resistance to corn borer feeding damage (assessed by feeding detached leaf discs to neonate larvae and scoring for feeding damage and larval mortality three days after infestation. The data in Table 6.3 reflect primary transgenic bioassay-positive transformation events. Plants derived from these transformation events may produce a range of seed counts, from zero to more than 300. Fertility and subsequent seed production significantly reduce the number of plants which go forward for field efficacy testing. Generally, only plants producing at least 50 seeds will progress to field trials. Otherwise, determination of segregation ratios is difficult.

Plant analysis

Laboratory analysis
Initial clones were selected by growing callus on herbicide-containing plates, indicating that the plant selectable marker had been incorporated into the genome. Clones that survived herbicide selection were further tested for the agronomic gene (*Bt*) using the polymerase chain reaction (PCR) followed by detached leaf feeding assays (described above in the transformation section).

Plants emerging from this cascade (as well as field efficacy testing, described below) are being evaluated for protein expression level and gene integration pattern. Protein levels will be determined for different plant tissues and life stages. Techniques employed include Southern blot analysis for copy number and restriction fragment length polymorphism (RFLP), as well as immuno-detection methods such as Western blot and enzyme-linked immunosorbent assay (ELISA) to measure expressed protein.

Field efficacy testing
Plants from five separate transformation events were tested in the field in 1996 for efficacy against first and second generation ECB. Several plants from each transformation event were evaluated at the whorl stage for first generation corn borer resistance, and those that demonstrated little or no feeding damage were re-infested at anthesis to mimic the natural second-generation infestation. Final scoring was performed after harvest by splitting whole stalks and measuring the actual length of the tunnel caused by corn borer larval feeding.

Mendelian segregation was also an important consideration at this stage, so that the future inheritance patterns of any candidate entering a breeding programme could be more readily predicted. All primary transgenics were crossed as males and as females, as fertility allowed, to an inbred line related to the B parent. At this screening level, the segregation ratio should therefore be 1:1. This was actually observed in the progeny of one plant from one event.

Events from the 1996 field trials are currently being analysed for copy number, rearrangements and protein expression levels. Southern blot analysis has shown copy number to vary between one and ten for transformation events

tested to date. A monoclonal antibody ELISA is under development and protein
levels of potential candidates are yet to be determined.

Currently our efforts are concentrated on the analysis of event C in Table
6.4, but the study of other events, particularly A, is still possible. Many more
events are being prepared for 1997 field efficacy testing, and emphasis has been
on improving seed set so that segregation ratios can be more accurately
determined.

Table 6.4. Summary of 1996 *cryV* field trial.

Event	Number of susceptible:number of resistant (by visual scoring of leaf damage)	*cryV* gene copy number (by Southern blot analysis)
A	8:252	2
B	39:4	nd
C	23:21	4
D	204:0	nd
E	138:3	>10

nd: not determined.

Introgression into elite germplasm

A combination of methods may be used to introgress the *cryV* transgene into
locally adapted germplasm. These include marker-assisted breeding using
RFLPs, simple sequence repeats (SSRs), and amplified fragment length
polymorphisms (AFLPs), as well as traditional breeding. Field efficacy testing
against the target species, Asian stem borer (ASB), must also be part of this
programme.

To date suitable candidates for transfer to Indonesia remain to be identified,
and further discussion is required with respect to incorporating any potentially
useful transformation event(s) into a breeding programme. Until recently,
biosafety constraints prevented the transfer of transgenic maize into Indonesia,
but regulations to overcome this problem have been developed. Additional
issues include intellectual property surrounding the transformation construct
and regulations which designate maize as a protected crop in Indonesia and
hence not subject to patent protection.

Transfer of Enabling Technologies to Indonesian Scientists via Training in the US

The first component of technology transfer involved training of visiting
scientists in transformation and related skills. The main training areas
accomplished were:

- maize transformation
- tissue culture of tropical germplasm
- insect bioassays – diet, transient callus, whole plant
- molecular characterization – Southern blot, PCR, ELISA, Western blot
- field evaluation and artificial infestation.

The following scientists all spent time at ICI/Garst, Slater and received training in one or more of the above areas:

- Ms Bahagiawati Amirhusin (July 1993–April 1994)
- Mr Saptowo Pardal (September 1994–September 1995)
- Mr Edy Listanto (January 1995–September 1995)
- Dr Firdaus Kasim (October 1994 and May 1995; two visits totalling almost 3 months).

In addition to practical training at Garst, Mr Pardal and Mr Listanto attended an international plant physiology meeting as well as local courses at Iowa State University. Mr Listanto also attended a technology transfer workshop in Indonesia. Two visiting scientists, Dr V. Sekar (India) and Ms Shireen Assem (Egypt), also spent time at Garst. Dr Sekar worked with CryV and CryIA(c) to investigate their respective binding sites in the European corn borer (ECB) gut and Ms Assem was trained in Whisker transformation technology.

The second component of technology transfer involves the forwarding of materials and products to Central Research Institute for Food Crops (CRIFC) and has so far included bioassay materials to conduct local assays against ASB (Table 6.1). In order to facilitate transfer of transformation constructs and *Bt* plants, an intellectual property review is in progress to determine the best route for continued research and commercial goals.

A further impediment to technology transfer is the status of maize as a protected crop in Indonesia. The Indonesian government at this time is actively addressing this issue.

Commercialization of Insect-resistant Germplasm Generated from the Project

There have been discussions regarding the best way forward with this objective, involving a potential commercial collaboration with Kaltimex-Jaya, Indonesia. It was determined that, at this time, there is no market for insect-resistant maize hybrids in Indonesia because most maize is open-pollinated. However, marketing in The Philippines is a possibility. Dr Achmad Fagi, Director of CRIFC, is exploring possibilities for hybrid maize production in Indonesia and this could also provide a commercial avenue. Commercialization through Pac Seeds, a sister company of Garst, was considered at the beginning of the programme and remains a possibility.

Discussion

As the programme nears completion it is time to evaluate progress against objectives and to reflect upon the accomplishments, deficiencies and problems encountered. The programme's objectives as outlined above were to produce transgenic insect-resistant maize, transfer-associated technology to CRIFC and explore commercial avenues for the products generated from the programme.

Undoubtedly, the technical aspects of the programme have been a success and all difficulties were overcome or circumvented. Transformation of different maize lines using both the Biolistic gun and Whiskers have been successfully completed with a variety of constructs, and many transformation events have been generated from which material for introgression into Indonesian germplasm could be selected after the 1997 field trial, no material has been selected to date.

Similarly, training of visiting scientists was successfully accomplished. Participants were offered opportunities to experience research and commercial operations in an industrial setting, as well as to develop a network foundation for the future.

Transfer of project-related materials and products has proven to be more difficult. Although some materials were transferred to CRIFC earlier in the programme, facilitating the important CryV activity studies against ASB, there has been no transfer of transgenic germplasm. Garst will securely maintain the germplasm pending future resolution of this issue.

Selection of suitable transformation events for introgression will continue after the completion of the field trial. It is probable that three-way crosses will need to be made before transferring the transgenic plants to CRIFC, and further discussions between Garst and CRIFC are required to determine how these will accomplished. An update on progress toward hybrid maize production in Indonesia will assist transfer of germplasm for both continued research and any potential commercialization.

There have been considerable constraints in the regulatory and intellectual property fields, which are major contributors to the lack of transfer at this time. Most of these issues have been mentioned above, but in summary include: crop protection and biosafety regulations in Indonesia; and ownership of construct components, enabling technologies and germplasm worldwide. Among the many issues for consideration are patents encompassing such technologies as transformation, codon optimization, gene ownership (*Bt*, selectable markers), transgenic plants and *Bt* maize. There are significant challenges to be met before the final transfer of these technologies can be made and benefits to Indonesia as a developing country can be assessed.

Acknowledgements

Jan Tippett (Hach Company) presented this information to the ABSP Project 'Biotechnology for a Better World' Conference at Monterey, California in April

1997. Upon Jan's departure from Garst Seed, Pam Robeff served as primary author in the submission of this chapter. Acknowledgements are due Saptowo Pardal (CRIFC), Edy Listanto (CRIFC), Kim Hagemann (Pioneer Hi-Bred), Bahagiawati Amirhusin (Purdue University), Kan Wang (Iowa State University), Martin Wilson (Stine Biotechnology), Bruce Held (Stine Biotechnology), and Jan Tippett for their contribution to the project, the information in this chapter and its authorship.

References

Ge, A.Z., Shivarova, N.I. and Dean, D.H. (1989) Location of the *Bombyx mori* specificity domain on a *Bacillus thuringiensis* delta endotoxin protein. *Proceedings of the National Academy of Sciences USA* 86, 4037–4041.

Schnepf, H.E. and Whiteley, H.R. (1985) Delineation of a toxin-encoding segment of a *Bacillus thuringiensis* crystal protein gene. *Journal of Biological Chemistry* 260, 6273–6780.

Tailor, R., Tippett, J., Gibb, G., Pells, S., Pike, D., Jordan, L. and Ely, S. (1992) Identification and characterization of a novel *Bacillus thuringiensis* delta endotoxin entomocidal to coleopteran and lepidopteran larvae. *Molecular Microbiology* 6(9), 1211–1217.

The Application of Biotechnology to Potato

7

Marc Ghislain, Maddalena Querci, Merideth Bonierbale, Ali Golmirzaie and Peter Gregory

International Potato Center (CIP), PO Box 1558, Lima, Peru

The application of biotechnology to potato has a long history marked by the development of successful strategies for the improvement of quality traits and pest and disease resistance. Both direct gene transfer and breeding with molecular markers, collectively referred to as molecular breeding, have been developed to improve potato. Direct gene transfer by *Agrobacterium tumefaciens*-mediated transformation has been successful in a number of cases. Successful examples include resistance to the Colorado potato beetle and potato tuber moth (PTM) mediated by crystal proteins of *Bacillus thuringiensis* (*Bt*), and partial protection against major potato viruses (potato potyvirus X (PVX), potato potyvirus Y (PVY), and potato leafroll luteovirus (PLRV)) mediated by viral sequences. Prospects for bacterial and fungal disease resistance are also encouraging.

Major agronomic traits are often governed by many genes that determine a continuous phenotypic variation in breeding populations. Environmental effects often complicate selection by conventional breeding methods. Genetic (DNA) markers for these quantitative trait loci (QTL) could greatly facilitate the identification and selection of improved genotypes with a minimum of unfavourable loci. These tools help to uncover the individual QTL contribution, intra- and interloci interactions and their inheritance, and isolate genes known only by their phenotype.

Molecular markers have been identified and, through collaborative research at the International Potato Center (CIP), are being used for map-based cloning of virus resistance genes, as well as for better access to genetic variation for late blight resistance and insect resistance mediated by glandular trichomes. These advances in molecular breeding for improved potatoes are of particular importance to CIP as

© CAB INTERNATIONAL 1998. *Agricultural Biotechnology in International Development* (eds C.L. Ives and B.M. Bedford)

complementary approaches to conventional breeding strategies and natural resources management. The deployment of new, improved potato varieties will decrease pesticide use in potato production and hence facilitate sustainable production systems suited to less-favoured potato producers in developing countries.

Introduction

Potato is the most important non-cereal food crop in the world. The crop represents roughly half of the world's annual output of all roots and tubers and is part of the diet of half a billion consumers in developing countries (FAO/CIP, 1995). Trends toward increasing the percentage of potato used for processing may lead to new income opportunities in developing countries.

Global potato production has nearly stagnated during the last 30 years at around 260–270 million tons and area planted declined from 22 to 18 Mha during the same period (FAO/CIP, 1995). The average yield increased by 25% from 12 to 15 t ha^{-1}. Differences between trends in developed countries and developing countries have been identified. Developed countries will continue to have a slow growth of potato production until the end of the century while developing countries will increase their share of the global potato output up to 34% by the year 2000.

Yields vary tremendously in potato production from lows of 6 t ha^{-1} in sub-Saharan Africa to 42 t ha^{-1} in The Netherlands. Schematically, three factors could raise yields in developing countries to the upper values: access to chemical fertilizers, pesticides and good-quality planting materials. Another noteworthy difference between developed and developing countries' potato trends is the negligible varietal change in developed countries, whereas improved varieties continue to play an important role in increasing yields in developing countries.

Today, potato production relies on the use of large quantities of toxic chemical pesticides to ensure stable yields. Pesticide overuse threatens not only the environment but also farmers' health, as pesticides are often handled inappropriately and are less regulated in developing countries. Lack of access to pesticides also has dramatic consequences on potato production, especially in areas where late blight disease is severe. Hence, pesticides play a pivotal role in the fragile economy of small potato producers due to their elevated purchase cost.

Potato Improvement for Pests and Diseases

Reduction of pesticide use in potato production is a first priority in the International Potato Center (CIP)'s potato improvement programmes. To that end, CIP uses a blend of conventional and molecular approaches to improve potato resistance to targeted pests and diseases. Molecular approaches are based on the utilization of defence and resistance genes from either exotic sources or from the *Solanum* germplasm.

Through its various institutional mechanisms to set research priorities, CIP's research agenda was debated in 1996, and formalized in a document entitled *CIP Midterm Plan* for the years 1998–2000 (CIP, 1997). This research agenda sets a strong focus of potato improvement projects for three constraints: late blight, viruses and potato tuber moth (PTM) (see Table 7.1). We briefly review applications of biotechnology to: (i) reduce these constraints by gene technology or genetic improvement; (ii) utilize the genetic diversity of the *Solanum* germplasm; (iii) transfer new genetic loci into potato varieties.

Potato pest and disease constraints

Late blight disease

Late blight is the main focus of our research as the disease poses a renewed threat to potato production by the new migratory wave of *Phytophthora infestans*, the causal agent of this fungal disease. The value of crop losses from late blight in developing countries is currently estimated at US $2.75 billion (French and Mackay, 1996).

Disease symptoms have been delayed by engineering the expression of several genes in potato, such as osmotins and glucose oxidase (*GO* gene). Osmotin displays membrane-disturbing properties and has been shown to inhibit hyphal growth *in vitro* and to cause sporangial lysis. Transgenic

Table 7.1. Targeted traits for potato improvement by molecular breeding methods ranked by priorities for CIP (VH=very high, H=high, M=moderate, L=low) sources of resistance:transgenes currently under trials and tuber-bearing *Solanum* species.

Constraint	Priority	Source of resistance:transgene	*Solanum* species
Late blight	VH	Osmotin, glucose oxidase, PR1 proteins, programmed cell death	*S. phureja, S. microdontum, S. verrucosum, S. tuberosum* subsp. *andigena*
Viruses	H	Coat protein, replicase	*S. tuberosum* subsp. *andigena, S. acaule, S. stoloniferum*
Bacterial wilt	M	Lysozyme, glucose oxidase, programmed cell death	*S. sparsipilum*
Insect pests	M	*Bacillus thuringiensis*, protease inhibitor	*S. berthaultii, S. sparsipilum*
Nematodes	L	Plantibodies, programmed cell death	*S. spegazzinii, S. tuberosum* subsp. *andigena, S. vernei*

potatoes have been produced with strong constitutive expression of tobacco, tomato and potato osmotins. Delayed disease symptoms were observed after inoculating detached leaves with *P. infestans* (Liu *et al.*, 1994; Zhu *et al.*, 1996). Several pathogenesis-related proteins of the *PR1* class, from tomato and tobacco, have been shown to inhibit germination of *P. infestans* zoospore (Niderman *et al.*, 1995). The *GO* gene from *Aspergillus niger* was inserted into the potato genome and led to a marked delay in appearance of late blight disease symptoms in potato leaves as a consequence of elevated levels of hydrogen peroxide (Wu *et al.*, 1995). Other promising transgenic approaches involve programmed cell death systems that mimic the hypersensitive reaction to fungal invasion. In one of the systems currently under development, transgenic potatoes containing a *P. infestans*-dependent cell-suicide system develop a quantitative resistance to *P. infestans* (Strittmatter *et al.*, 1995). These programmed cell death systems represent a general approach for pathogen resistance.

Recent breeding for late blight resistance has favoured use of polygenic race-non-specific resistance over *R* genes (conditioning race-specific hypersensitive responses). This type of resistance is apparently more durable with respect to new races of the pathogen. For several years, CIP breeders have conducted recurrent selection using populations of *Solanum tuberosum* subsp. *andigena*, native to South America, to improve quantitative resistance to late blight in the *tuberosum* germplasm. Beyond the *tuberosum* boundaries, another source of high levels of quantitative resistance has been identified in the native cultivated, diploid potato *S. phureja* (Cañizares and Forbes, 1995; Trognitz *et al.*, 1996). Molecular markers have been recently correlated with late blight resistance segregation as putative quantitative trait loci (QTL) in diploid crosses from *S. phureja* (Ghislain *et al.*, 1997). Once confirmed, these QTL will be introgressed into the cultivated tetraploid potato following a marker-assisted breeding scheme combined with tetraploidization via *2n* gametes produced by the donor species.

Potato virus diseases
Potato viruses are economically important not only as a direct cause of severe crop losses but also as a barrier to seed trade due to phytosanitary requirements imposed by most countries. Among the three main potato viruses, potato potyvirus X (PVX) causes only mild symptoms when infecting plants alone, but when it occurs together with potato potyvirus Y (PVY) it causes a synergistic interaction that results in significant crop losses. Infections by potato leaf roll luteovirus (PLRV) or PVY cause the two major potato viral diseases. At CIP, resistance to control virus diseases has a high priority because of their worldwide importance.

Viral sequences (e.g. coat protein and replicase) have often provided cross-protection to the corresponding virus when integrated into the potato genome. Coat-protein-mediated protection has been successful in developing PVX resistance (Hoekema *et al.*, 1989; Jongedijk *et al.*, 1992), and combined resistances to PVY and PVX in potato variety Russet Burbank (Kaniewski *et al.*, 1990; Lawson *et al.*, 1990). Several potato varieties with combined transgenic resistance to PVX and PVY have been released in Europe, Mexico and the United States. The transfer of replicase genes into potato, coding for replication-related protein of these viruses, has also produced significant levels of resistance to PVX (Braun and Hemenway, 1992), PVY (Audy *et al.*, 1994) and PLRV in variety Russet Burbank. The latter led to a variety release in 1995, by the US-based Monsanto Company.

At CIP, we work with genes for extreme resistance to PVX and PVY identified in the *Solanum* germplasm by conventional and molecular approaches. The dominant genes *Ry* and *Rx* from *S. tuberosum* subsp. *andigena* have been incorporated into breeding populations at high frequencies. The genetic mapping of these genes conferring extreme resistance to PVX and PVY has been achieved (Ritter *et al.*, 1991; Hämäläinen *et al.*, 1997; Brigneti *et al.*, 1997). Molecular cloning of *Rx* from *S. acaule* and *Ry* from *S. stoloniferum* are currently underway through a collaborative research project between CIP and the Sainsbury laboratory, UK. In both cases, molecular markers have been identified in the vicinity of the dominant genes. Flanking markers for *Rx* are at 0.17 cM and at 0.06 cM (Bendahmane *et al.*, 1997), and for *Ry* two amplified fragment length polymorphism (AFLP) markers (M17 and M6) are at less than 0.3 cM (D.C. Baulcombe, 1997, personal communication).

Genetic resistance to PLRV always has been of high interest to breeders. Only partial resistance has been encountered despite several germplasm screening efforts. Building up durable resistance to PLRV has been hampered by the multifactorial and probably multigenic nature of the resistance. Different resistance mechanisms (infection, replication, antixenosis and antibiosis to the vector) are conferring, individually, only moderate levels of resistance. In addition, previous studies at CIP have shown that PVX and/or PVY reduce the durability and level of resistance against PLRV for which no 'immunity' genes have been found. A better understanding of the precise mechanism by which extreme resistance to PVX or PVY operate in potato could make it possible to engineer a broad-spectrum and durable resistance to virus diseases.

The insect pests

The PTM (*Phthorimaea operculella*) is the most damaging potato insect pest in stores and fields of developing countries. PTM resistance has been obtained by transferring a *Bacillus thuringiensis* (*Bt*) gene that codes for an insecticidal protein into potato. At least two crystal proteins have proved to be effective against *P. operculella*: CryIA(b) and CryIB (Jansens *et al.*, 1995). At CIP, we have collaborated with Plant Genetic Systems, Belgium, to transfer a modified *cryIA(b)* gene with high levels of expression into various breeding clones and varieties with different attributes and adapted for different agroecologies (see Table 7.2). High levels of resistance were obtained in both foliage and tubers. Tuber resistance has been also shown to last for more than 4 months. Protease inhibitors can be transferred to potato in combination with *Bt* genes to ensure durability of the engineered PTM resistance.

Table 7.2. Transgenic potatoes with *crylA(b)* gene for potato tuber moth (PTM) resistance developed at CIP.

Clones/varieties	Attributes	Agroecologies
Costanera (LT-8)	Immune to PVX and PVY, processing qualities, earliness, adapted to short days	Lowland tropics
Sangema	Hypersensitive to PVY, moderately resistant to late blight, adapted to long and short days	Humid tropics
Cruza 148	Moderately resistant to late blight, tolerant to bacterial wilt, earliness, short days	Lowland and humid tropics
Achirana-INTA	Moderately resistant to late blight, resistant to PLRV, adapted to long and short days	Mid-elevation tropics
Revolución	Salt tolerant, adapted to short days	Highlands (Peru)
María Tambeña	Processing qualities, resistant to leafminer flies, adapted to short days	Lowland tropics and coastal area

Tuber resistance to PTM has been found in *S. sparsipilum* and was used at CIP to enhance resistance to PTM in diploid breeding populations (Ortiz *et al.*, 1990). Other insects such as aphids may be effectively controlled by glandular trichomes from wild species such as *S. berthaultii*. This trait was characterized both biochemically and morphologically and at the genome level by genetic mapping (Bonierbale *et al.*, 1994). As the insect pests in potato were successfully managed by integrated approaches in the case of the PTM, moderate priority is given at CIP to developing host-plant resistance to insect pests.

Utilization of potato germplasm

Potato germplasm consists of the genus *Solanum* which embodies over 200 tuber-bearing species and represents a wide array of adaptation to different agroecologies. This genetic diversity will continue to be the focus of most of our research efforts to identify and use new genetic loci to defeat pests and diseases. Biotechnology applications offer efficiency and precision in characterization and utilization of genetic resources. The developing world that hosts the largest reservoir of genetic resources will largely benefit from the appropriate use of biotechnology to develop new products and crops.

The key question is how to effectively apply biotechnology to better exploit this germplasm to develop new products and better crops. Two interconnected routes are conceivable once specific genotypes are selected from new germplasm sources for the trait of interest. Newly identified genotypes carrying a trait of interest can be used directly to isolate genes. This is particularly needed if modification of gene expression is desirable, e.g. higher levels, new targeting for protein accumulation or novel induction patterns. Isolated genes can be transferred directly into potato via *Ag-tumefaciens*-mediated transformation. The other route, valid when genes are unknown or difficult to access, involves molecular markers to tag and monitor these genes in breeding. Methods to analyse the segregation of DNA markers with the trait of interest have been developed in potato. These 'tags' or DNA markers associated with genes governing the trait of interest, can assist breeding by providing genotypic proof of introgression without progeny testing and by helping to more rapidly eliminate unwanted segments of the donor genome.

Methods to transfer new genetic loci

Direct gene transfer in potato

Genetic engineering in potato already has a long history, from the production of the first transgenic potatoes about 15 years ago to new commercial products developed in the past 2 years. Gene transfer into potato was achieved nearly a decade ago, early in the development of gene transfer technology via *agroinfection* (An *et al.*, 1986; De Block, 1988). Potato was one of the first crops transformed genetically. Because of its ease of transformation, potato has been often used as a model species to test the expression of foreign proteins in plants.

Genetic engineering is increasingly adopted in potato improvement programmes because of the worldwide importance of this food crop, the relative ease of potato transformation, its clonal mode of multiplication and some of the inherent complications of traditional potato breeding. The widespread use of genetic engineering for potato improvement will, however, depend largely on the availability of well-adapted potato cultivars, consumer acceptance and the successful marketing of transgenic food crops.

The most widely used system to transfer foreign genes into potato is derived from the natural DNA transfer system of *Agrobacterium* cells (*agroinfection*; for a review, see De Block, 1993; Zupan and Zambryski, 1995). It is an efficient, easy and inexpensive system for gene transfer and is therefore well adapted for developing-country laboratories, which do not always have the capacity to purchase sophisticated and costly equipment. The agroinfection protocol is amenable to the large-scale production of transgenic lines which is needed to select the best-performing transgenic line. Transformation of organelles, such as the chloroplast, can also be achieved via the Biolistic Gun™ approach using a chloroplast-selectable marker (Daniell, 1993; McBride *et al.*, 1995). This organelle transformation offers new possibilities for engineering crop plants and provides maternal inheritance of the transgene.

Transferring an intact foreign gene into a plant genome does not necessarily imply transgene expression. The level of expression can vary significantly among the same transgenic lines presumably because of the positional effect of an integrated transgene (Peach and Velten, 1991). Numerous lines must be produced to select for the best-performing line. This apparent disadvantage can be beneficial as breeders often seek genetic variation. In some cases, however, when the genotype is recalcitrant to genetic transformation and as a result only a few transgenic lines are obtained, the position effect should be minimized. Possible solutions are under development with the use of the scaffold attachment region to stabilize gene expression (Allen *et al.*, 1993, 1996) or the targeting of the transgene to specific high-expressing genomic sites by homologous recombination (Ohl *et al.*, 1994).

Several phenomena that alter the expression of transgenes following their genetic integration have been observed in transgenic crop management and can be grouped into either cosuppression or gene silencing. Cosuppression of transgene expression occurs as a coordinated decrease in gene expression of two transgenes or a transgene and a homologous endogenous gene in a single plant. Gene silencing via inactivation of the transgene can occur between homologous genes by at least one mechanism – the methylation of either the transgene or endogenous gene. The importance of these modifications of transgene expression to agricultural practices should not be overestimated in the case of potatoes because: potatoes with a low copy number (preferably one) of transgenes will be most acceptable; additionally, low homology to endogenous potato sequences will be the most judicious choice in transgene design; and above all, clonal propagation of the potato is an advantage as the transgenic potato can be maintained as a hemizygous crop plant. Nevertheless, the phenomena of gene silencing will become more important when genetic resistance in plants is pyramided. Indeed, the extensive use of the 35S promoter and of mainly one selectable marker gene in all gene constructs may favour methylation of homologous sequences and hence reduce the expected overall expression of resistance genes. More promoters and selectable markers are needed to cotransform plants and build up quantitative traits via genetic engineering.

Transgenic potato research eventually aims at the release of new varieties with added traits that pose low risk for human or animal consumption and to the environment. Field trials are an obligatory step in the development of new transgenic potato varieties. We have already conducted such field trials for three consecutive years at the CIP experimental station of San Ramón using transgenic lines with *Bt* genes. We have identified good transgenic lines with persistence of the original attributes and PTM resistance. This work follows Peruvian national regulations on field trials of genetically modified organisms. Hybridization with related or wild species of potato is possible in the Andean region and hence we have taken drastic measures to avoid rare but possible gene flow into other potato varieties or *Solanum* species. Flower buds are removed every day during the flowering period, transgenic foliage is destroyed after harvest and monitoring of the field over the following cropping seasons is performed. Following recommendations of a regional workshop at Iguazú, Argentina in 1995 (BAC/IICA 1995), the fitness of the transgene in the testing and cultivation area should be determined. If a convincing fitness exists, CIP recommends the use of male sterile varieties. All these precautions were taken in CIP field trials in the mid-elevation station of San Ramón in Peru where no wild species are present.

Molecular genetics and potato breeding

The application of marker-assisted breeding in potato is a new challenge. The Irish potato belongs to an out-breeding tetraploid species: breeding is slow because of the large number of characters that must be included in a tetraploid crop subject to inbreeding depression. This and its clonal propagation has made potato a highly heterozygous crop. Classical potato genetics is still poorly developed because of these limitations. Hence, most of the genetics in potato is learned from studies made with diploid potatoes. Ploidy levels can be interchanged in potato as some diploid potatoes produce 2n gametes and therefore can be brought to the tetraploid level. Molecular genetics studies have produced high-density potato genetic maps (Tanksley *et al.*, 1992). The conservation of marker order between potato and tomato, with the exception of five large inversions of chromosome segments, is an advantage as tomato genetics is further advanced. Molecular markers are now being used to map several traits from native and wild species with the goal of introgressing these traits into cultivated potatoes. We have developed genetic maps for late blight and insect resistance from interspecific diploid populations.

The breeding scheme at CIP to improve late blight resistance in potato essentially follows two routes. One is classical breeding and genetics that make use of tetraploid germplasm followed by recurrent and mass selection methods. The other is a molecular approach involving genetic mapping at the diploid level of loci that determine the resistance to late blight in *Solanum* species and the selective genotyping of phenotypic extremes. Both approaches overlap partially in that the same material is shared. The exploitation of *Solanum* germplasm has led to the identification of a valuable source of resistance to late blight and of several molecular markers associated with this polygenic resistance character. Two interspecific hybrid populations were developed at CIP for this purpose from the *S. phureja* species. Molecular markers, random amplified polymorphic DNA (RAPD), microsatellites and AFLP were scored at CIP with the collaboration of the Scottish Crop Research Institute and the Centro de Investigaciones en Ciencias Veterinarias molecular biology laboratory in Castelar, Argentina. We have already identified several markers associated with late blight resistance. Most of them belong to independent linkage groups suggesting that several loci constitute the genetic architecture of this trait. The genetic maps we are developing in our diploid *S. phureja* populations will eventually be compared with the potato map positions of resistance and defence genes and of QTL for late blight resistance (see Fig. 7.1). Markers flanking the QTL which account for a large variation in the resistance will be used to introgress this trait in cultivated potato. A first stage will be to breed at the diploid level with the help of markers for indirect selection. The production of 2n gametes will lead to the tetraploid level and the eventual development of improved potato varieties.

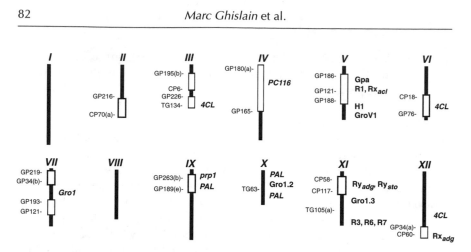

Fig. 7.1. Composite potato map of resistance and defence genes: selected DNA markers on the left side of each chromosome flank QTL for late blight resistance (boxes; Leonards-Schippers *et al.*, 1994); defence related genes are *4CL* (4-coumarate CoA ligase), *PC116* and *prpI* both differentially expressed by *Phytophthora infestans*, *PAL* (phenylalanine ammonia-lyase) (Gebhardt *et al.*, 1991), resistance genes for *Phytophthora infestans* R1 (Leonards-Schippers *et al.*, 1992), R3 (El-Kharbotly *et al.*, 1994), R6 and R7 (El-Kharbotly *et al.*, 1996), extreme resistance gene to viruses, Rx_{acl} and Rx_{adg} (Ritter *et al.*, 1991), Ry_{adg} (Hämäläinen *et al.*, 1997), Ry_{sto} (Brigneti *et al.*, 1997), resistance genes to nematodes *GroI* (Barone *et al.*, 1990), *GroI.2* and *I.3* (Kreike *et al.*, 1993), *GroVI* (Jacobs *et al.*, 1996), *H1* (Pineda *et al.*, 1993; Gebhardt *et al.*, 1993) and *Gpa* (Kreike *et al.*, 1994).

A valuable type of insect resistance from the wild Bolivian species *S. berthaultii* is associated with the presence of trichomes on its foliage. This resistance is polygenic and several components of this resistance have been identified and placed on the potato genetic map (Bonierbale *et al.*, 1994). At Cornell University, where this work was developed, this trait has been partially introgressed into the cultivated potato. Two types of trichomes are involved in this resistance – the A type and the B type – each with specific biochemical components. The efforts at Cornell to breed for tuber yield and resistance mediated by both types of trichomes were hindered by the association between type B trichome biochemistry and lateness and thus poor tuber production under temperate conditions. Plants producing tubers lacked sugar droplets on their type B trichomes and plants with type B trichome products did not produce tubers early enough in the growing season. Molecular genetic mapping in *S. berthaultii* × *tuberosum* hybrid populations demonstrated genetic linkage between lateness and type B trichome properties. Indeed, important loci controlling both characters mapped to chromosome 5 but at different map positions. Molecular markers could be useful to break this linkage by selecting recombinants between these loci and observing their type B properties as well as their lateness.

Evaluation of *Solanum* germplasm results also in the identification of single genes determining high levels of resistance (Table 7.3). The addition of single traits to existing potato varieties is not easy by conventional means due to

Table 7.3. Natural resistance genes from *Solanum* tuber-bearing germplasm targeted for mapping.

Resistance	Gene	Pest and pathogen species
Fungal		
S. demissum	*R1 to R11*	*Phytophthora infestans*
Virus		
S. tuberosum. subsp. *andigena*	Rx_{adg}	PVX
S. acaule	Rx_{acl}	PVX
S. stoloniferum	Ry_{sto}	PVY
S. tuberosum. subsp. *andigena*	Ry_{adg}	PVY
Nematode		
S. spegazzinii	*Gro1*	*Globodera rostochiensis* (Ro1)
S. tuberosum subsp. *andigena*	*H1*	*Globodera rostochiensis* (Ro1, Ro4)
S. vernei	*GroV1*	*Globodera rostochiensis* (Ro1)
S. spegazzinii	*Gpa*	*Globodera pallida* (Ro1)

potato's heterozygosity. Several genes have been identified in potato that confer immunity or hypersensitive reaction to PVX and PVY viruses. Nematode resistance genes have also been identified in *Solanum* species. The first six genes are currently being cloned by different laboratories: the Sainsbury Laboratory in Great Britain, for the virus resistance genes; and the Max Planck Institute in Cologne, Germany, for late blight and nematode resistance genes.

Both efforts proceed via a fine-mapping of the target gene followed by map-based cloning methods. Recent developments in genomic library vectors allow the cloning of large inserts (100–150 kb) in a binary plasmid which can replicate in *E. coli* and *A. tumefaciens*. The plasmid carries all features of a plant transformation vector and hence allows the direct transfer of the large DNA insert into potato without subcloning steps. The molecular isolation of these resistance genes will allow their direct transfer into susceptible potato varieties without changing their genetic backgrounds. The cloned genes will also allow the engineering of new resistance types by manipulation of genic components (promoters, enhancers, coding sequence). Eventually, a broad-based resistance to pathogens may be possible in potato by applying these molecular breeding methods.

Prospects for the Application of Biotechnology in Potato Production

Genetic engineering for potato improvement has led to substantial achievements since the first generation of transgenic potatoes more than a decade ago. Healthy transgenic potatoes have been produced and the process is now amenable to large-scale production of hundreds of transformed plants per transgene. The agroinfection system is particularly well suited to this purpose in developing countries. A growing pool of cloned genes is becoming available to enhance pest and disease resistance and to improve abiotic stress tolerance and tuber quality. Genes can now be engineered for stable expression, for regulation of expression in both time (developmental regulation) and space (tissue specificity), and, soon, for insertion at specific sites in the genome. Foreign genes can be tailored so that proteins encoded by them are accumulated in almost all subcellular compartments or are secreted. Abiotic and biotic control of gene expression is progressing via the development of numerous chimeric promoters in model transgenes and plants. This area of research has developed so much that a comprehensive list of possibilities for controlling and targeting gene expression is no longer feasible. Much research remains to be done to optimize and refine these processes.

The challenge to build durable pest and disease resistance by direct gene transfer will be to produce oligo-transgenic plants in order to pyramid resistance genes and to modify biosynthetic pathways of defence compounds as well as to introduce new ones in potato. Recent developments in engineering disease resistance in plants reveal that the combination of several transgenes in a single genotype led to a synergistic effect on disease resistance (Zhu *et al.*, 1994; Jach *et al.*, 1995; Lorito *et al.*, 1996). Therefore, an oligo-transgenic approach has to be pursued to pyramid genes for resistance. Transgenic approaches should complement natural host-plant resistance by the direct gene transfer to locally adapted varieties with the highest endogenous resistance available.

Whole-plant approaches, using the phenotypic expression of resistance genes, have recently been demonstrated through the use of molecular technology to clone several resistance genes (Martin *et al.*, 1993; Jones *et al.*, 1994; Bent *et al.*, 1994; Whitham *et al.*, 1994). All these resistance genes confer a qualitative resistance that is easy to score for the presence versus absence of the resistance gene in a segregating progeny. Other types of resistance are inherited quantitatively (quantitative trait) and methods for identification of QTL have been developed (for a review, see Tanksley, 1993). QTL analyses allow the identification of resistance-related genes that account for a significant amount of the phenotypic variation observed in a segregating population for the trait of interest. DNA markers have allowed the exploitation of wild and native *Solanum* germplasm, but so far have not allowed the development of new varieties *per se*. The application of marker-assisted selection in potato therefore remains a major challenge, mainly because of the potato's genetic constitution.

The growing importance of the transgenic approach in potato crop improvement makes it urgent to develop appropriate public awareness activities and reduce risks associated with transgenic potatoes. Commercial applications of potato genetic engineering have so far been restricted to developed countries, but developing countries will also likely elaborate improved locally adapted varieties and products for local or regional markets using biotechnology. These molecular techniques and products can be efficiently transferred by international agricultural research centres to national agricultural research systems of developing countries. Farmers and small-scale industries in developing countries will be able to generate higher incomes by developing new products at lower costs. The overall research efforts in applying sophisticated technology to potato improvement aim to develop sustainable production systems that will improve the well-being of potato farmers worldwide.

References and Further Reading

Allen, G.C., Hall, G.E. Jr, Childs, L.C., Weissinger, A.K., Spiker, S. and Thompson, W.F. (1993) Scaffold attachment regions increase reporter gene expression in stably transformed plant cells. *Plant Cell* 5, 603–613.

Allen, G.C., Hall, G. Jr, Michalowski, S., Newman, W., Spiker, S., Weissinger, A.K. and Thompson, W.F. (1996) High-level transgene expression in plant cells: effects of a strong scaffold attachment region from tobacco. *Plant Cell* 8, 899–913.

An, G., Watson, B.D. and Chiang, C.C. (1986) Transformation of tobacco, tomato, potato, and *Arabidopsis thaliana* using a binary Ti vector system. *Plant Physiology* 81, 301–305.

Audy, P., Palukaitis, P., Slack, S.A. and Zaitlin, M. (1994) Replicase-mediated resistance to potato virus Y in transgenic tobacco plants. *Molecular Plant–Microbe Interactions* 7(1), 15–22.

BAC/IICA (1995) Environmental concerns with transgenic plants in centers of diversity: Potato as a model. *Proceedings of a regional workshop, Parque Nacional Iguazú, Argentina*, BAC/IICA.

Barone A., Ritter, E., Schachtschabel, U., Debener, T., Salamini, F. and Gebhardt, C. (1990) Localization by restriction fragment length polymorphism mapping in potato of a major dominant gene conferring resistance to the potato cyst nematode, *Globodera rostochiensis*. *Molecular and General Genetics* 224, 177–182.

Bendahmane, A., Kanyuka, K. and Baulcombe, D.C. (1997) High resolution genetical and physical mapping of the *Rx* gene for extreme resistance to potato virus X in tetraploid potato. *Theoretical Applied Genetics* 95, 153–162.

Bent, A.F., Kunkel, B.N., Dahlbeck, D., Brown, K.L., Schmidt, R., Giraudat, J., Leung, J. and Staskawicz, B.J. (1994) *RPS2* of *Arabidopsis thaliana*: A leucine-rich repeat class of plant disease resistance genes. *Science* 265, 1856–1860.

Bonierbale, M.W., Plaisted, R.L., Pineda, O. and Tanksley, S.D. (1994) QTL analysis of trichome-mediated insect resistance in potato. *Theoretical and Applied Genetics* 87, 973–987.

Braun, C.J. and Hemenway, C.L. (1992) Expression of amino-terminal portions or full-length viral replicase genes in transgenic plants confers resistance to potato virus X infection. *Plant Cell* 4, 735–744.

Brigneti, G., Garcia-Mas, J. and Baulcombe, D.C. (1997) Molecular mapping of the potato virus Y resistance gene Rysto in potato. *Theoretical and Applied Genetics* 94, 198–203.

Cañizares, C.A. and Forbes, G.A. (1995) Foliage resistance to *Phytophthora infestans* (Mont.) de Bary in the Ecuadorian national collection of *Solanum phureja* subsp. *phureja* Juz. & Buk. *Potato Research* 38, 3–10.

CIP (1997) *Medium-term Plan 1998–2000*. International Potato Center, Peru.

Daniell, H. (1993) Foreign gene expression in chloroplasts of higher plants mediated by Tungsten particle bombardment. *Method Engineering* 217, 536–556.

De Block, M. (1988) Genotype-independent leaf disc transformation of potato (*Solanum tuberosum*) using *Agrobacterium tumefaciens*. *Theoretical and Applied Genetics* 76, 767–774.

De Block, M. (1993) The cell biology of plant transformation: current state, problems, prospects and the implications for plant breeding. *Euphytica* 71, 1–14.

El-Kharbotly, A., Leonards-Schippers, C., Huigen, D.J., Jacobsen, E., Pereira, A., Stiekema, W.J., Salamini, F. and Gebhardt, C. (1994) Segregation analysis and RFLP mapping of the *R1* and *R3* alleles conferring race-specific resistance to *Phytophthora infestans* in progeny of dihaploid potato parents. *Molecular and General Genetics* 242, 749–754.

El-Kharbotly, A., Palomino-Sánchez, C., Salamini, F., Jacobsen, E. and Gebhardt, C. (1996) *R6* and *R7* alleles of potato conferring race-specific resistance to *Phytophthora infestans* (Mont.) de Bary identified genetic loci clustering with the *R3* locus on chromosome XI. *Theoretical and Applied Genetics* 92, 880–884.

FAO/CIP (1995) *Potatoes in the 1990s: situation and prospects of the world potato economy*. FAO, Rome.

French, E.R. and Mackay, G.R. (1996) *Enhancing the global late blight network. Report of the Project Design Meeting on the Global Initiative on Late Blight*. CIP, Lima, Peru.

Gebhardt, C. (1994) RFLP mapping in potato of qualitative and quantitative genetic loci conferring resistance to potato pathogens. *American Potato Journal* 71, 339–345.

Gebhardt, C., Ritter, E., Barone, A., Debener, T., Walkemeier, B., Schachtschabel, U., Kaufmann, H., Thompson, R.D., Bonierbale, M.W., Ganal, M.W., Tanksley, S.D. and Salamini, F. (1991) RFLP maps of potato and their alignment with the homoeologous tomato genome. *Theoretical and Applied Genetics* 83, 49–57.

Gebhardt, C., Mugniery, D., Ritter, E., Salamini, F. and Bonnel, E. (1993) Identification of RFLP markers closely linked to the H1 gene conferring resistance to *Globodera rostochiensis* in potato. *Theoretical and Applied Genetics* 85, 541–544.

Ghislain, M., Trognitz, B., Rosario Herrera, M., del, Hurtado, A. and Portal, L. (1997) *DNA markers for the introgression of late blight resistance in potato*. CIP program report 1995–1996, International Potato Center, Peru.

Hämäläinen, J.H., Watanabe, K.N., Valkonen, J.P.T., Arihara, A., Plaisted, R.L., Pehu, E., Miller, L. and Slack, S.A. (1997) Mapping and marker-assisted selection for a gene for extreme resistance to potato virus Y. *Theoretical and Applied Genetics* 94, 192.

Hamilton, C.M., Frary, A., Lewis, C. and Tanksley, S.D. (1996) Stable transfer of intact high molecular weight DNA into plant chromosomes. *Proceedings of the National Academy of Sciences USA* 93, 9975–9979.

Hoekema, A., Huisman, M.J., Molendijk, L., van den Elzen, P.J.M. and Cornelissen, B.J.C. (1989) The genetic engineering of two commercial potato cultivars for resistance to potato virus X. *Bio/Technology* 7, 273–278.

Jach, G., Görnhardt, B., Mundy, J., Logemann, J., Pinsdorf, E., Leah, R., Schell, J. and Maas, C. (1995) Enhanced quantitative resistance against fungal disease by combinatorial expression of different barley antifungal proteins in transgenic tobacco. *Plant Journal* 8(1), 97–109.

Jacobs, J.M.E., van Eck Herman, J., Horsman, K., Arens P.F.P., Verkerk-Bakker, B., Jacobsen, E., Pereira, A. and Stiekema, W.J. (1996) Mapping of resistance to the potato cyst nematode *Globodera rostochiensis* from the wild potato species *Solanum vernei*. *Molecular Breeding* 2, 51–60.

Jansens, S., Cornelissen, M., De Clercq, R., Reynaerts, A. and Peferoen, M. (1995) *Phthorimaea operculella* (Lepidoptera: Gelechiidae) resistance in potato by expression of the *Bacillus thuringiensis* CryIA(b) insecticidal crystal protein. *Journal of Economic Entomology* 88(5), 1469–1476.

Jones, D.A., Thomas, C.M., Hammond-Kosack, K.E., Balint-Kurti, P.J. and Jones, J.D.G. (1994) Isolation of the tomato *Cf-9* gene for resistance to *Cladosporium fulvum* by transposon tagging. *Science* 266, 789–793.

Jongedijk, E., de Schutter, A.A.J.M., Stolte, T., van den Elzen, P.J.M. and Cornelissen, B.J.C. (1992) Increased resistance to potato virus X and preservation of cultivar properties in transgenic potato under field conditions. *Bio/Technology* 10, 422–429.

Kaniewski, W., Lawson, C., Sammons, B., Haley, L., Hart, J., Delannay, X. and Tumer, N.E. (1990) Field resistance of transgenic Russet Burbank potato to effects of infection by potato virus X and potato virus Y. *Bio/Technology* 8, 750–754.

Kreike, C.M., de Koning, J.R.A., Vinke, J.H., van Ooijen, J.W., Gebhardt, C. and Stiekema, W.J. (1993) Mapping of loci involved in quantitatively inherited resistance to the potato cyst nematode *Globodera rostochiensis* pathotype Ro1. *Theoretical and Applied Genetics* 87, 464–470.

Kreike, C.M., de Koning, J.R.A., Vinke, J.H., van Ooijen, J.W. and Stiekema, W.J. (1994) Quantitatively-inherited resistance to *Globodera pallida* is dominated by one major locus in *Solanum spegazzinii. Theoretical and Applied Genetics* 88, 764–769.

Lawson, C., Kaniewski, W., Haley, L., Rozman, R., Newell, C., Sanders, P. and Tumer, N.E. (1990) Engineering resistance to mixed virus infection in a commercial potato cultivar: resistance to potato virus X and potato virus Y in transgenic Russet Burbank. *Bio/Technology* 8, 127–134.

Leonards-Schippers, C.H., Gieffers, W., Salamini, F. and Gebhardt, C.H. (1992) The R1 gene conferring race-specific resistance to *Phytophthora infestans* in potato is located on potato chromosome V. *Molecular and General Genetics* 233, 278–283.

Leonards-Schippers, C., Gieffers, W., Salamini, F. and Gebhardt, C., (1994) Quantitative resistance to *Phytophthora infestans* in potato: A case study for QTL mapping in an allogamous plant species. *Genetics* 137, 67–77.

Liu, D., Raghothama, K.G., Hasegawa, P.M. and Bressan, R.A. (1994) Osmotin over-expression in potato delays development of disease symptoms. *Plant Molecular Biology* 91, 1888–1892.

Lorito, M., Woo, S.L., D'Ambrosio, M.D., Harman, G.E., Hayes, C.K., Kubicek, C.P. and Scala, F. (1996) Synergistic interaction between cell wall degrading enzymes and membrane affecting compounds. *Molecular Plant–Microbe Interactions* 9(3), 206–213.

Martin, G.B., Brommonschenkel, S.H., Chunwongse, J., Frary, A., Ganal, M.W., Spivey, R., Wu, T., Earle, E.D. and Tanksley, S.D. (1993) Map-based cloning of a protein kinase gene conferring disease resistance in tomato. *Science* 262, 1432–1436.

McBride, K.E., Svab, Z., Schaaf, D.J., Hogan, P.S., Stalker, D.M. and Maliga, P. (1995) Amplification of a chimeric *Bacillus* gene in chloroplasts leads to an extraordinary level of an insecticidal protein in tobacco. *Bio/Technology* 13, 362–365.

Niderman, T., Genetet, I., Bruyère, T., Gees, R., Stintzi, A., Legrand, M., Fritig, B. and Mösinger, E. (1995) Pathogenesis-related PR-1 proteins are antifungal. Isolation and characterization of three 14-kilodalton proteins of tomato and of a basic PR-1 of tobacco with inhibitory activity against *Phytophthora infestans. Plant Physiology* 108, 17–27.

Ohl, S., Offringa, R., van den Elzen, P.J.M. and Hooykas, J.J. (1994) In: Paszkowski, J. (ed.) *Homologous Recombination and Gene Silencing in Plants.* Kluwer Academic Publishers, Dordrecht, The Netherlands, pp. 191–217.

Ortiz, R., Iwanaga, M., Raman, K.V. and Palacios, M. (1990) Breeding for resistance to potato tuber moth, *Phthorimaea operculella* (Zeller), in diploid potatoes. *Euphytica* 50, 119–125.

Peach, C. and Velten, J. (1991) Transgene expression variability (position effect) of *CAT* and *GUS* reporter genes driven by linked divergent T-DNA promoters. *Plant Molecular Biology* 17, 49–60.

Pineda, O., Bonierbale, M.W., Plaisted, R.L., Brodie, B.B. and Tanksley, S.D. (1993) Identification of RFLP markers linked to the *H1* gene conferring resistance to the potato cyst nematode *Globodera rostochiensis. Genome* 36, 152–156.

Ritter, E., Debener, T., Barone, A., Salamini, F. and Gebhardt, C. (1991) RFLP mapping on potato chromosomes of two genes controlling extreme resistance to potato polyvirus X (PVX). *Molecular and General Genetics* 227, 81–85.

Strittmatter, G., Janssens, J., Opsomer, C. and Botterman, J. (1995) Inhibition of fungal disease development in plants by engineering controlled cell death. *Bio/Technology* 13, 1085–1089.

Tanksley, S.D. (1993) Mapping polygenes. *Annual Review of Genetics* 27, 205–233.

Tanksley, S.D., Ganal, M.W., Prince, J.P., Vincente, M.C. de, Bonierbale, M.W., Broun, P., Fulton, T.M., Giovannoni, J.J., Grandillo, S., Martin, G.B., Messeguer, R., Miller, J.C., Miller, L., Paterson, A.H., Pineda, O., Röder, M.S., Wing, R.A., Wu, W. and Young, N.D. (1992) High density molecular linkage maps of the tomato and potato genomes. *Genetics* 132, 1141–1160.

Trognitz, B.R., Ghislain, M., Crissman, C. and Hardy, B. (1996) Breeding potatoes with durable resistance to late blight using novel sources of resistance and nonconventional methods of selection. *CIP Circular* 22(1), 6–9.

Whitham, S., Dinesh-Kumar, S.P., Choi, D., Hehl. R., Corr, C. and Baker, B. (1994) The product of the tobacco mosaic virus resistance gene N: similarity to toll and the interleukin 1 receptor. *Cell* 78, 1101–1115.

Wu, G., Shortt, B.J., Lawrence, E.B., Levine, E.B., Fitzsimmons, K.C. and Shah, D.M. (1995) Disease resistance conferred by expression of a gene encoding H_2O_2-generating glucose oxidase in transgenic potato plants. *Plant Cell* 7, 1357–1368.

Zhu, B., Chen, T.H.H. and Li, P.H. (1996) Analysis of late-blight disease resistance and freezing tolerance in transgenic potato plants expressing sense and antisense genes for an osmotin-like protein. *Planta* 198, 70–77.

Zhu, Q., Maher, E.A., Masoud, S., Dixon, R.A. and Lamb, C.J. (1994) Enhanced protection against fungal attack by constitutive co-expression of chitinase and glucanase genes in transgenic tobacco. *Bio/Technology* 12, 807–812.

Zupan, J.R. and Zambryski, P. (1995) Transfer of T-DNA from *Agrobacterium* to the plant cell. *Plant Physiology* 107, 1041–1047.

Development of Virus-resistant Sweetpotato

Maud Hinchee

Monsanto Company, 700 Chesterfield Parkway North, St Louis, MO 63198, USA

Sweetpotato is a subsistence crop which provides high calorific and nutritional value throughout the developing world. Due to its ability to withstand environmental stress, it is often a crop which can provide needed food during times of extreme drought. Despite its importance in Eastern Africa, sweetpotato yields can be very low due to virus diseases and insect infestations.

A team composed of scientists from the Kenya Agricultural Research Institute (KARI), Monsanto and the ABSP project have worked together to develop reproducible transformation procedures for the development of sweetpotatoes containing genes for virus resistance. Working in collaboration with Roger Beachy, The Scripps Research Institute, and Jim Moyer, North Carolina State University, plant transformation vectors containing the coat protein (*cp*) gene of the sweet potato feathery mottle virus (SPFMV, strain rc) were developed.

Nearly 200 independent transformation events of a sweetpotato variety grown in Kenya were produced which contained the SPFMV *cp* gene. The plants are currently undergoing evaluation for resistance against the rc strain of the virus. Encouraging results have been obtained and efforts will be undertaken to test resistance to the most virulent SPFMV isolates in Kenya. Growth chamber virus resistance assays will be conducted at Monsanto in 1998 to identify transgenic plants with high levels of resistance. This will be followed by field tests in Kenya in 1999.

Introduction

Sweetpotato is the fifth most important crop behind wheat, rice, maize and barley. It is grown as a staple food in some regions, as a food supplement in other regions and as food security during times of drought in most countries. Sweetpotato forms an important year-round food base, as a source of calories, vitamins and minerals. Its full impact as a food source can be realized only in its absence, as most of the crop is eaten before reaching the market. Kenya is currently producing about one million tonnes of sweetpotato annually (KARI Annual Report, 1990) which feeds a substantial population of poor farmers, especially women and children. It is a crop that a small stakeholder farmer can rely upon to feed his family when other crops fail, and it is very important in staving off famine in many regions of Kenya and Africa.

Sweetpotato can be a very productive crop with a world average yield of 14 t ha^{-1}. Yields of 18 t ha^{-1} have been realized in China; however, Africa's production is significantly lower (6 t ha^{-1}). Pests and diseases are the primary causes of this yield reduction. Viral diseases in Africa cause reduction of potential harvests by 20–80%. Control of sweetpotato viral diseases has been difficult.

The insects that act as vectors for the transmission of sweetpotato viruses are abundant during the sweetpotato growing season. Since sweetpotato is vegetatively propagated from cuttings, the viruses are unavoidably propagated from one year to the next. Farmers exchange planting stocks as a cultural practice, and the viruses are further perpetuated and disseminated from one farmer to another. Pesticide control for insects which spread the Kenyan sweetpotato viruses is somewhat ineffective, and the frequency at which insecticidal sprays are required for effective control makes this method unaffordable by resource-poor African farmers. Attempts have been made to plant virus-free sweetpotato stock, but the rate of reinfection of the clean material is too rapid within the first season (50% reinfection) to be economically feasible or practical (F. Wambugu, 1991, personal communication).

A possibly feasible and long-term avenue for control of sweetpotato viral diseases by the resource-poor farmers may be provided using biotechnology. Virus resistance has been achieved against several different viruses using a coat-protein-mediated approach. Monsanto, an international agribusiness company based in St Louis, Missouri, in conjunction with Dr Roger Beachy, during his tenure at Washington University in St Louis, has demonstrated that transgenic plants, which contain a gene for a virus coat protein, can be protected against subsequent infection by the virus which was the source of the coat protein. Solanaceous plants which show resistance in the field to tobacco mosiac virus (TMV), tomato mosiac virus (ToMV), cucumber mosaic virus (CMV), potato potyvirus X (PVX), potato potyvirus Y (PVY) and potato leaf roll luteovirus (PRLV) have been produced. The coat protein approach towards resistance to sweetpotato viruses was therefore considered likely to be successful.

The impact of virus resistant sweetpotatoes on African food security would be significant. If yield benefits of 15% could be realized, the total gain from sweetpotatoes in Africa alone would be 1.8 million t year^{-1}. It is estimated that the extra 1.8 million t of sweetpotatoes would supply the dietary needs for about 10 million people with no additional costs, inputs or acreage associated with its production.

Research Collaboration

Scientists at Monsanto feel that the virus resistance technology developed at Monsanto for commercial crops such as potato and tomato has the potential to reduce world hunger by increasing yields of subsistence crops such as sweetpotato. Towards this end, Monsanto entered into a pioneering programme in 1991 for technology transfer to the developing world. The programme, as originally conceived by Dr Ernest Jaworski and Dr Robert Horsch at Monsanto and Dr Joel Cohen at the US Agency for International Development (USAID), envisioned a partnership between researchers from Africa and from Monsanto for the development of virus-resistant sweetpotato using coat protein technology, which would be readily available to the resource-poor farmers. USAID and Monsanto committed to jointly sponsoring an African scientist to come to Monsanto to develop the transformation technology for sweetpotato, produce virus resistant plants, and facilitate the transfer of the transformation and virus-resistance technology, and the resulting transgenic plants, back to his or her home country for local evaluation and continued development.

The search for an African scientist produced Dr Florence Wambugu, a research scientist responsible for developing disease resistance strategies for sweetpotato in the Root and Tuber Program of the Kenya Agricultural Research Institute (KARI) in Nairobi, Kenya. Dr Cyrus Ndiritu, the Director for KARI, and Dr John Wafula, Biotechnology Program Director for KARI, supported the temporary assignment of Dr Wambugu to Monsanto. Dr Wambugu came to Monsanto in December 1991, with a wealth of knowledge concerning sweetpotato viruses and sweetpotato regeneration and micropropagation techniques as this was the basis for her PhD thesis. The match between Dr Wambugu's in-depth knowledge of sweetpotato and its diseases with Monsanto's general competency in plant transformation and molecular biology for virus resistance has proved to be a fertile relationship.

Dr Wambugu, together with Monsanto molecular biologist Dr Nilgun Tumer, Dr Roger Beachy and Dr Jim Moyer of North Carolina State University, identified the first target of the sweetpotato virus resistance project as the sweet potato feathery mottle virus (SPFMV). This virus does not cause major yield losses by itself, yet is frequently found in combination with one or several other viruses, resulting in a synergistic increase in disease severity caused by the other viruses. SPFMV is therefore implicated in many of the most devastating viral diseases of sweetpotato. It is hoped that by controlling this virus, deleterious symptoms caused in conjunction with other viruses may also be controlled.

The development of plant transformation vectors was the result of the efforts of several different laboratories. Dr Moyer provided the coat protein (*cp*) clone of the SPFMV strain rc, which is most common universally. Dr Beachy's laboratory (now at The Scripps Research Institute) introduced the coat protein clone into a plant expression gene cassette and provided it for Monsanto. Dr Wambugu, working with Monsanto scientists Dr Nilgun Tumer and Dr Fred Perlak, introduced the expression cassette, containing a *35S* promoter driving the SPFMV rc *cp* gene, into optimized plant transformation binary vectors for *Agrobacterium*-mediated plant transformation. The binary vector also contained two other genes: one which contained the *NOS* promoter driving the *npt* II 128 (a version of the neomycin phosphotransferase gene) selectable marker gene, and another which contained the *FMV* promoter driving the intron-containing *GUS* (β-glucuronidase) scorable marker gene. The resulting vectors were pMON10574 and pMON10575 which differed only in the respective orientation of the coat protein gene relative to the selectable marker gene. These vectors were introduced into *Nicotiana bentheminana* and expression of the SPFMV coat protein gene was confirmed by western analysis.

Dr Wambugu's work at Monsanto during 1992–1994, focused on developing the basis for a sweetpotato transformation system. She worked with a KARI scientist, Ms Charity Macharia, to have virus-free samples of eight different locally-adapted African sweetpotato genotypes shipped to Monsanto. The African varieties could be used in the laboratory, but were not allowed to be environmentally released since they had not undergone US Department of Agriculture (USDA) quarantine. These genotypes, as well as eight US genotypes, were assessed for their responsiveness to several different micropropagation and regeneration methods. Dr Wambugu's own PhD research contributed directly to the development of a regeneration protocol which worked for several genotypes, but most notably with a genotype designated CPT560.

Much more basic research towards the development of the regeneration and transformation methodology for sweetpotato had to be done at Monsanto than had been expected, as previously publicly described methods did not perform as hoped in the Monsanto laboratories. Dr John Wafula and Dr Cyrus Ndiritu of KARI continued to support the project by allowing their top scientists to come to Monsanto to continue research towards virus resistance in sweetpotato. A second scientist from KARI, Mr Daniel Maingi, was sponsored in 1994, to support and accelerate the adaption of *Agrobacterium* transformation techniques to the established regeneration protocol developed by Dr Wambugu. Direct USAID support for the Monsanto sweetpotato virus resistance project was withdrawn in 1994, due to budgetary constraints. Monsanto continued support of the project, and KARI stepped forward to support the work of Mr Daniel Maingi through the Mid-America International Agriculture Consortium (MIAC) training programme administered by Dr Charles Campbell at the University of Missouri. USAID continued supporting the project indirectly, through its sponsorship of the MIAC programme at KARI and through collaboration with the Agricultural Biotechnology for Sustainable Productivity (ABSP) project based at Michigan State University (MSU).

Dr Wambugu left the project to pursue a career opportunity as Director of the AfriCenter office of the International Service for the Acquisition of Agri-Biotech Applications (ISAAA). She continues to support the project as a key consultant in the area of sweetpotato virology and epidemiology. At Monsanto, Mr Maingi continued development of a reliable sweetpotato transformation method. Dr Jeffrey Lowe, a postdoctoral researcher sponsored by the ABSP project, who had a charter to develop sweetpotato transformation technology for Indonesia, joined Mr Maingi in the effort to develop a highly efficient system. Dr Lowe focused his attention on modifying both the Monsanto protocol and published protocols to ensure that a broad range of sweetpotato genotypes would be amenable to transformation. Mr Maingi continued to focus on optimization and reliability of the CPT560 transformation protocol.

The transformation and regeneration protocol is based on using explants from micropropagated plants of sweetpotato. Some 3 to 4 weeks after sub-culturing stem segments with axillary buds on micropropagation medium, the plantlets provided the best young leaf material for transformation and regeneration. The leaves were precultured for a day, inoculated with *Agrobacterium*, cocultured for 3 days and then placed immediately on selective callus induction medium containing paromomycin. The paromomycin selected callus was placed on shoot induction medium, with or without paromomycin selection, and regenerated shoots were elongated then placed on micropropagation medium for asexual multiplication and confirmation of transformation.

The transformation protocol which Mr Maingi developed between 1994 and 1996, ultimately produced 200 transgenic sweetpotato plants (the majority obtained from the genotype CPT560) containing either pMON10574 or pMON10575 (containing the SPFMV *cp* gene). Each transgenic plant was shown to express either the *npt* II and/or *GUS* gene by enzyme-linked immunosorbent assay (ELISA). A limited number of lines were further confirmed to contain the *cp* gene by Southern analysis. The sweetpotato transformation and regeneration procedure was confirmed at this point to be reliable and reproducible, and had reached a point at which it could be easily provided to other researchers.

The Monsanto sweetpotato team continued their sweetpotato transformation training with the ABSP project through efforts coordinated by Dr Catherine Ives and Mr Bruce Bedford. Another ABSP-sponsored KARI scientist, Ms Charity Macharia, joined Mr Maingi for 6 months training in sweetpotato transformation and to assist in the production of transgenic plants. She returned to Kenya in mid-1996, to establish the sweetpotato regeneration and transformation system at KARI, laying the cornerstone for successful technology transfer. At approximately the same time, an Indonesian scientist, Ms Minantyorini, from the Central Research Institute for Food Crops, Indonesia, was sponsored by ABSP to visit the laboratory at Monsanto for several months to train in the sweetpotato transformation protocol with Ms Macharia and Mr Maingi.

With the availability of a large number of transgenic sweetpotato plants, it became a high priority to develop reliable methods for screening the plants for their resistance to SPFMV infection. KARI scientists Dr Anne Wangai, Dr Duncan Kirubi and Ms Charity Macharia, and Monsanto scientist Ms Maria Kaniewska, with the advice and support of another Monsanto scientist, Dr Wojciech Kaniewski, began the painstaking task of developing a reliable screen for SPFMV resistance. First, Ms Kaniewksa worked on this task with Ms

Macharia and a student intern from Costa Rica, Mr Miquel Rendon. They found that the SPFMV rc strain is not easily transmitted by mechanical means, and evaluated two alternative methods of virus transmission: (i) grafting transgenic scions to SPFMV root stock, and (ii) feeding with viliferous green peach aphids. Both methods worked effectively, but aphid transmission was chosen as the most efficient method for infection. Dr Wangai, followed by Dr Kirubi, worked to achieve infection rates in control plants in the range of ≥70%.

One complication discovered when screening the infected plants for the virus using the ELISA technique was that the SPFMV virus was not distributed uniformly throughout the stem, petiole and leaf tissue of the plant. The irregular distribution of SPFMV in the plants led to significant sampling errors when determining resistance. An improved method for detecting the spread of the virus in infected transgenic plants was developed which relies on tissue printing. Results to date are highly suggestive that the SPFMV *cp* gene from the rc strain can provide resistance when challenged with the rc strain of the SPFMV virus. Up to five lines, from 28 evaluated, were shown in preliminary testing to be negative for the presence of SPFMV rc infection and spread after aphid transmission, while control plants without the *cp* gene were infected at levels greater than 70%. This is a very exciting result considering that approximately 170 plants are in the process of being screened, and more candidates demonstrating resistance are expected to be found.

Currently, evaluation continues of all the transgenic CPT560 sweetpotato lines for their level of resistance to the rc strain of the SPFMV. In addition, Monsanto is supporting KARI in collecting field isolates of SPFMV from infected sweetpotatoes in Kenyan fields. Ms Charity Macharia, with the support of Dr John Wafula and the advice of Dr Florence Wambugu, has collected isolates which will be used in virus challenge experiments at Monsanto in St Louis. Transgenic plants which have shown resistance to the rc strain will be challenged with the Kenyan field isolates, and the transgenic lines which perform best against the Kenyan isolates will be sent to Kenya for further evaluation. Prior to shipment, these plants will be fully characterized by Southern blot for their insert copy number and by Western blot and quantitative ELISA for their transgene expression (SPFMV strain rc *cp*, *GUS*, and *npt* II). It is expected that the plants will be sent to Kenya in 1998 and possibily tested that year or the following year in the field.

The ABSP project has been working to faciliate the transfer of the SPFMV resistant sweetpotatoes by taking responsibility for drafting the notification to USAID regarding the shipment of transgenic material from the United States to Kenya. Dr Patricia Traynor, an ABSP biosafety consultant of Virginia Tech has played an essential role in developing this document.

The KARI team of Dr Duncan Kirubi, Ms Charity Macharia and Ms Fatuma Omari will continue the work on the sweetpotato project in Kenya. This team, with other KARI employees, will be responsible for the biosafety approvals for testing the transgenic sweetpotatoes in the field in Kenya. They will be reponsible for ensuring that adequate resources and personnel are in place for successful greenhouse and field evaluation of the SPFMV-resistant sweet-potatoes. The KARI sweetpotato team is also expected to continue additional work on improving the technology for Kenya. This work takes the form of additional transformations for SPFMV resistance in other locally adapted sweetpotato varieties and further characterization of the performance of these transgenic plants against SPFMV.

This project is just one example of what is possible when a public institution from a developing country forms a partnership with a biotechnology company. It serves as a model for how multiple institutions can work together as a team towards a common goal of increasing crop yields for subsistence growers. USAID, through MIAC and ABSP, ISAAA, KARI and Monsanto all were essential contributors to the technical progress made at Monsanto by Kenyan scientists. It is also a model for how scientists from a developing country can take a leadership role in initiating and eventually managing biotechnology projects for locally important crops. In addition, the project can be viewed as a model for future endeavours in which commercial technology can be adapted for use by subsistence growers. The customers in this case are small stakeholders in need of readily accessible tools to help them increase their yields and food security.

Acknowledgements

Maud Hinchee[3] served as lead author on this chapter and presented this information to the ABSP Project 'Biotechnology for a Better World' Conference at Monterey, California in April 1997. Acknowledgements are due Daniel Maingi[1], Duncan Kirubi[1], Florence Wambugu[2], Charity Macharia[1], Jeff Lowe[3], Maria Kaniewska[3], Anne Wangai[1], Wojciech Kaniewski[3] and Robert Horsch[3] for their contribution to the project, to the information in this chapter and its authorship.

[1] Kenya Agricultural Research Institute, PO Box 57811, Nairobi, Kenya.
[2] ISAAA AfriCenter Office, Naivasha Road, PO Box 25171, Nairobi, Kenya.
[3] Monsanto, 700 Chesterfield Parkway North, St Louis, Missouri, USA.

The Application of Biotechnology to Rice

9

Gurdev S. Khush and Darshan S. Brar

*International Rice Research Institute (IRRI), PO Box 933,
1099 Manila, The Philippines*

Modern varieties have had a dramatic impact on the increase in rice production. World rice production doubled in a 25-year period, from 257 million tonnes in 1966, to 520 million tonnes in 1990; and most of the rice-growing countries in Asia, where 92% of rice is grown, became self-sufficient. However, the population of rice consumers continues to increase at an alarming rate and the demand for rice is likely to call for increased production by the end of this century. To meet the demand for rice in the next century, rice varieties with higher yield potential and yield stability are needed.

Recent breakthroughs in cellular and molecular biology have provided tools that can increase the efficiency of breeding methods and allow unconventional approaches to rice improvement. These tools are being used to develop rice varieties which will meet the challenge of increased rice production. Some of the advances in rice biotechnology include: successful regeneration from protoplasts of japonica and indica rice; production of transgenic rice lines through protoplast manipulation and biolistic approaches; development of a comprehensive molecular map consisting of more than 2000 DNA markers; tagging of genes for economic importance with restriction fragment length polymorphism (RFLP) and random amplified polymorphic DNA (RAPD) markers; and molecular characterization of pathogen populations and production of interspecific hybrids through embryo rescue.

Anther culture and molecular-aided selection are being used to increase the efficiency of conventional rice breeding. Molecular characterization of pathogen populations is helpful in determining gene deployment strategies. Wide hybridization and transformation are being employed to broaden the gene pool of rice.

© CAB INTERNATIONAL 1998. *Agricultural Biotechnology in International Development*
(eds C.L. Ives and B.M. Bedford)

Introduction

Major advances have been made in increasing rice production as a result of large-scale adoption of modern high-yielding varieties and improved production technologies. The world rice production has more than doubled from 257 million tonnes in 1966, to 560 million tonnes in 1996. High-yielding varieties were mainly developed through the application of classical Mendelian principles and conventional plant breeding methods. To meet the need of growing human populations, we will have to produce 70% more rice by the year 2025. To meet this challenge, we need varieties with higher yield potential. Moreover, several biotic and abiotic stresses continue to limit rice productivity. To overcome these constraints to rice production, varieties must have multiple resistance to diseases and insects, and tolerance of abiotic stresses.

Plant breeding consists of two phases: the evolutionary phase, where variable populations are created; and the evaluationary phase, where superior genotypes are selected on the basis of evaluation. Recent advances in cellular and molecular biology have provided scientific tools to increase the efficiency of both phases. Some of the exciting developments in biotechnology include: (i) successful plant regeneration from protoplasts of several japonica and indica cultivars; (ii) production of somatic hybrids and cybrids through protoplast fusion; (iii) production of transgenic plants carrying agronomically important genes; (iv) development of a comprehensive molecular map consisting of more than 2000 DNA markers; (v) tagging, via linkage with molecular markers, several genes governing resistance to major diseases and insects; (vi) mapping of quantitative trait loci (QTL) governing several traits of low heritability; (vii) development of polymerase chain reaction (PCR)-based markers; (viii) development of protocols for marker-aided selection; (ix) determination of synteny relationships between genomes of rice and other cereals; (x) construction of bacterial artificial chromosome (BAC) libraries for physical mapping of the rice genome and their utilization in map-based cloning of genes; and (xi) transfer of useful genes from wild into cultivated rice across crossability barriers.

Evaluation Phase: Increasing Selection Efficiency

Anther culture

As early as 1968, Niizeki and Oono reported the production of haploids from anther cultures of rice (Niizeki and Oono, 1968). Since then, the anther culture technique has been greatly refined. The effect of the genotype, the physiological state of the donor plant, the stage of pollen development, the pretreatment of anthers, and the media composition for callus induction and efficiency of plant regeneration from anther culture have been extensively investigated (Raina, 1989). As a result, it is now possible to produce haploids from the anther cultures of many japonica and indica rices, although the frequency of regenerated plants is relatively lower in indicas.

Anther culture is important in developing true breeding lines in the immediate generation from any segregating population, thereby shortening the breeding cycle of new varieties. In conventional breeding, after a cross is made between desired parents, the segregating populations from F_2 to F_7 are grown to ultimately develop true breeding homozygous lines; anther culture-derived lines lack heterozygosity and can be multiplied and evaluated in the immediate generation. The selection efficiency with doubled haploid (DH) lines is higher, especially when dominance variation is significant (Snape 1982). Early generation lines (F_3–F_5) show phenotypic differences because of additive and dominance variance. In contrast, there is no dominance variance among DH lines and heritability of the trait is high. DH lines are also useful for developing mapping populations for molecular analysis. Seeds of such populations can be distributed to workers in other laboratories, and populations can be grown repeatedly in many different environments, greatly facilitating the additional mapping of both DNA markers and genetic loci controlling traits of agronomic importance. In rice, mapping populations (DH lines) produced through anther culture of IR64 × Azucena, an indica/japonica cross, and CT9993 × IR62266 are being used in the molecular mapping of genes including quantitative trait loci.

Use of anther culture in the development of rice varieties has been reported in several countries, including China, Japan, the Republic of Korea and the US. Most of the anther culture-derived varieties are japonicas. Indica rices are planted on 90% of land under rice cultivation, but they are generally regarded as recalcitrant, not only to anther culture but also to other tissue culture techniques. At the International Rice Research Institute (IRRI), the anther culture technique is being employed to obtain DH lines from many crosses for different rice-breeding objectives. To date, about 8000 DH lines have been regenerated. On average, about 20–50 DH lines are produced from one cross. This efficiency is being further increased to reach the target of 100 lines per cross. Several superior DH lines have been selected from crosses for salinity tolerance (Zapata *et al.*, 1991). One of the anther culture derived lines, IR51500-AC-11-1 has been named as a variety (PSBRC50) in The Philippines. Many DH lines produced through anther culture are now being used as parents in breeding programmes.

Molecular-marker-aided selection

Numerous genes of economic importance, such as those for disease and insect resistance, are repeatedly transferred from one varietal background to another by plant breeders. Most genes behave in a dominant and recessive manner and require time-consuming efforts to transfer. Sometimes the screening procedures are cumbersome and expensive, and require large field space. If such genes can be tagged by tight linkage with DNA or isoenzyme markers, time and money can be saved in transferring these genes from one varietal background to another. The presence or absence of the associated molecular marker would indicate, at a very early stage, the presence or absence of the desired target gene. Codominance of the associated molecular marker allows all the possible genotypes to be identified in any breeding scheme, even if the gene for economic traits cannot be scored directly. A molecular marker very closely linked to the target gene can act as a 'tag' which can be used for indirect selection of the gene(s) in a breeding programme.

Molecular maps

The availability of a comprehensive molecular genetic map for rice comprising more than 2000 DNA markers has been a major advance in rice genetics. A molecular genetic map of rice, based on restriction fragment length polymorphisms (RFLPs), was developed at Cornell University, Ithaca, USA, in collaboration with IRRI (McCouch *et al.*, 1988). This map was based on an indica × japonica F_2 population and mapped sequences were cloned from a genomic library derived from the indica variety IR36. A second RFLP map was based on a different indica × japonica cross (Saito *et al.*, 1991). Causse *et al.* (1994) developed the map, comprising 726 markers. The mapping population was derived from the backcross between the cultivated rice *Oryza sativa* and a wild species, *O. longistaminata*. More recently, a comprehensive molecular genetic map consisting of 1383 DNA markers has been developed under the Rice Genome Research Program in Japan (Kurata *et al.*, 1994b). The markers, distributed along 1575 cM on 12 linkage groups, comprise 883 cDNAs, 265 genomic DNAs, 147 random amplified polymorphic DNAs (RAPDs) and 88 other DNAs. Recently, Singh *et al.* (1996) mapped centromeres on the molecular genetic map of rice and determined the correct orientation of linkage groups.

Gene tagging

Two of the most serious and widespread diseases in rice production are rice blast caused by the fungus *Pyricularia oryzae*, and bacterial blight caused by *Xanthomonas oryzae* pv. *oryzae*. Development of varieties with durable resistance to these diseases is the focus of a coordinated effort at IRRI using molecular marker technology. Early gene mapping efforts have been directed at tagging several single genes conferring resistance to these pathogens. As a result, several genes for bacterial blight and for blast resistance have been tagged with molecular markers (Table 9.1).

Resistance to blast has long posed a serious challenge in rice improvement. Resistance, particularly when based on single major genes, has generally been overcome in one or a few years in blast prone environments. Some cultivars, however, have shown longer lasting or durable resistance. In several cases, durable resistance to blast is believed to be associated with quantitative or polygenic inheritance. Under these conditions, there would be little or no gain in fitness for a pathogen variant that could overcome only a fraction of the polygenes. Many rice improvement programmes now aim at incorporating quantitative or partial resistance into rice varieties.

An RFLP mapping study was conducted to understand the genetics of blast resistance in Moroberekan, a traditional West African upland cultivar considered to have durable resistance (Wang *et al.*, 1994). A population of F_7 recombinant inbred lines was used for this analysis. One gene conferring complete resistance was identified on chromosome four and was designated *Pi-5(t)*. Both interval analysis and linear regression have identified nine regions of the genome with quantitative effects on blast resistance. These were considered to be putative QTL for blast resistance. It was interesting that three of the loci found to be associated with partial resistance in this study had been previously identified as being linked to genes for complete resistance. This raises the possibility that the same loci conditioning complete resistance under some circumstances may act as QTL under others. Analyses of single resistance genes have also been carried out using near isogenic lines (NILs). The NILs have been used for mapping of the genes *Pi-2(t)* and *Pi-4(t)*.

Efforts to detect markers closely linked to bacterial blight resistance genes have similarly taken advantage of the availability of ten sets of NILs. These lines each contain a single gene for resistance and, in the case of *Xa-21*, the gene has been introgressed from the wild relative of rice, *O. longistaminata* (Khush *et al.*, 1990). Segregating populations have been used to confirm co-segregation between the RFLP markers and *Xa-1*, *Xa-2*, *Xa-3*, *Xa-4*, *xa-5*, *Xa-10*, *xa-13* and *Xa-21*.

To test the utility of markers as selection tools and to study the effect of gene combinations in a particular genetic background, pairs of genes were combined by marker assisted selection (MAS) and confirmed when possible by phenotypic selection. These markers provide immediate information on plants which carry a target gene and whether the gene exists in a homozygous or heterozygous state. Markers also allow accurate identification of genotypes in which two or more genes have been combined in one individual, but where the phenotype of the individual is indistinguishable from those carrying fewer genes. On the basis of this capability, the hypothesis is being tested that polygenic forms of resistance can be effectively developed by combining genes that were originally identified as conferring qualitative or single gene resistance. Unexpected interactions between genes that have been well characterized individually can also be evaluated using this approach.

Genes for aroma, wide compatibility, fertility restoration and thermo-sensitive genetic male sterility, brown plant hopper resistance, and tolerance of tungro and submergence have been tagged with molecular markers (Table 9.1). PCR-based markers have been developed for some of these traits to facilitate selection.

Nair *et al.* (1995, 1996) developed PCR-based markers for *Gm-2* and *Gm-4(t)* which are useful in MAS for developing gall midge resistant varieties. The emphasis now is on tagging genes of economic importance by developing PCR-based markers. This will enhance the efficiency of MAS and pyramiding of useful genes for tolerance to various biotic and abiotic stresses.

Table 9.1. Some examples of mapping genes of agronomic importance with molecular markers in rice.

Gene	Trait	Chromosome	Linked marker and distance		Reference
Pi-1	Blast resistance	11	Npb181	3.5 cM	Yu, 1991
Pi-2(t)	Blast resistance	6	RG64	2.1 cM	Yu et al., 1991
					Hittalmani et al., 1995
Pi-4	Blast resistance	12	RG869	15.3 cM	Yu et al., 1991
Pi-ta	Blast resistance	12	RZ397	3.3 cM	Yu et al., 1991
Pi-5 (t)	Blast resistance	4	RG498	5–10 cM	Wang et al., 1994
			RG788		
Pi-6 (t)	Blast resistance	12	RG869	20 cM	Yu, 1991
Pi-7 (t)	Blast resistance	11	RG103	5–10 cM	Wang et al., 1994
Pi-9 (t)	Blast resistance	6	RG16	–	R. Nelson (per. com.)
Pi-10 (t)	Blast resistance	5	RRF6, RRH18	–	Naqvi et al., 1995
Pi-11 (t)	Blast resistance	8	BP127	2.4 cM	Zhu et al., 1992
Pi-b	Blast resistance	2	RZ123	–	Miyamoto et al., 1996
Xa-1	Bacterial blight resistance	4	Npb235	3.3 cM	Yoshimura et al., 1992
			Npb197	7.2 cM	
Xa-2	Bacterial blight resistance	4	Npb235	3.4 cM	Yoshimura et al., 1992
			Npb197	9.4 cM	
Xa-3	Bacterial blight resistance	11	Npb181	2.3 cM	Yoshimura et al., 1992
			Npb78	3.5 cM	
Xa-4	Bacterial blight resistance	11	Npb181	1.7 cM	Yoshimura et al., 1992, 1995
			Npb78	1.7 cM	
xa-5	Bacterial blight resistance	5	RG556	0–1 cM	McCouch et al., 1991
Xa-10	Bacterial blight resistance	11	$OP07_{2000}$	5.3 cM	Yoshimura et al., 1995
xa-13	Bacterial blight resistance	8	RZ390	0 cM	Yoshimura et al., 1995
			RG136	3.8 cM	Zhang et al., 1996

Gene	Trait	Chr.	Marker	Distance	Reference
Xa-21	Bacterial blight resistance	11	pTA818 pTA248 RG103	0–1 cM	Ronald *et al.*, 1992
RTSV	Rice tungro spherical virus resistance	4	RZ262	5.5 cM	Sebastian *et al.*, 1996
Bph-1	Brown plant hopper resistance	12	XNpb 248	–	Hirabayashi and Ogawa, 1995
Bph-10 (t)	Brown plant hopper resistance	12	RG457	3.68 cM	Ishii *et al.*, 1994
ef	Early flowering	10	CDO98	9.96 cM	Ishii *et al.*, 1994
fgr	Fragrance	8	RG28	4.5 cM	Ahn *et al.*, 1992
Wph-1	Whitebacked plant hopper resistance	7	–	–	McCouch, 1990
Gm-2	Gall midge resistance	4	RG329 RG476	1.3 cM 3.4 cM	Mohan *et al.*, 1994
Rf-3	Fertility restorer	1	RG532	0–2 cM	Zhang *et al.*, 1997
S-5	Wide compatibility	6	RG213	4.4 cM	Yanagihara *et al.*, 1995
Se-1	Photoperiod sensitivity	6	RG64	0	Mackill *et al.*, 1993
Se-3	Photoperiod sensitivity	6	A19	5–10 cM	Maheswaran, 1995
sdg(t)	Semidwarf	5	RZ182	4.3 cM	Liang *et al.*, 1994
sd-1	Semidwarf	1	RG109	0.8 cM	Cho *et al.*, 1994
tms-1	Thermosensitive male sterility	8	–	–	Wang *et al.*, 1995b
tms-3(t)	Thermosensitive male sterility	6	$OPAC3_{640}$	–	Subudhi *et al.*, 1997
PMS 1	Photoperiod sensitivity male sterility	7	RG477	4.3 cM	Zhang *et al.*, 1993
PMS 2	Photoperiod sensitivity male sterility	3	RG191	–	Zhang *et al.*, 1993
Sub-1(t)	Submergence tolerance	9	RZ698	–	Nandi *et al.*, 1997

QTL mapping

Although a number of important characters are determined by loci having major effects on phenotype, most economically important traits such as yield, quality and tolerance of various abiotic stresses (drought, salinity, submergence, etc.) are of a quantitative nature. Genetic differences affecting such traits (within and between populations) are controlled by a relatively large number of loci, each of which can make a small positive or negative contribution to the final phenotypic value of the traits. The genes governing such traits, called polygenes or minor genes, also follow Mendelian inheritance but are greatly influenced by the environment. Biometrical procedures involving special experimental designs and data analyses are used to study genetics of quantitative traits.

The advent of molecular markers has made it possible to map the QTL which have large genotypic effects on phenotype. These molecular markers provide methods to transform QTL into Mendelian or quasi-Mendelian entities that can be manipulated in classical breeding programmes.

Wang *et al.* (1994) probed 281 recombinant inbred lines (RIL) with 156 RFLP markers. The RIL were evaluated for qualitative resistance using five isolates of blast pathogen, and for quantitative resistance using a single isolate in polycyclic tests in which multiple infection cycles were allowed to proceed. Two genes, *Pi-5(t)* and *Pi-7(t)*, were mapped on chromosomes 4 and 11 respectively. Nine QTLs having quantitative effect on blast resistance to isolate *Po6−6* were identified (Table 9.2).

Table 9.2. Some examples of QTL mapping in rice.

Trait	QTLs	Reference
Blast resistance	9	Wang *et al.*, 1994
Root–shoot ratio	3	Champoux *et al.*, 1995
Root dry weight	9	Champoux *et al.*, 1995
Coleoptile length	2	Redona and Mackill, 1996
Root length	5	Redona and Mackill, 1996
Mesocotyl length	5	Redona and Mackill, 1996
Plant height	6	Xiao *et al.*, 1996b
Days to heading	3	Xiao *et al.*, 1996b
Days to maturity	2	Xiao *et al.*, 1996b
Panicle length	2	Xiao *et al.*, 1996b
Panicles per plant	1	Xiao *et al.*, 1996b
Spikelets per panicle	4	Xiao *et al.*, 1996b
1000 grain weight	3	Xiao *et al.*, 1996b
Grains per plant	3	Xiao *et al.*, 1996b
Grain yield	2	Xiao *et al.*, 1996b
Total root weight	5	Yadav *et al.*, 1997
Root thickness	4	Yadav *et al.*, 1997
Maximum root length	6	Yadav *et al.*, 1997
Submergence tolerance	4	Nandi *et al.*, 1997

Redona and Mackill (1996) identified two QTL controlling root length, and five each influencing coleoptile and mesocotyl length. Nandi *et al.* (1997) used amplified fragment length polymorphism (AFLP) markers and identified four QTL for submergence tolerance on chromosomes 6, 7, 11 and 12 of rice. In addition, a major gene, *Sub-1(t)*, for submergence tolerance was localized on chromosome 9. Xiao *et al.* (1996b) analysed QTL governing various agronomic traits. Some of these traits were controlled by one to six QTL (Table 9.2). Xiao *et al.* (1996a) analysed BC_2 test cross families from the interspecific cross (*O. sativa* × *O. rufipogon*) and found that *O. rufipogon* alleles at marker loci RM5 on chromosome 1, and RG256 on chromosome 2, were associated with enhanced yield potential. The phenotypic advantage of the lines carrying *O. rufipogon* alleles at these loci corresponded to 18% and 17% yield increase over the *O. sativa* parent. These results indicate that molecular markers can be used to identify QTL from wild species responsible for transgressive segregation. Comparative mapping among different cereals has also increased the efficiency of orthologous QTL mapping.

Future research should focus on: (i) identification of QTL which could be exploited in different environments; (ii) exploitation of complementary QTL to isolate transgressive segregants, particularly from interspecific crosses; (iii) identification of orthologous QTL among different species, as the conservation of such QTL among species may provide new opportunities for manipulation of economic traits; (iv) high resolution of QTL to help determine whether QTL are single genes or clusters of tightly linked genes and whether overdominance plays a significant role in heterosis; and (v) cloning of QTL, based on high resolution mapping to usher in a new era in molecular quantitative genetics.

Pyramiding genes for bacterial blight (BB) resistance

The development of saturated molecular maps, the possibility of finding tight linkage of target genes with molecular markers such as RFLPs and conversion of these markers to PCR-based markers have provided new opportunities to use MAS in rice breeding. Recently, we have developed protocols for PCR-based MAS in rice (Zheng *et al.*, 1995). In MAS, individuals carrying target genes are selected in a segregating population based on tightly linked markers rather than on their phenotypes. Thus, the populations can be screened at an early seedling stage and in various environments. MAS can overcome interference from interactions between alleles of a locus or of different loci. MAS increases the efficiency and accuracy of selection, especially for traits which are difficult to phenotype. We have successfully used MAS for pyramiding genes (*Xa-4, xa-5, xa-13, Xa-21*) for bacterial blight resistance (Huang *et al.*, 1997). Breeding lines with two, three or four BB resistance genes have been developed. The pyramided lines showed a wider spectrum and higher level of resistance than lines with only a single gene. Pyramided lines carrying four BB resistance genes are being used in MAS to transfer these genes into elite breeding lines of a new rice plant type.

Yoshimura *et al.* (1995) selected lines carrying *Xa-4* + *xa-5* and *Xa-4* + *Xa-10* using RFLP and RAPD markers linked to the BB resistant genes. These lines were evaluated for reaction to eight strains of BB, representing eight pathotypes and three genetic lineages. Lines carrying *Xa-4* + *xa-5* were more resistant to isolates of race 4 than were either of the parental lines. Such pyramided lines with different gene combinations are useful for developing varieties with durable resistance.

In order to improve the durability of resistance of future rice varieties, we are using molecular marker analysis to identify pathogen populations with wide diversity, which are then employed for screening and for developing resistant genotypes. We are also using deployment strategies based on understanding of pathogen population genetics and on the genetic basis of durable resistance (Leung *et al.*, 1993).

The high-density molecular genetic map of rice is also of great value in understanding genome organization and in map-based cloning of agriculturally important genes. Rice serves as a model for genome research in monocots because of its small genome size, excellent germplasm collection, good genetic stocks and relatively well-developed molecular genetic maps. Development of molecular genetic maps has been of great value in understanding the homoeologous relationships between the genomes of various crop plants. Ahn *et al.* (1993) found extensive homoeologies in a number of regions of the genomes of wheat, rice and maize. Kurata *et al.* (1994a) analysed synteny between rice and wheat and found that many wheat chromosomes contained homoeologous genes and genomic DNA fragments in a similar order to that found in rice. Comparative genome mapping in rice, maize, wheat, barley and sorghum is proceeding rapidly, and the rice research community is in an excellent position to take advantage of the possibilities for exchanging materials and information on these crops as opportunities arise. Based on comparative mapping, the species with smaller genome size, such as rice, may be used to accelerate map-based cloning of orthologous genes. The small genome size of rice, about 40 times smaller than that of wheat and only 2.5 times larger than that of *Arabidopsis thaliana*, makes rice an excellent candidate for isolation of genes through chromosome walking and map-based cloning. The possibilities of cloning rice genes based on map position have been greatly enhanced with the development of bacterial artificial chromosome (BAC) libraries. Wang *et al.* (1995a) used BAC library and identified clones linked to the *Xa-21* bacterial blight resistance gene in rice. Song *et al.* (1995) isolated the *Xa-21* gene by positional cloning and used this gene in rice transformation. At IRRI, a BAC library has been constructed from IR64 genomic DNA (Yang *et al.* 1997). The library consists of 18,432 clones. We used 31 RFLP markers on chromosome 4 to screen the library by colony hybridization. Positive clones were analysed to generate 29 contigs. These contigs are serving as landmarks for physical mapping of chromosome 4, and as starting points for chromosome walking towards map-based cloning of disease resistance genes.

Evolutionary Phase: Broadening Gene Pool of Rice Cultivars

Wide hybridization

The genus *Oryza*, to which cultivated rice belongs, has 22 wild species. Diploid *O. sativa* ($2n = 24$) has the AA genome and is the predominantly cultivated species, while *O. glaberrima*, also with the AA genome, is cultivated in a limited area in West Africa. The wild species of rice have either $2n = 24$ or 48 chromosomes with AA, BB, BBCC, CC, CCDD, EE, FF, GG and HHJJ genomes. Genetic variability for some traits, such as resistance to tungro, sheath blight and yellow stem borer, and tolerance of abiotic stresses, is limited in the cultivated germplasm. Wild species offer useful sources of genes for rice improvement.

Several barriers are encountered in transferring useful genes from wild species to cultivated rice (Sitch, 1990). The barrier most commonly encountered is lack of crossability because of chromosomal and genic differences. Biotechnology tools such as embryo rescue and protoplast fusion have become available to overcome the crossability barriers, and several interspecific hybrids have been produced. Molecular techniques have been employed in the precise monitoring of alien gene introgression and in the transfer of useful genes to cultivated crop plants.

Production of interspecific hybrids through embryo rescue
Abortion of embryos at different stages of development is a characteristic feature of wide crosses. Hybrids have been produced through embryo rescue between elite breeding lines or varieties of rice and several accessions of 11 wild species representing BBCC, CC, CCDD, EE, FF, GG and HHJJ genomes (Brar *et al.*, 1991). These hybrids have been produced to transfer useful genes for resistance to brown plant hopper (BPH), bacterial blight, blast, sheath blight, tungro virus, and yellow stem borer. Additional hybrids involving rice and AA genome wild species have been produced for the diversification of cytoplasmic male sterility and for the transfer of tolerance to acid sulphate conditions. In most interspecific crosses, the F_1 hybrids are completely sterile. Progenies are advanced through embryo rescue in subsequent backcrosses to the respective recurrent rice parents until plants with $2n = 24$ and 25 chromosomes become available. Monosomic alien addition lines (MAALs) have been produced involving alien chromosomes of five wild species. These MAALs serve as an additional source of alien genetic variation.

Gene transfer from wild species to rice
A number of useful genes have been transferred from wild species to rice (Jena and Khush 1990; Khush *et al.*, 1990; Brar and Khush, 1997). Khush (1977) transferred the gene for grassy stunt virus resistance from *O. nivara* to rice through backcrossing. Grassy stunt virus resistance has been bred into several rice cultivars. Recently, useful genes were transferred from other wild species

(Table 9.3). Jena and Khush (1990) transferred genes for resistance to three Philippine biotypes of BPH from *O. officinalis* into the elite breeding line IR31917-45-3-2. Some of the derived lines have also shown resistance to BPH populations in Bangladesh and India. These lines are free from the undesirable features of wild species such as grain shattering and poor plant type, and are on par with the recurrent parent for grain yield. They are being used as parents in rice breeding at IRRI and in other national programmes. Four breeding lines resistant to BPH derived from *O. officinalis* have been released as varieties in Vietnam. Genes for blast, bacterial blight and BPH resistance have been transferred to rice from other wild species: *O. minuta*; *O. latifolia*; *O. australiensis*; and *O. brachyantha* (Table 9.3).

Cytoplasmic diversification

Most of the commercial hybrids of indica rice are based on the wild abortive (WA) source of cytoplasmic male sterility (CMS). More than 95% of the rice hybrids grown in China have WA cytosterile cytoplasm. Such cytoplasmic uniformity increases the vulnerability of hybrid rice to diseases and insects. To overcome this problem, diversification of the cytoplasmic male sterility source is essential. We crossed 45 accessions of *O. perennis* and four accessions of *O. rufipogon* as the female parents with the widely grown varieties IR54 and IR64. Both IR54 and IR64 can restore the fertility of CMS lines possessing WA cytoplasm.

Of all the backcross derivatives, one line with the cytoplasm of *O. perennis* (Acc. 104823) and the nucleus of IR64 was found to be stable for complete pollen sterility. The newly developed CMS line has been designated IR66707A (Dalmacio *et al.*, 1995). Crosses of IR66707A with six restorers of WA cytoplasm also show almost complete pollen sterility, indicating that this source of CMS is different from that of WA cytoplasm. Southern hybridization of IR66707A, *O. perennis* and IR66707B with eight mitochondrial DNA-specific probes was carried out. Of 40 combinations, 18 showed a monomorphic pattern, while in 22 polymorphic combinations the banding patterns of IR66707A and *O. perennis* were identical. The results indicated that IR66707A has the same mitochondrial genome as the donor *O. perennis* and that CMS may not be caused by any major rearrangement or modification of mitochondrial (mt)DNA.

Characterization of alien genetic variation through RFLP markers

Both isoenzyme and RFLP markers detect extensive polymorphism between rice and wild species and have proved useful as genetic markers in the characterization of alien genetic variation. Introgression has been detected for isoenzyme loci from several wild species. RFLP analysis of the introgression lines derived from *O. sativa* × *O. officinalis* showed introgression of the chromosome segments in 11 of the 12 chromosomes of *O. officinalis* (Jena *et al.*, 1992). Using molecular markers, introgression of small chromosome segments has been detected from chromosomes 10 and 12 of *O. australiensis* into rice (Ishii *et al.*,

Table 9.3. Genes of wild *Oryza* species transferred into cultivated rice.

| Trait transferred to *O. sativa* (AA genome) | Donor *Oryza* species | | |
	Wild species	Genome	Accession number
Grassy stunt resistance	*O. nivara*	AA	101508
Bacterial blight resistance	*O. longistaminata*	AA	–
	O. officinalis	CC	100896
	O. minuta	BBCC	101141
	O. latifolia	CCDD	100914
	O. australiensis	EE	100882
	O. brachyantha	FF	101232
Blast resistance	*O. minuta*	BBCC	101141
Brown plant hopper resistance	*O. officinalis*	CC	100896
	O. minuta	BBCC	101141
	O. latifolia	CCDD	100914
	O. australiensis	EE	100882
	O. granulata[a]	GG	100879
Whitebacked plant hopper resistance	*O. officinalis*	CC	100896
Cytoplasmic male sterility	*O. sativa* f. *spontanea*	AA	–
	O. perennis	AA	104823
	O. glumaepatula	AA	100969
Yellow stem borer resistance	*O. brachyantha*[a]	FF	101232
	O. ridleyi [b]	HHJJ	100821
Sheath blight resistance	*O. minuta*[a]	BBCC	101141
Tungro tolerance	*O. rufipogon*[a]	AA	105908
	O. rufipogon[a]	AA	105909
	O. officinalis[b]	CC	105220
Increased elongation ability	*O. rufipogon*[a]	AA	CB751
Tolerance of acid sulphate soils	*O. rufipogon*[a]	AA	106412
	O. rufipogon[a]	AA	106423

[a] Material under test.
[b] Advance backcross progenies (introgression lines) being produced.

1994). We are now using RFLPs to determine introgression of the chromosome segments from distantly related genomes of wild species such as *O. brachyantha* and *O. granulata* into rice (Brar and Khush 1997). Under an IRRI–Japan shuttle project, *in situ* hybridization techniques are being employed to precisely detect introgression of the chromosome segments from wild species into rice (Fukui *et al.*, 1994).

Tagging alien genes with RFLP markers

Alien genes introgressed for resistance to BPH, bacterial blight and blast have been tagged with molecular markers. RFLP analysis was carried out on the introgression line IR65482-4-136-2-2 with resistance to three BPH biotypes derived from *O. sativa* × *O. australiensis*. We surveyed the recurrent parent, the wild species (donor) and the introgressed line for RFLP polymorphism (Ishii *et al.*, 1994). Fourteen probes, previously mapped to chromosome 12, were found to be polymorphic between the recurrent parent and the wild species. Thirteen probes did not detect any introgression. Only RG457 detected introgression from *O. australiensis*. Cosegregation for BPH reaction and RG457 was studied in the F_2 population. The results showed that the gene for resistance to BPH is linked to molecular marker RG457, with a crossover value of $3.68 \pm 1.29\%$. Such tight linkage should facilitate selection for BPH resistance during the transfer of this resistance to other elite breeding lines of rice. Analysis of this introgressed line indicates that the mechanism of alien gene transfer is genetic recombination rather than substitution of whole or large segments of the chromosome of wild species.

Somaclonal variation

Somaclonal variation refers to the variation arising through tissue culture in regenerated plants and their progenies. Somaclonal variation has been reported in various plant species and occurs for a series of agronomic traits such as disease resistance, plant height, tiller number and maturity as well as for various biochemical traits. The technique consists of growing callus or cell suspension cultures for several cycles and regenerating plants from such long-term cultures. The regenerated plants and their progenies are evaluated in order to identify individuals with a new phenotype. Some useful somaclonal variants, including those for disease resistance and male sterile lines, have been isolated in rice. Heszky and Simon-Kiss (1992) tested several somaclonal variants of anther culture origin. Of these, one was released as a variety named Dama. This variety is resistant to *Pyricularia* and has good cooking quality. Similarly, Ogura and Shimamoto (1991) identified useful somaclonal variants from protoplast regenerated progenies of Koshihikari, and a new variety, Hatsuyume, was released. This variety is late by 1 week, shorter in height, lodging resistant and has a 10% higher grain yield than the mother variety, Koshihikari.

Somatic hybridization

Somatic hybridization involves the isolation, culture and fusion of protoplasts from different species and the regeneration of somatic hybrid plants. Since the first demonstration of plant regeneration from mesophyll protoplasts of tobacco, protoplasts have been successfully cultured and regenerated into plants from more than 300 plant species. Yamada *et al.* (1985) were the first to successfully regenerate plants from rice protoplasts. Since then, many laboratories have regenerated plants from protoplasts of several japonica and indica cultivars. Hayashi *et al.* (1988) produced somatic hybrids between rice and four wild species of *Oryza*.

One of the most important applications of somatic hybridization is the production of cytoplasmic hybrids (cybrids). In cybridization, the nuclear genome of one parent is combined with the organelles of the second parent in one step. For example, transfer of CMS to elite breeding lines requires five to seven backcrosses. The donor–recipient protoplast fusion method has made it possible to transfer CMS within several months. Yang *et al.* (1989) produced cybrid rice plants through asymmetric hybridization by electrofusing the gamma-irradiated protoplasts of A-58 CMS and the iodoacetamine treated protoplasts of the fertile cultivar Fujiminori. The technique has been successfully used to transfer CMS from the indica rice Chinsurah Boro II into japonica cultivars (Kyozuka *et al.*, 1989).

Transformation

Introduction of alien genes from bacteria, viruses, fungi, animals and, of course, plants into crop species allows plant breeders to accomplish breeding objectives considered impossible a decade ago. Transformation of rice is now possible through several techniques, e.g. electroporation, polyethylene glycol (PEG)-induced uptake of DNA into protoplasts and microprojectile bombardment. More recently *Agrobacterium*-mediated transformation has become feasible. Transgenic plants have been produced in japonica as well as in indica rices.

Protocols for protoplast culture and DNA transformation in rice have been reviewed by Hodges *et al.* (1991). Several laboratories have produced transgenic rices mainly through protoplast-mediated DNA transformation but also via microprojectile bombardment. Recently, a major breakthrough was made in *Agrobacterium*-mediated transformation of rice (Hiei *et al.*, 1994). Where a large number of fertile transgenic rice plants were obtained, the efficiency of transformation was similar to that obtained by the methods used routinely for the transformation of dicot plants through *Agrobacterium*. Molecular and genetic analyses of transformants in the R_0, R_1 and R_2 generations showed stable integration, expression and inheritance of transgenes.

Genes for rice transformation

Major efforts are under way to isolate genes for disease and insect resistance and abiotic stress tolerance, and to introduce them into rice in order to increase the level of host resistance. Several laboratories have produced transgenic plants both in japonica and indica rices. Recently, transgenic rices carrying agronomically important genes for resistance to stem borer, virus tolerance, resistance to fungal pathogens and herbicide tolerance have also been produced (Table 9.4).

Table 9.4. Some examples of transgenic rice plants carrying agronomically important genes.

Transgene	Gene transfer method	Useful trait	Reference
bar	Microprojectile bombardment	Tolerance of basta (phosphinothricin)	Cao *et al.*, 1992
bar	PEG-mediated	Tolerance of herbicide	Datta *et al.*, 1992
Coat protein gene	Protoplast electroporation	Tolerance of stripe virus	Hayakawa *et al.*, 1992
Chitinase	PEG-mediated	Sheath blight resistance	Lin *et al.*, 1995
crylA(b)	Protoplast electroporation	Resistance to striped stem borer	Fujimoto *et al.*, 1993
crylA(b)	Particle bombardment	Tolerance of yellow stem borer and striped stem borer	Wuhn *et al.*, 1996
crylA(c)	Particle bombardment	Resistance to yellow stem borer	Nayak *et al.*, 1997
crylA(b)	Particle bombardment	Resistance to yellow and striped stem borers	Ghareyazie *et al.*, 1997
Corn cystatin (CC)	Protoplast electroporation	Insecticidal activity for *Sitophilus zeamais*	Irie *et al.*, 1996

GENES FOR INSECT RESISTANCE. As early as 1987, genes coding for toxins from *Bacillus thuringiensis* (*Bt*) were transferred to tomato, tobacco and potato where they provided protection against lepidopteran insects. A major target for *Bt* deployment in transgenic rice is the yellow stem borer, *Scirpophaga incertulas* (Walker). The pest is widespread in Asia and has the potential to cause substantial crop losses. Improved rice cultivars are either susceptible to the insect or have only moderate levels of resistance. Thus, *Bt* transgenic rice has much appeal for controlling the yellow stem borer. Fujimoto *et al.* (1993) introduced a truncated endotoxin gene, *cry1A(b)*, of *B. thuringiensis*. The coding sequence was modified on the basis of the codon usage of rice genes. Transgenic rice plants efficiently expressed the modified *cry1A(b)* gene at both the mRNA and protein levels. Transgenic plants in the R_2 generation expressing the *cry1A(b)* protein had increased resistance to striped stem borer and leaf folder.

Wuhn *et al.* (1996) introduced the *cry1A(b)* gene into IR58 through particle bombardment. The transgenic plants in the R_0, R_1 and R_2 generations showed a significant insecticidal effect on several lepidopterous insect pests. Feeding studies showed up to 100% mortality for yellow stem borer and striped stem borer. Recently, Nayak *et al.* (1997) transformed IR64 through particle bombardment using the *cry1A(c)* gene placed under the control of the maize ubiquitin 1 promoter, along with the first intron of the maize *ubiquitin 1* gene, and the *nos* terminator. Six independent transgenic lines showing high expression of insecticidal crystal protein were identified. The transferred synthetic *cry1A(c)* gene was stably expressed in the T_2 of these lines and the transgenic rice plants were highly toxic to yellow stem borer larvae, resulting in little feeding damage.

Another category of genes for transforming rice for insect resistance is the inhibitors of insect digestive enzymes. The storage tissues of most plants contain chemicals that limit predation by insects and other herbivores. Some are the protease inhibitors of insect digestive enzymes, and the genes coding for them are sources for insect resistance. Chen (1991) has purified and characterized five starch digestive enzymes (α-amylase isoenzymes) from the gut of three major storage pests (the rice weevil, the red flour beetle and the yellow mealworm). Hilder *et al.* (1987) have purified a putative inhibitor from cowpea, which is a typical serine protease inhibitor with broad spectrum effectiveness against insect pests. The cowpea gene (*CpTi*) was transferred to tobacco, where it conferred resistance to tobacco bud worm and other insect pests. This gene is now being tested for its effectiveness against BPH, a very serious pest of rice. Additional inhibitors from other plants such as the mungbean trypsin inhibitor, potato proteinase inhibitors I and II, arrowhead proteinase inhibitors, and towel gourd trypsin inhibitors are being evaluated against the digestive enzymes of rice insects.

Irie *et al.* (1996) transformed the japonica rice cultivar Nipponbare by introducing corn cystatin (CC) cDNA through electroporation of protoplasts. The transgenic plants showed strong inhibitory activity against rice proteinases that occur in the gut of the insect pest, *Sitophilus zeamais*.

We are attempting to transform rice with the soybean trypsin inhibitor (*SBTi*) gene. Hygromycin-resistant plants have been regenerated that were cotransformed with the *SBTi*. Rice plants transformed with the *SBTi* are likely to be toxic to yellow stem borer and they may delay the breakdown of *Bt* toxins within the larval midgut and thereby enhance the action of *Bt* toxins.

Other genes that might confer enhanced insect resistance are genes encoding α-amylase inhibitors, lectins and chitinases. α-amylase inhibitors might be toxic to insects through interference in the digestion of dietary carbohydrates. It is widely assumed that the proteinase inhibitors and α-amylase inhibitors that accumulate in wounded tissue are produced as a normal defence against insects. Many lectins are toxic to insects, presumably through some deleterious interaction with intestinal glycoproteins. Chitinases might be toxic to insects if they are able to degrade the chitin layer of the peritropic membrane, which protects the insect's gut epithelium.

GENES FOR DISEASE RESISTANCE. Several viral, fungal and bacterial diseases attack rice and cause serious yield losses. Sources of resistance to some diseases (blast and bacterial blight) have been identified within cultivated rice germplasm, and improved germplasm with resistance has been developed. However, sources of resistance to sheath blight are not available and only a few donors for resistance to tungro disease, caused by two virus particles, are known.

A highly successful strategy, termed coat-protein (CP)-mediated protection, has been employed against certain viral diseases such as tobacco mosaic virus in tobacco and tomato. When expressed in the transgenic crop, a chimeric gene made by combining a strong promoter with the virus gene encoding the capsid protein results in the accumulation of capsid protein in plant cells. Such plants are resistant to infection by the virus from which the gene stems. While the mechanism of resistance is not understood, it appears that disassembly of the virus is inhibited and that this is due to the accumulated capsid protein and not to the mRNA.

CP genes for the two component viruses that cause tungro disease have been cloned (Hay *et al.*, 1991) and efforts are under way to express these genes in rice plants. A coat protein gene for rice stripe virus was introduced into two japonica varieties by electroporation of the protoplasts (Hayakawa *et al.*, 1992). The resultant transgenic plants expressed the CP at high levels (up to 0.5% of the total soluble proteins) and exhibited a significant level of resistance to virus infection, and this resistance was inherited. In addition to CP cross-protection in RNA viruses, other mechanisms for transgenic virus resistance have been reported. Of particular relevance to transgenic tungro resistance is the report that the DNA virus (tomato golden mosaic virus) has been controlled in transgenic tobacco expressing the antisense product of the viral *AL1* gene. The AL1 protein is involved in the replication of viral DNA. It may be possible to exploit a similar antisense approach in controlling rice tungro bacilliform virus, a DNA virus.

Sheath blight of rice is caused by the pathogen *Rhizoctonia solani*, which has a wide host range. Transgenic tobacco and canola plants with enhanced resistance to *R. solani* have been obtained by introducing the bean chitinase gene under the control of the CaMV35S promoter (Broglie *et al.*, 1991). Similarly, Logemann *et al.*, (1992) transformed tobacco plants with a barley gene encoding a ribosome inactivating protein under the control of the wound inducible *wun-2* promoter from potato. We are exploring the possibility of transforming rice with similar gene constructs to develop resistance to sheath blight.

Chitinases and glucanases degrade the major structural polysaccharides of the fungal cell wall. These enzymes have both a binding domain and a catalytic domain for their respective polysaccharides. Alone or in combination, they attack the growing hyphal tip and are potent inhibitors of fungal growth. About six chitinase genes have been identified in rice and are being manipulated to increase the level of resistance to fungal diseases (Zhu and Lamb, 1991). Lin *et al.* (1995) introduced a 1.1 kb rice genomic DNA fragment containing a chitinase gene through PEG-mediated transformation. The presence of this chimeric chitinase gene in T_0 and T_1 transgenic plants was detected by Southern blot analysis. Western blot analysis of transgenic plants and progeny revealed the presence of two proteins that reacted with the chitinase antibody.

Progeny from the chitinase positive plants were tested for resistance to sheath blight pathogen. The degree of resistance correlated with the level of chitinase expression.

GENES FOR ABIOTIC STRESS TOLERANCE. Transgenic mechanisms for resistance to abiotic stresses in rice may also be feasible. Tarczynski *et al.* (1993) have reported the use of the mannitol-1-phosphate dehydrogenase gene, *mtl D*, from *Escherichia coli* to provide salt tolerance in transgenic tobacco. The mechanism of resistance involves the accumulation of mannitol in plant tissues. Given the importance of osmolytes in drought tolerance, such mechanisms of salt tolerance may also provide some protection against drought. Other abiotic stresses for which transgenic mechanisms of protection have been reported are heavy metal ions and oxidative stress.

GENES FOR NUCLEAR MALE STERILITY AND FERTILITY RESTORATION. Nuclear male sterility has been engineered in tobacco and oilseed rape by expression of a bacterial ribonuclease (barnase) in the tapetal cells that provide the nutrition to developing microspores (Mariani *et al.*, 1990). Tapetum specific expression of the ribonuclease was achieved through the use of a tapetum specific promoter from tobacco. Restoration of male fertility was achieved by crossing transgenic plants containing the barnase gene with transgenic plants expressing the bacterial gene encoding the barnase inhibitor barstar, again under the control of the tapetum specific promoter (Mariani *et al.*, 1992). Sufficient barstar was produced in some crosses to block barnase activity. Such a restoration mechanism of nuclear male sterility would be valuable in rice and would permit diversification of limited sources of cytoplasmic male sterility for hybrid rice production.

Summary

Major advances have been made in increasing rice production as a result of large-scale adoption of modern high-yielding varieties and improved production technologies. World rice production more than doubled from 257 million tonnes in 1966, to 560 million tonnes in 1996. To meet the growing need of future rice consumers, we will have to produce 70% more rice by 2025. Thus, we need varieties with higher yield potential, multiple resistance to diseases and insects, and tolerance of abiotic stresses. Recent breakthroughs in cellular and molecular biology have provided new tools to increase the efficiency of breeding methods and follow unconventional approaches to rice improvement. Anther culture is routinely used to obtain doubled haploid populations. A molecular genetic map of more than 2000 DNA markers has been prepared and genes of economic importance as well as QTL for traits of low heritability have been tagged with molecular markers. Molecular-marker-aided selection is being employed to move genes from one varietal background to the other and for pyramiding genes into some cultivars. Regeneration from protoplasts of many indica and japonica varieties has been obtained. This allows introduction of novel genes into elite germplasm through transformation. Other methods of transformation such as use of biolistic gun and *Agrobacterium* have been employed to introduce genes for disease and insect resistance. Similarly, wide hybridization has been used to transfer genes from wild species into elite germplasm. All of these biotechnology techniques are being widely employed for rice improvement.

References

Ahn, S., Bollich, C.N. and Tanksley, S.D. (1992) RFLP tagging of a gene for aroma in rice. *Theoretical and Applied Genetics* 84, 825–828.

Ahn, S., Anderson, J.A., Sorrells, M.E. and Tanksley, S.D. (1993) Homoeologous relationships of rice, wheat and maize chromosomes. *Molecular and General Genetics* 241, 483–490.

Brar, D.S. and Khush, G.S. (1997) Alien introgression in rice. *Plant Molecular Biology* 35, 35–47.

Brar, D.S., Elloran, R. and Khush, G.S. (1991) Interspecific hybrids produced through embryo rescue between cultivated and eight wild species of rice. *Rice Genetics Newsletter* 8, 91–93.

Broglie, K., Chet, I., Holliday, M., Cressman, R., Biddle, P., Knowlton, S., Mauvals, C.J. and Brogile R. (1991) Transgenic plants with enhanced resistance to fungal pathogen *Rhizoctonia solani*. *Science* 254, 1194–1197.

Cao, J., Duan, X., McElroy, D. and Wu, R. (1992) Regeneration of herbicide resistant transgenic rice plants following microprojectile-mediated transformation of suspension culture cells. *Plant Cell Reports* 11, 586–591.

Causse, M.A., Fulton, T.M., Cho, Y.G., Ahn, S.N., Chungwongse, J., Wu, K., Xiao, J., Yu, Z., Ronald, P.C., Harrington, S.E., Second, G., McCouch, S.R. and Tanksley S.D. (1994) Saturated molecular map of the rice genome based on interspecific backcross population. *Genetics* 138, 1251–1274.

Champoux, M.C., Wang, G., Sarkarung, S., McKill, D.J., O'Toole, J.C., Huang, N. and McCouch, S.R. (1995) Locating genes associated with root morphology and drought avoidance in rice via linkage to RFLP markers. *Theoretical and Applied Genetics* 90, 969–981.

Chen, M. (1991) Characterization of insect alpha-amylases and cloning and expression of *oryzacystanin* cDNA in *E. coli*. PhD thesis, Kansas State University, Manhattan, Kansas, USA.

Cho, Y.G., Eun, M.Y., McCouch, S.R. and Chae, Y.A. (1994) The semidwarf gene, *sd-1*, of rice (*Oryza sativa* L.) II Molecular mapping and marker-assisted selection. *Theoretical and Applied Genetics* 89, 54–59.

Dalmacio, R., Brar, D.S., Ishii, T., Sitch, L.A., Virmani, S.S. and Khush, G.S. (1995) Identification and transfer of a new cytoplasmic male sterility source from *Oryza perennis* into indica rice (*O. sativa*). *Euphytica* 82, 221–225.

Datta, S.K., Datta, K., Soltanifar, N., Donn, G. and Potrykus, I. (1992) Herbicide-resistant indica rice plants from IRRI breeding line IR72 after PEG-mediated transformation of protoplasts. *Plant Molecular Biology* 20, 619–629.

Fujimoto, H., Itoh, K., Yamamoto, M., Kyozuka, J. and Shimamoto, K. (1993) Insect resistant rice generated by introduction of a modified δ-endotoxin gene of *Bacillus thuringiensis*. *Bio/Technology* 11, 1151–1155.

Fukui, K., Ohmido, N. and Khush, G.S. (1994) Variability in rDNA loci in the genus *Oryza* detected through fluorescence *in situ* hybridization. *Theoretical and Applied Genetics* 87, 893–899.

Ghareyazie, B., Alinia, F., Menguito, C.A., Rubia, L., de Palma, J.M., Liwanag, E.A., Cohen, M.B., Khush, G.S. and Bennett, J. (1997) Enhanced resistance to two stem-borers in an aromatic rice containing a synthetic *cry1A(b)* gene. *Molecular Breeding* 3, 401–414.

Hay, J.M., Jones, M.C., Blakebrough, M., Dasgupta, I., Davies, J.W. and Hull, R. (1991) An analysis of the sequence of an infectious clone of rice tungro bacilliform virus, a plant pararetrovirus. *Nucleic Acids Research* 19, 2615–2621.

Hayakawa, T., Zhu, Y., Itoh, K. and Kimura, Y. (1992) Genetically engineered rice resistant to rice stripe virus, an insect-transmitted virus. *Proceedings of the National Academy of Sciences, USA* 89, 9865–9869.

Hayashi, Y., Kyozuka, J. and Shimamoto, K. (1988) Hybrids of rice (*Oryza sativa* L.) and wild *Oryza* species obtained by cell fusion. *Molecular and General Genetics* 214, 6–10.

Heszky, L.E. and Simon-Kiss, I. (1992) 'DAMA', the first plant variety of biotechnology origin in Hungary, registered in 1992. *Hungarian Agricultural Research* 1, 30–32.

Hiei, Y., Ohta, S., Komari, T. and Kumashiro, T. (1994) Efficient transformation of rice (*Oryza sativa* L.) mediated by *Agrobacterium* and sequence analysis of the boundaries of the T-DNA. *Plant Journal* 6, 271–282.

Hilder, V.A., Gatehouse, A.M.R., Sheerman, S.E., Baker, R.F. and Boulter, D. (1987) A novel mechanism of insect resistance engineered into tobacco. *Nature* 330, 160–163.

Hirabayashi, H. and Ogawa, T. (1995) RFLP mapping of *Bph-1* (brown planthopper resistance gene) in rice. *Breeding Science* 45, 369–371.

Hittalmani, S., Foolad, M.R., Mew, T.V., Rodriguez, R.L. and Huang, N. (1995) Development of PCR-based marker to identify rice blast resistance gene, *Pi-2(t)* in a segregating population. *Theoretical and Applied Genetics* 91, 9–14.

Hodges, T.K., Peng, J., Lyznik, A. and Koetje, D.S. (1991) Transformation and regeneration of rice protoplasts. In: Khush, G.S. and Toenniessen, G. (eds) *Rice Biotechnology*. CAB International, Wallingford, UK, pp. 157–164.

Huang, N., Angeles, E.R., Domingo, J., Magpantay, G., Singh, S., Zhang, G., Kumaravadivel, N., Bennett, J. and Khush, G.S. (1997) Pyramiding of bacterial blight resistance genes in rice: marker assisted selection using RFLP and PCR. *Theoretical and Applied Genetics* 95, 313–320.

Irie, K., Hosoyama, H., Takeuchi, T., Iwabuchi, K., Watanabe, H., Abe, M., Abe, K. and Arai S. (1996) Transgenic rice established to express corn cystatin exhibits strong inhibitory activity against insect gut proteinases. *Plant Molecular Biology* 30, 149–157.

Ishii, T., Brar, D.S., Multani, D.S. and Khush, G.S. (1994) Molecular tagging of genes for brown planthopper resistance and earliness introgressed from *Oryza australiensis* into cultivated rice, *O. sativa*. *Genome* 37, 217–221.

Jena, K.K. and Khush, G.S. (1990) Introgression of genes from *Oryza officinalis* Well ex Watt to cultivated rice, *O. sativa* L. *Theoretical and Applied Genetics* 87, 737–745.

Jena, K.K., Khush, G.S. and Kochert, G. (1992) RFLP analysis of rice (*Oryza sativa* L.) introgression lines. *Theoretical and Applied Genetics* 84, 608–616.

Khush, G.S. (1977) Disease and insect resistance in rice. *Advances in Agronomy* 29, 265–341.

Khush, G.S., Bacalangco, E. and Ogawa, T. (1990) A new gene for resistance to bacterial blight from *O. longistaminata. Rice Genetics Newsletter* 7, 121–122.

Kurata, N., Moore, G., Nagamura, Y., Footo, T., Yano, M., Minobe, Y. and Gale M. (1994a) Conservation of genome structure between rice and wheat. *Bio/Technology* 12, 276–278.

Kurata, N., Nagamura, Y., Yamamoto, K., Harushima, Y., Sue, N., Wu, J., Antonio, B.A., Shomura, A., Shimizu, T., Lin, S.Y., Inoue, T., Fukuda, A., Shimano, T., Kuboki, Y., Toyama, T., Miyamoto, Y., Kirihara, T., Hayasaka, K., Miyao, A., Nonna, L., Zhong, H.S., Tamura, Y., Wang, Z.X., Momma, T., Umehara, Y., Yano, M., Sasaki, T. and Minobe Y. (1994b) A 300 kilobase interval genetic map of rice including 883 expressed sequences. *Nature Genetics* 8, 365–372.

Kyozuka, J., Kaneda, T. and Shimamoto, K. (1989) Production of cytoplasmic male sterile rice (*Oryza sativa* L.) by cell fusion. *Bio/Technology* 7, 1171–1174.

Leung, H.R., Nelson, J. and Leach, J.E. (1993) Population structure of plant pathogens fungi and bacteria. *Advances in Plant Pathology* 10, 158–205.

Liang, C.Z., Gu, M.H., Pan, X.B., Liang, G.H. and Zhu, L.H. (1994) RFLP tagging of a new scmidwarf gene in rice. *Theoretical and Applied Genetics* 88, 898–900.

Lin, W., Anuratha, C.S., Datta, K., Potrykus, I., Muthukrishnan, S. and Datta, S.K. (1995) Genetic engineering of rice for resistance to sheath blight. *Bio/Technology* 13, 686–691.

Logemann, J., Jach, G., Tommerup, H., Mundy, J. and Scheel, J. (1992) Expression of a barley ribosome-inactivating protein leads to increased fungal protection in transgenic tobacco plants. *Bio/Technology* 10, 305–308.

Mackill, D.J., Salam, M.A., Wang, Z.Y. and Tanksley, S.D. (1993) A major photoperiod-sensitivity gene tagged with RFLP and isozyme markers in rice. *Theoretical and Applied Genetics* 85, 536–540.

Maheswaran, M. (1995) Identification of quantitative trait loci for days to flowering and photoperiod sensitivity in rice (*Oryza sativa* L.). PhD thesis, Tamil Nadu Agricultural University, Coimbatore, India.

Mariani, C., De Beuckeleer, M., Truettner, J., Leemans, J. and Goldberg, R.B. (1990) Induction of male sterility in plants by a chimaeric ribonuclease gene. *Nature* 347, 737–741.

Mariani, C., Gosselle, V., De Beuckeleer, M., De Block, M., Goldberg, R.B., De Greef, W. and Leemans, J. (1992) A chimaeric ribonuclease-inhibitor gene restores fertility to male sterile plants. *Nature* 357, 384–387.

McCouch, S.R. (1990) Construction and applications of a molecular linkage map of rice based on restriction fragment length polymorphism (RFLP). PhD thesis, Cornell University, Ithaca, New York, USA.

McCouch, S.R., Abenes, M.L., Angeles, R., Khush, G.S. and Tanksley, S.D. (1991) Molecular tagging of a recessive gene *xa-5*, for resistance to bacterial blight of rice. *Rice Genetics Newsletter* 8, 143–145.

McCouch, S.R., Kochert, G., Yu, Z-H., Wang, Z-Y., Khush, G.S., Coffman, D.R. and Tanksley, S.D. (1988) Molecular mapping of rice chromosomes. *Theoretical and Applied Genetics* 76, 815–829.

Miyamoto, M., Ando, I., Rybka, K., Kodama, O. and Kawasaki, S. (1996) High resolution mapping of the indica-derived blast resistance gene I. *Pi-b. Molecular Plant–Microbe Interactions* 9, 6–13.

Mohan, M., Nair, S., Bentur, J.S., Rao, U.P. and Bennett, J. (1994) RFLP and RAPD mapping of the rice *Gm2* gene that confers resistance to biotype 1 of gall midge (*Orseolia oryzae*). *Theoretical and Applied Genetics* 87, 782–788.

Nair, S., Bentur, J.S., Rao, U.P. and Mohan, M. (1995) DNA markers tightly linked to a gall midge resistance gene (*Gm2*) are potentially useful for marker-aided selection in rice breeding. *Theoretical and Applied Genetics* 91, 68–73.

Nair, S., Kumar, A., Srivastava, M.N. and Mohan, M. (1996) PCR-based DNA markers linked to a gall midge resistance gene *Gm4t* has potential for marker aided selection in rice. *Theoretical and Applied Genetics* 92, 660–665.

Nandi, S., Subudhi, P.K., Senadhira, D., Manigbas, N.L., Sen-Mandi, S. and Huang, N. (1997) Mapping QTLs for submergence tolerance in rice by AFLP analysis and selective genotyping. *Molecular and General Genetics* 255, 1–8.

Naqvi, N.I., Bonman, J.M., Mackill, D.J., Nelson, R.J. and Chattoo, B.B. (1995) Identification of RAPD markers linked to a major blast resistance gene in rice. *Molecular Breeding* 1, 341–348.

Nayak, P., Basu, D., Das, S., Basu, A., Ghosh, D., Ramakrishnan, N.A., Ghosh, M. and Sen, S.K. (1997) Transgenic elite indica rice plants expressing CryIAc δ-endotoxin of *Bacillus thuringiensis* are resistant against yellow stemborer (*Scirophaga incertulas*). *Proceedings of the National Academy of Sciences, USA* 94, 2111–2116.

Niizeki, H. and Oono, K. (1968) Induction of haploid rice plants from anther culture. *Proceedings of the Japan Academy* 44, 554–557.

Ogura, H. and Shimamoto, K. (1991) Field performance of protoplast derived rice plants and the release of a new variety. In: Bajaj, Y.P.S. (ed.). *Biotechnology in Agriculture and Forestry*, Vol. 14. Springer-Verlag, Berlin, pp. 269–282.

Raina, S.K. (1989) Tissue culture in rice improvement: status and potential. *Advances in Agronomy* 42, 339–398.

Redona, E.D. and Mackill, D.J. (1996). Molecular mapping of quantitative trait loci in japonica rice. *Genome* 39, 395–403.

Ronald, P.C., Albano, B., Tabien, R., Abenes, L., Wu, K., McCouch, S.R. and Tanksley, S.D. (1992) Genetic and physical analysis of the rice bacterial blight resistance locus, *Xa-21*. *Molecular and General Genetics* 236, 113–120.

Saito, A., Yano. M., Kishimoto, N., Nakagahra, M., Yoshimura, A., Saito, K., Kuhura, S., Ukai, Y., Kawase, M., Nagamine, T., Yoshimura, S., Ideta, O., Ohsawa, R., Hayano, Y., Iwata, N. and Sugiura, M. (1991) Linkage map of restriction fragment length polymorphism loci in rice. *Japanese Journal of Breeding* 41, 665–670.

Sebastian, L.S., Ikeda, R., Huang, N., Imbe, T., Coffman, W.R. and McCouch, S.R. (1996) Molecular mapping of resistance to rice tungro spherical virus and green leafhopper. *Phytopathology* 86, 25–30.

Singh, K., Ishii, T., Parco, A., Huang, N., Brar, D.S. and Khush, G.S. (1996) Centromere mapping and orientation of the molecular linkage map of rice (*Oryza sativa* L.). *Proceedings of the National Academy of Sciences, USA* 93, 6163–6168.

Sitch, L.A. (1990) Incompatibility barriers operating in crosses of *Oryza sativa* with related species and genera. In: Gustafson, J.P. (ed.) *Gene Manipulation in Plant Improvement*. II. Plenum Press, New York, USA, pp. 77–94.

Snape, J.W. (1982) The use of doubled haploids in plant breeding. In: *Induced Variability in Plant Breeding*, Centre for Agricultural Publishing and Documentations, Wageningen, The Netherlands, pp. 52–58.

Song, W-H., Wang, G-L., Chen, L-L., Kim, H-S., Pi, L-Y., Holsten, T., Gardner, J., Wang, B., Zhai, W-X., Zhu, L-H., Fauquet, C. and Ronald, P. (1995) A receptor kinase-like protein encoded by the rice disease resistance gene, *Xa-21*. *Science* 270, 1804–1806.

Subudhi, P.K., Borkakati, R.P., Virmani, S.S. and Huang, N. (1997) Molecular mapping of a thermosensitive genetic male sterility in rice using bulk segregant analysis. *Genome* 40, 188–194.

Tarczynski, M.C., Jensen, R.G. and Bohnert, H.J. (1993) Stress protection of transgenic tobacco by production of the osmolyte mannitol. *Science* 259, 508–510.

Wang, G.L., Mackill, D.J., Bonman, J.M., McCouch S.R. and Nelson, R.J. (1994) RFLP mapping of genes conferring complete and partial resistant to blast in a durably resistant rice cultivar. *Genetics* 136, 1421–1434.

Wang, G.L., Holsten, T.E., Song, W.Y., Wang, H.P. and Ronald, P.C. (1995a) Construction of a rice bacterial artificial chromosome library and identification of clones linked to *Xa-21* disease resistance locus. *Plant Journal* 7, 725–733.

Wang, B., Xu, W., Wang, J., Wu, W., Zheng, H., Yang, Z., Ray, J.D. and Nguyen, H.T. (1995b) Tagging and mapping the thermosensitive genic male sterile gene in rice (*Oryza sativa* L.) with molecular markers. *Theoretical and Applied Genetics* 91, 1111–1114.

Wuhn, J., Kloti, A., Burkhardt, P.K., Ghosh Biswas, G.C., Launis, K., Iglesias, V.A. and Potrykus I. (1996) Transgenic indica rice breeding line IR58 expressing a synthetic *cry1A(b)* gene from *Bacillus thuringiensis* provides effective insect pest control. *Bio/Technology* 14, 171–176.

Xiao, J., Grandillo, S., Ahn, S.A., McCouch, S.R., Tanksley, S.D., Li, J. and Yuan, L. (1996a) Genes from wild rice improve yield. *Nature* 384, 223–224.

Xiao, J., Li, J., Yuan, L. and Tanksley, S.D. (1996b) Identification of QTLs affecting traits of agronomic importance in a recombinant inbred population derived from a subspecific rice cross. *Theoretical and Applied Genetics* 92, 230–244.

Yadav, R., Courtois, B., Huang, N. and McLaren, G. (1997) Mapping genes controlling root morphology and root distribution in a doubled-haploid population of rice. *Theoretical and Applied Genetics* 94, 619–622.

Yamada, Y., Yang Z-Q. and Tang, D-T. (1985) Regeneration of rice plants from protoplasts. *Rice Genetics Newsletter* 2, 94–95.

Yanagihara, S., McCouch, S.R., Ishikawa, K., Ogi, Y., Maruyama, K. and Ikehashi, H. (1995) Molecular analysis of the inheritance of the S-5 locus, conferring wide compatibility in indica/japonica hybrids of rice (*O. sativa* L.). *Theoretical and Applied Genetics* 90, 182–188.

Yang, D., Parco, A., Nandi, S., Subudhi, P., Zhu, Y., Wang, G. and Huang N. (1997) Construction of a bacterial artificial chromosome (BAC) library and identification of overlapping BAC clones with chromosome 4 specific RFLP markers in rice. *Theoretical and Applied Genetics* 95, 1147–1154.

Yang, Z-Q., Shikanai, T., Mori, M. and Yamada, Y. (1989) Plant regeneration from cytoplasmic hybrids of rice (*Oryza sativa* L.). *Theoretical and Applied Genetics* 77, 305–310.

Yoshimura, S., Yoshimura, A., Saito, A., Kishimoto, N., Kawase, M., Yano, M., Nakagahra, M., Ogawa, T. and Iwata, N. (1992) RFLP analysis of introgressed chromosomal segments in three near-isogenic lines of rice bacterial blight resistance genes, *Xa-1*, *Xa-3* and *Xa-4*. *Japanese Journal of Genetics* 67, 29–37.

Yoshimura, S., Yoshimura, A., Iwata, N., McCouch, S.R., Abenes, M.L., Baraoidan, M.R., Mew, T.W. and Nelson, R.J. (1995) Tagging and combining bacterial blight resistance genes in rice using RAPD and RFLP markers. *Molecular Breeding* 1, 375–387.

Yu, Z.H. (1991) Molecular mapping of rice (*Oryza sativa* L.) genes via linkage to restriction fragment length polymorphism (RFLP) markers. PhD thesis, Cornell University, Ithaca, New York, USA.

Yu, Z.H., Mackill, D.J., Bonman, J.M. and Tanksley, S.D. (1991) Tagging genes for blast resistance in rice via linkage to RFLP markers. *Theoretical and Applied Genetics* 81, 471–476.

Zapata, F.J., Alejar, M.S., Torrizo, L.B., Nover, A.U., Singh, V.P. and Senadhira, D. (1991) Field performance of anther culture derived lines from F_1 crosses of indica rices under saline and non-saline conditions. *Theoretical and Applied Genetics* 83, 6–11.

Zhang, Q.F., Shen, B.S., Dai, X.K., Mei, M.H., Saghai Maroof, M.A. and Li, Z.B. (1993) An RFLP-based genetic analysis of photoperiod sensitive male sterility in rice. *Rice Genetics Newsletter* 10, 94–97.

Zhang, G., Angeles, E.R., Abenes, M.L.P., Khush, G.S. and Huang, N. (1996) RAPD and RFLP mapping of the bacterial blight resistance gene *xa-13* in rice. *Theoretical and Applied Genetics* 93, 65–70.

Zhang, G., Bharaj, T.S., Lu, Y., Virmani, S.S. and Huang, N. (1997) Mapping of the *Rf-3* nuclear fertility-restoring gene for WA cytoplasmic male sterility in rice using RAPD and RFLP markers. *Theoretical and Applied Genetics* 94, 27–33.

Zheng, K., Huang, N., Bennett, J. and Khush, G.S. (1995) PCR-based marker-assisted selection in rice breeding. *IRRI Discussion Paper Series No. 12*, International Rice Research Institute, Manila, The Philippines, pp. 24.

Zhu, Q. and Lamb, C.J. (1991) Isolation and characterization of a rice gene encoding a basic chitinase. *Molecular and General Genetics* 226, 289–296.

Zhu, L., Chen, Y., Ling, Z., Xu, Z. and Xu, J. (1992) Identification of molecular markers linked to a blast resistance gene in rice. *Proceeding Asia-Pacific Conference Agricultural Biotechnology*. August 20–24, 1992, Beijing, China, pp. 213.

The Application of Biotechnology to Non-traditional Crops

III

10

Current Advances in the Biotechnology of Banana

Oscar Arias

Agribiotecnología de Costa Rica, S.A., PO Box 100–4003, Alajuela, Costa Rica

Bananas are the world's largest fruit crop. In 1997, banana production was approximately 58 million tonnes and plantain production was approximately 30 million tonnes. It has not received as much intensive research and development (R&D) effort as many other smaller crops. Production is exclusively based on selections of naturally occurring clones.

Modern biotechnology could, and in fact in some cases already does, contribute to the overall improvement of bananas. The uses of biotechnology to provide more sustainable banana production are apparent in at least five areas: (i) plant propagation; (ii) plant improvement through the development of new genotypes with resistance to or tolerance of the main pests and diseases; (iii) tolerance of low temperatures (below 16°C) and drought; (iv) improved fruit quality and storability; (v) intra- and international germplasm transfer.

Through continuous efforts during the last 10 years, Agribiotecnología de Costa Rica (ACR) has proved the advantage of micropropagated banana planting material in Central and South America. The need for high-quality planting material at a fair price for growers was the basis for starting a collaborative project between ACR and DNA Plant Technology Corporation, sponsored by the United States Agency for International Development (USAID) through the Agricultural Biotechnology for Sustainable Productivity (ABSP) project, with the general goal of implementing more efficient micropropagation methods for tropical crops. Banana micropropagation methods were adapted to bioreactor vessels using liquid media.

The improvement of *Musa* through biotechnology carries a potentially tremendous social impact in terms of the production of staple food for domestic consumption in many developing countries. This is

© CAB INTERNATIONAL 1998. *Agricultural Biotechnology in International Development*
(eds C.L. Ives and B.M. Bedford)

especially true in the equatorial belt, where an estimated 400 million people derive more than one-quarter of their food energy requirement from bananas and plantains, which are at present devastated by Black Sigatoka disease.

Introduction

Banana and plantain are members of the *Musaceae* family. Most cultivars are products of evolution in the *Eumusa* series, genus *Musa*. Cultivars of the *Eumusa* series had their origins in two wild species *Musa acuminata* Colla and *Musa balbisiana* Colla. All bananas grown for export originate from *M. acuminata*, whereas some bananas favoured in certain areas for domestic consumption are hybrids between these two parental species (Simmonds and Shepherd, 1955).

Banana, plantain and cooking banana are staple foods for nearly 400 million people in developing countries. Banana has been ranked first among fruit crops with an annual world production of approximately 58 million tonnes as of 1997 (FAOSTAT, 1997). Plantains have an annual world production of approximately 30 million tonnes as of 1997 (FAOSTAT, 1997). Many inhabitants in the tropics of all continents make their living directly or indirectly from *Musa* as a source of food or cash export. The economic impact of banana production in banana-exporting countries in Latin America and the Caribbean is significant, not only as a source of hard currency but as a generator of work and income to more than 700,000 families and also to national economies through taxes.

New market opportunities, especially in Europe, have induced Latin American countries to increase the area of land under banana production. Between 1985 and 1996, the area planted increased by 134,358 ha. The largest increases occurred in Ecuador (79,140 ha), Costa Rica (30,239 ha) and Colombia (19,069 ha). In the remaining countries, the increase was not very significant as shown in Table 10.1. In Costa Rica, banana is the most important agricultural export. It is farmed on about 52,000 ha. In 1996, it generated an income of US $693.6 million, with 49% of Costa Rica's banana export going to the United States, 48% to the European Economic Community (EEC) and 3% to the rest of Europe.

The objective of this chapter is to show the impact that biotechnology has and can have on banana production. With the recent advances made in biotechnology, potential increased production and improved quality of banana and plantain soon could become a reality.

Banana Biotechnology

Bananas are one of the least genetically improved plants, when compared with other major food crops. Production is exclusively based on banana cultivars

Table 10.1. Area (ha) of cultivated banana in Latin America.

Country	1985	1996	Increase
Colombia	24,191	43,260	19,069
Costa Rica	20,285	50,254	30,239
Ecuador	48,000	127,140	79,140
Guatemala	7,688	14,197	6,509
Honduras	17,590	18,743	843
Mexico	12,000	8,900	−3,100
Nicaragua	2,761	1,780	−981
Panamá	13,541	14,772	1,231
Venezuela	100	1,508	1,408
Total	146,156	280,554	134,358

occurring in nature. There are at least five areas where modern biotechnology could, and in some cases already does, contribute to overall improvement of banana. These areas are: (i) plant propagation; (ii) plant improvement through the development of new genotypes with resistance to or tolerance of pests and diseases; (iii) tolerance of low temperature (below 16°C) and drought; (iv) improved fruit quality and storability; and (v) intranational and international germplasm transfer.

Micropropagation has been successfully applied for the rapid propagation of high-quality and disease-free planting material and has been used commercially since 1980 in Taiwan, Israel, Australia and The Philippines (Novak, 1992). It has been used for the rapid multiplication and distribution of planting material on a large scale, replacing the traditional sources of planting material (suckers and bits). Micropropagation has also greatly facilitated the international exchange of banana germplasm, because plants can be transferred between regions, countries or continents in a closed, sterile environment, free of diseases.

Since 1985, Agribiotecnología de Costa Rica (ACR) has been producing bananas through tissue culture for internal markets and for export to Central and South America and the Caribbean. ACR's experience with tissue culture planting material has shown that tissue-cultured bananas yield about 30–35% higher than conventional plants and do not require the use of nematicides, at least during the 5 years following planting. This results in substantial savings as nematicides cost approximately US $500 ha^{-1} year^{-1}. In addition, the crop is highly uniform when compared with plants derived from other propagation methods. Over the years it has been well established that *Musa* species and other cultivated monocotyledonous plants are recalcitrant for tissue culture and somatic cell manipulations, which results in a certain degree of inefficiency when micropropagated. This biological inconvenience, and the need for high-quality and disease-free planting material at a reasonable price for the growers, was the basis for starting a collaborative project between ACR and DNA Plant Technology Corporation (DNAP), sponsored by the US Agency for International

Development (USAID) through the Agricultural Biotechnology for Sustainable Productivity (ABSP) Project. The general goal of the collaboration was the implementation of more efficient micropropagation methods for tropical crops through the use of bioreactor vessels.

After several trials, micropropagation protocols were adapted to bioreactor vessels in liquid culture. Several thousand banana plants were produced from bioreactor cultures, but three major problems appeared:

1. An endogenous bacteria found in shoot tips appeared 2–3 weeks after explant inoculation (this contamination also has been a serious limitation for callus induction);

2. Apical dominance inhibited clump subdivision, producing a reduced number of well-developed plants and a significant number of small-sized plants in different stages of development, mostly without roots;

3. Screenings of the bioreactor-derived plants in the greenhouse and nursery showed an increased rate of mutation (15%). This rate of mutation is high when compared with traditional tissue culture (usually 5% offtypes). Field evaluations of the remaining plants showed that 100% of them were true-to-type.

Based on research reports made by Novak (1992) and Escalant *et al.* (1994), we were encouraged to use somatic embryogenesis as an alternative method for banana micropropagation. To avoid the endogenous contamination problem during callus induction, young male flowers were used as a source of explant and Escalant's modified methodology was adopted. The procedure was time consuming and the plants obtained from the differentiation of somatic embryos showed a 60–80% rate of mutation, primarily dwarf plants.

The Ideal Transgenic Banana Plant

Conventional banana agriculture involves highly specialized production systems that use high inputs of fertilizer, pesticides and a variety of other materials that are a source of environmental contamination. High banana yields during the second part of this century are the result of two technological changes: (i) the use of high levels of chemicals; and (ii) the use of different varieties, from Gros Michel to Valery and then to Gran Nain, the clone most used commercially (Table 10.2).

Maintaining efficient production is essential to the banana industry, especially in countries like Costa Rica where production costs are high. The development of management strategies that embrace the concepts of sustained soil productivity, reduced use of pesticides, reduced environmental impact and improved economic stability are important. In general, the major constraint to banana production is disease. The spread of Black Sigatoka *(Mycosphaerella fijiensis)* to western Africa and the American tropics within the last 25 years has made the disease the most economically important problem of banana and

Table 10.2. Evolution of banana production in Central America and the Caribbean.

Year	Clone	Yield (t ha^{-1})	Chemical input
1870–1960	Gros Michel	10–20	Low
1960–1980	Valery	40–50	High
1980–Present	Gran Nain	60–80	High

Low: < US $2000 ha^{-1} year^{-1}.
High: > US $2000 ha^{-1} year^{-1}.

plantain. In Costa Rica, chemical control costs are US $78 million annually ($1500 ha^{-1} year^{-1}).

Development of disease-resistant bananas by conventional breeding has been hampered by long generation times, triploidy and the sterility of most edible cultivars. Nevertheless, the Fundación Hondureña de Investigación Agrícola (FHIA, 1996), has produced seven hybrids with resistance to or tolerance of different diseases, particularly Black Sigatoka (Table 10.3). At least four of these hybrids have shown potential for commercial use and have been selected for further regional tests.

Another, more difficult, route to obtaining new banana cultivars is transformation using somatic embryos, single cells or protoplasts. The use of genetic engineering techniques is extremely promising as demonstrated in other crops such as potato and tomato. The recent success achieved with banana protoplasts (Panis *et al.*, 1993; Sagi *et al.*, 1995) has opened up the possibility of DNA or gene transfer into isolated *Musa* protoplasts prior to their regeneration into whole plants. These alternative methods of banana breeding offer greater potential than the conventional route because the long seed-to-seed cycle is eliminated, genetic diversity can be increased, costs are reduced and genetic fingerprinting techniques can differentiate and identify new cultivars.

Table 10.3. Disease resistance of banana hybrids produced by FHIA.

Hybrid	Type	Black Sigatoka (*M. fijiensis*)	Moko (*P. solanacearum*)	Panama disease (*F. oxysporum* f.sp. *cubense*)	Fruit green life
FHIA-01 AAAB	Dessert banana	√	√	√	Long
FHIA-03 AABB	Cooking banana	√	√	√	Short
FHIA-18 AAAB	Dessert banana	√	√	√	Medium
FHIA 21 AAAB	Plantain	√	√	√	Short

Based on its horticultural characteristics, the Gran Nain cultivar may be the ideal commercial banana. Efforts in plant transformation need to be focused on the improvement of Gran Nain. The characteristics of the ideal banana are described in Table 10.4: a plant with short stature, so they are not easily blown down by winds; shorter ratooning; a high potential yield (at least $83 \, t \, ha^{-1}$ $year^{-1}$); tolerance of lower temperatures (below 16°C); drought resistance; improved fruit quality; storability; and resistance to pests and diseases. Gran Nain is not resistant to Black Sigatoka, race 4 of fusarium, burrowing nematode or bunchy top virus. The improvement of these characteristics remains a challenge in banana transformation (Robinson, 1996).

Conclusions

The scientific advances recently achieved in banana biotechnology may make new methods for banana micropropagation feasible, allowing for more efficient

Table 10.4. Characteristics of the ideal transgenic banana plant.

Characteristics	Gran Nain	Ideal transgenic plant
Genomic composition	AAA	AAA or AAAA
Plant phenology		
Height (m)	3.3–3.5	2.5 or less
Ratooning index	1.5	1.5 or more
Annual yield (t ha^{-1})	73.6–83.2	83 or more
Harvest index	Regular	Better than Gran Nain
Cold tolerance (below 16°C)	None	Better than Gran Nain
Drought tolerance	Low	Better than Gran Nain
Fruit		
Finger length (cm)	20	20 or more
Transport quality	Good	Equal or better
Storability	Good	Equal or better
Ripening	Gas control	Natural control
Pest and disease resistance		
Black Sigatoka (*M. fijiensis*)	Susceptible	Resistant
Race 1 of *Fusarium oxysporum*	Resistant	Resistant
Race 4 of *Fusarium oxysporum*	Susceptible	Resistant
Moko (*Pseudomonas solanacearum*)	Susceptible	Resistant
Burrowing nematode (*Radopholus similis*)	Susceptible	Resistant
Banana weevil borer (*Cosmopolites sordidus*)	Susceptible	Resistant
Banana bunchy top virus (BBTV)	Susceptible	Resistant
Cucumber mosaic virus (CMV)	Susceptible	Resistant
Banana streak virus (BSV)	Susceptible	Resistant

propagation and reducing the price of the plant delivered to farmers. Transformation coupled with micropropagation may create new varieties of plants which can be introduced to farmers. From an economic point of view, resistance to or tolerance of the major pests and diseases would be the prime target for the application of contemporary biotechnology to banana.

Of the world's banana and plantain production, 90% is consumed for subsistence in developing countries, making access to *Musa* science and technology an important issue. If developing countries are to participate in and benefit from the application of contemporary biotechnology, individual countries should establish collaborative and informational links with other producing countries, international agencies and research and development groups in industrialized countries in order to secure access to *Musa* biotechnology. The ABSP Project has foreseen this situation and has already established joint projects to promote understanding and dissemination of new technology.

References

Escalant, J.V., Teisson, C. and Cote, F. (1994) Amplified somatic embryogenesis from male flowers of triploid banana and plantain cultivars (*Musa* spp.). In vitro *Cell Development Biology* 30, pp. 18–186.

FAOSTAT (1997) http://apps.fao.org/lim500/nph-wrap.pl?Production.Crops.Primary&Domain=SUA.

Fundación Hondureña de Investigación Agrícola (FHIA) (1996) *Strengthening FHIA's Breeding Program for Development of Disease-Resistant Bananas and Plantains for Domestic Consumption.* Annual report 1995, La Lima, Honduras, pp. 25.

Novak, F.J. (1992) *Musa* (Banana and plantain). In: Hammerschag, F.A. and Litz, R.E. (eds) *Biotechnology of Perennial Fruit Crops.* CAB International, Wallingford, UK, pp. 449–488.

Panis, B., vanWauwe, A. and Swennen, R. (1993) Plant regeneration through direct somatic embryogenesis from protoplasts of banana (*Musa* spp.). *Plant Cell Reports* 12, 403–407.

Robinson, J.C. (1996) *Bananas and Plantains.* CAB International, Wallingford, UK.

Sagi, L., Panis, B., Remy, S., Schoofs, H., De Smet, K., Swennen, R. and Cammue, P.A. (1995) Genetic transformation of banana and plantain (*Musa* spp.) via particle bombardment. *Biotechnology* 13, 481–485.

Simmonds, N.W. and Shepherd, K. (1955) The taxonomy and origins of the cultivated bananas. *Journal of the Linnean Society of Botany, London* 55, 302–312.

The Application of Biotechnology to Date Palm

Mohamed Aaouine

Domaine Agricole El Bassatine, BP 299, Meknes, Morocco

The date palm is a strategic crop for the entire region of North Africa and the Middle East. It supports the life of more than 8 million people in North Africa (Mauritania, Morocco, Algeria, Tunisia and Libya), and makes up 20–75% of the income of the growers. Besides providing a steady source of food and income to local farmers, the tree plays a key role in the multiple-stage farming system by providing protection to fruit trees and other associated crops beneath. Oasis agriculture depends heavily on this providential tree that makes life possible in the desert as it preserves the ecosystem and maintains a favourable microclimate for agronomic activities surrounding human subsistence. Livestock and agronomic crops as well as fruits, vegetables and dyes, such as henna (*Lawsonia inermis*) and saffron (*Crocus sativus*) are integrated with the date palm culture in a unique ecosystem in the oases.

Without any question, Bayoud disease is the most serious enemy of this crop throughout the world. The disease constitutes a plague to Saharan agriculture; its devastating effect unbalanced the ecology of several areas and posed serious human, social and economic problems. Bayoud disease was first discovered in Morocco in 1870 and has already killed more than 10 million trees there. It moved to Algeria where it killed 3 million date palms and is threatening the date palm industry in Tunisia and Mauritania. The problem is aggravated by the fact that the pathogen attacks the most vigorous and productive trees of the best commercial cultivars. Thus, the income of the farms is severely affected, as is biodiversity, with the best genotypes lost forever.

All known measures have been tried with only very limited success and genetic resistance is the sole alternative left to combat the disease. However, breeding for Bayoud resistance is hindered by the long generation time required for date palms, by the dioecious and highly heterozygous nature of the tree, by the limited knowledge concerning pathogen–host interactions, by the complexity of the genetic make up of the date palm and by insufficient information regarding the genetics of disease resistance and of fruit quality.

Biotechnology has played a key role in safeguarding the Moroccan date palm industry. Indeed, tissue culture was successfully used for large-scale propagation of the endangered, sensitive cultivar selections combining disease tolerance and fruit quality. The regeneration system established in Morocco, when combined with genetic transformation techniques, will have significant positive technological, economical, sociological and environmental impacts. Furthermore, tissue-culture-derived plants have been successfully established on soil with 15 g l^{-1} salinity. Thus, if Bayoud-resistant genotypes are obtained, a land reclamation process can be seriously considered.

Introduction

The date palm (*Phoenix dactylifera* L.) is an arborescent, dioecious, highly heterozygous, monocotyledonous plant with a very slow growth rate and a late reproductive phase. It is an ancient crop and one of the five fruit trees among the seven species mentioned in the Bible (Zohary and Spiegel-Roy, 1975) and the Koran. Date palm was domesticated and brought under cultivation in prehistoric times (4000 BC in Iraq) and is presently grown in countries of the Middle East, Asia, North and South Africa and North and Latin America.

In 1996, world production reached 4,491,504 t with 307,348 t for export (US $302,662,000) and 338,411 t for import (US $273,647,000) (FAOSTAT, 1997). Some 75% of world production is consumed in the date-growing countries, 10% is exported to developed countries and 15% to developing countries. The date palm supports the life of more than eight million people in North Africa (Algeria, Libya, Mauritania, Morocco and Tunisia), provides work for large numbers of them since it requires 200 labour days per hectare per year, and makes up 20–75% of their income, depending on the quality of the dates and the productivity of the associated crops and livestock.

The date palm is a strategic crop for the entire region of North Africa and the Middle East. Beside providing a steady source of income to local farmers, the tree plays a key role in the multiple-stage farming system by providing protection to fruit trees and other associated crops which grow beneath. Oasitic agriculture depends heavily on this providential tree that makes life possible in the desert as it preserves the ecosystem and maintains a favourable micro-climate for agronomic activities surrounding human subsistence in the oases. Livestock and agronomic crops (wheat, barley, lucerne) as well as fruits, vegetables and dyes, such as henna (*Lawsonia inermis*) and saffron (*Crocus sativus*), are integrated with the date palm culture in a unique ecosystem in many isolated areas of the Middle East and North Africa. The loss of the date palm would certainly result in the loss of these companion crops which cannot withstand the high light intensities and temperatures of the desert. Furthermore, the trees are grown mainly on marginal areas and play a key role in the ecology by providing a viable means to control desertification in arid lands.

In Morocco, there are approximately four million date palm trees, grown on 84,500 ha, from which more than one million people derive more than 50% of their income, allowing them to continue to live in a hostile environment (Toutain, 1967). The annual production in Morocco is a little over 100,000 t, of which about 80% is used locally in the production zone and about 20% in the rest of the country; very little, if any, is exported.

Uses of Date Palm and Date Palm Products

While the date is often considered a dessert fruit, in many countries, particularly in the Saharan regions of Morocco, it is a staple for a considerable part of the population. The date is a high-energy, nutritious fruit with sugar content varying from 70% to 90% of dry weight. Dates provides 0.8–1.3 kJ 100 g^{-1} (Toutain, 1967) which is more than most other fruits, cooked rice, wheat bread and meat. It is also one of the greatest producers of food per hectare. A well-managed date palm orchard can produce 0.109 million kJ ha^{-1}, the equivalent of 8 t of cereals. The date is also rich in sodium (Na) (10 mg kg^{-1}), calcium (Ca) (590 mg kg^{-1}), iron (Fe) (20–30 mg kg^{-1}), copper (Cu), zinc (Zn), manganese (Mn), potassium (K) (648 mg kg^{-1}), magnesium (Mg) (600 mg kg^{-1}), chlorine (Cl), phosphorus (P) (630 mg kg^{-1}) and sulphur (S) (Toutain, 1967), contains vitamins A, B_1, B_2 and B_7 (total of 0.36 mg 100 g^{-1}), proteins (2–3%), carbohydrates (64%), fibre (2.5–4%), lipids (0.2%) and no cholesterol.

The date is not only the bread of the Saharan people but it is also their hard currency. The inhabitants exchange part of their production for other products such as sugar, tea, cooking oil and clothing. The pit of the fruit is used to feed animals and to fatten very weak camels. The pit can provide 5.4 kJ kg^{-1} and contains 5.8% protein, 9% lipid, 16.2% fibre, 62% sugar and 1.12% ash (Toutain, 1967; Toutain 1973).

In desert areas, the pruning wood and leaves are as important as the fruit itself. One hectare of date palm can produce 4–5 t of pruning wood used for heating or construction. The leaves are used as windbreaks to prevent sand movement, as decorations for houses during feasts and for producing articles such as hats, baskets and ropes. The sap can be used to make palm wine and the terminal bud is considered a delicacy.

Date Production and Consumption

The annual per capita consumption in date-producing countries is 150–185 kg. In Morocco, the annual consumption in the date-growing areas is about

30–40 kg per person compared with about 2 kg in the rest of the country and only 100 g in developed countries (Toutain, 1973). Presently, national production is insufficient to meet local demand and Morocco imports dates from other Arab countries, particularly during the fasting month of Ramadan when date consumption reaches 100–300 g day^{-1} person^{-1} and reaches 400–1500 g on the day of Achoura (Toutain, 1973).

Date imports are increasing with the increase in population and the decrease of the date palm population caused by the ravages of Bayoud disease (from the Arabic word 'abiad' meaning white, in reference to the whitening of the fronds of diseased trees) caused by *Fusarium oxysporum* f. sp. *albidinis* Malençon. Increased imports result in the loss of badly needed hard currency.

The decimation of the date palm has led to increased desertification and to the degradation of the Steppes. This is due to the excessive use of the surrounding Steppes as a source of firewood, no longer available from palm groves, and to overgrazing by animals that no longer find enough fodder in the devastated oases. This also results in migration of people to large cities, thus increasing the already large numbers of unemployed in urban areas. This becomes even more serious when water becomes scarce or absent as in the Draa Valley of southern Morocco, where more than 2500 families have migrated to the north of the country or abroad and more than 7600 temporarily left the area to find seasonal work elsewhere.

In reality, Bayoud disease constitutes a plague to Saharan agriculture and poses serious human, social and economic problems. Although date palm culture is limited to the southeastern parts of Morocco, the crop is of national importance and its production affects the entire country.

Bayoud Disease

Introduction

The date palm is very important for the ecology, economy and sociology of the Sahara environment. It is irreplaceable in the irrigated desert lands. Bayoud disease is without question the most serious enemy of this crop in all the date-growing areas of the world.

The disease was first reported in the Draa Valley, north of Zagora, Morocco in about 1870 (Malençon, 1934c; Pereau-Leroy, 1958) and by 1958 it had killed ten million palms or about two-thirds of the Moroccan date palm population, including the most vigorous and productive palms of the best commercial cultivars (Pereau-Leroy, 1958). Resistant genotypes are horticulturally worthless. Because of this, the income of farmers is severely affected as well as the biodiversity of the date palm. Two genotypes (Idrar and Berni) have been lost forever because of the disease and Mejhoul, one of the most prized date cultivars in Morocco and abroad, which was an important commercial item in European markets since the 17th century, has practically disappeared. The equally important Bou Feggous cultivar is decreasing every year due to the disease and is presently found only in non-irrigated marginal areas where Bayoud disease is not very serious. Furthermore, the best commercial cultivars of Mauritania, Algeria, Tunisia, Libya, Iraq and Saudi Arabia have all been shown to be very sensitive to the disease.

The soil-borne fungus enters the roots and invades the trunk, the leaf bases and finally the terminal bud (Bulit *et al.*, 1967). It has spread both eastward and westward in Morocco. To the west, the last groves were infected in 1954 and 1960 (Malençon, 1934a,b; Toutain, 1965). To the east, the disease advanced relentlessly to affect not only other date-growing areas of Morocco but also western and central parts of Algeria. Despite the prophylactic measures and regular eradication attempts undertaken by Algeria (Kada and Dubost, 1975; Kellou and Dubost, 1975), three million trees have been destroyed and the disease is advancing eastward to the Deglet Nour growing areas of Algeria and Tunisia.

Each year 6% of the susceptible cultivars of excellent quality fruit are destroyed compared with only 1.5% for the moderately resistant cultivars which produce medium- or low-quality fruit (Toutain and Louvet, 1972). However, during a rainy season and in intensively irrigated oases, the loss rate may reach 10–15%. The trees are usually killed 2–5 years after symptoms appear. Consequently, areas formerly with 300–400 palms ha^{-1} have been reduced to only 5–10 trees and as a result, Morocco, which once was exporting dates, must now import them.

Epidemiology

The spread of Bayoud disease from tree to tree occurs through root contact (Malençon, 1949; Toutain, 1970; Brochard and Dubost, 1970; Toutain and Louvet, 1977). The spread is rapid in irrigated orchards but slow in non-irrigated orchards where tree production and vigour are also reduced. High salt content in both soil and water (5 g l^{-1}) does not prevent or slow down the spread of the disease (Toutain, 1974; Toutain and Louvet, 1977). The spread of the disease from oasis to oasis is ascribed to the transfer of infected plant material and soil, and articles made of diseased tissues such as saddle packs, baskets and ropes (Toutain, 1965). The pathogen can persist for at least several weeks in infected rachis (Pereau-Leroy, 1958). When such articles are discarded and placed in a moist place, the fungus may resume growth to become established in the soil and infect palms. The disease spread has followed the major roads of the caravan merchants established in the traditional exchange of products between oases (Pereau-Leroy, 1958). With the development of modern means of transport, the risk of its progression to further countries throughout the Middle East is very high and remains of great concern to all date-growing countries.

More recently, the causal organism of Bayoud disease was isolated in Italy and France from *Phoenix canariensis* L., another palm species widely used as an ornamental tree all over the world (Mercier and Louvet, 1973). The symptoms are identical to those of the date palm. Similarly, *Sabal* sp. and *Washingtonia robusta* are also sensitive to Bayoud disease.

The strategy for dealing with this disease has been well-defined (Bulit *et al.*, 1967; Kellou and Dubost, 1975; Louvet and Bulit, 1970; Louvet *et al.*, 1970; Louvet and Toutain, 1973; Snyder and Watson, 1974). The first objective is to stop or at least slow down the spread of the disease to the Deglet-Nour-producing areas of Algeria and Tunisia. A second objective, of prime importance to Morocco, is to replant areas where date palms have been destroyed with resistant cultivars having the quality of Deglet Nour or Mejhoul.

To achieve the first objective, prophylactic measures (Toutain, 1965; Munier, 1955; Carpenter and Elmer, 1978) including the use of systemic fungicides, disease eradication, cultural control, and extension and grower education have been attempted. However, while these efforts did slow down the spread of the disease, they did not stop it (Toutain and Louvet, 1977). The problem for Morocco, where all date-growing areas except the marginal zone of Marrakech are infected, is to grow date palm in the presence of the fungus. The only practical way of doing this is the use of genetic resistance.

Developing Resistance in Date Palm

The future of the date palm industry in Morocco will depend on the success of developing Bayoud-resistant genotypes. Two approaches can be used: (i) adoption of a selection and breeding programme; and (ii) the use of biotechnology.

Breeding for resistance

All date-growing areas in Morocco, except the marginal zone of Marrakech, are infected, thus efforts must be placed on growing date palm in the presence of the fungus. This can be accomplished only through the use of genetic resistance. Historically, breeding for Bayoud resistance was attempted and has been under way since 1963 (Louvet and Toutain, 1973). The programme involves screening cultivars for tolerance and actively hybridizing and selecting potentially resistant cultivars with adequate economic performance. Presently, approximately 50% of the date palm population of Morocco is seedlings (about two million trees). Resistant cultivars have been observed in heavily infected areas both in Morocco and Algeria (Pereau-Leroy, 1958; Toutain, 1968; Toutain and Louvet, 1972; Toutain and Louvet, 1977). Unfortunately, the selected cultivars are of marginally acceptable quality and are sold only in the local markets.

The breeding programme also involved crosses with, among others, high-quality males from Indio, California, USA. Unfortunately, all of the resistant genotypes developed are horticulturally worthless or of low quality. Furthermore, the breeding programme was handicapped by: (i) the limited number of selected genotypes, which in most cases were represented by single individuals; (ii) the low quality contributed by the resistant selections; (iii) the dioecious and highly heterozygous nature of the tree; (iv) the limited knowledge concerning the pathogen–host interactions; (v) the complexity of the genetic make-up of the date palm; (vi) insufficient information regarding the genetics of the disease resistance and of fruit quality; and (vii) the long generation time required for dates and their dioecious habit. It takes 3 months before one can test for resistance of young seedlings, two more months to be able to select resistant seedlings in the nursery, 5 years to test their actual resistance in heavily infected areas and 7–10 years to determine their sex (usually there are 50% males and 50% females), fruit quality and yield potential (Toutain and Louvet, 1977). The situation is rendered more complex by the lack of an effective means to delay the spread of the disease and the lack of an effective method of clonal propagation – only eight to ten offshoots are produced each year during the 5–8 years of juvenility. At the present time, no selection has been obtained which incorporates resistance and acceptable quality.

As a precautionary measure, in case Bayoud disease develops in other countries, an international programme of testing valuable non-Moroccan cultivars was initiated using two cultivars from Algeria, four each from Tunisia and Mauritania and six each from Libya and Iraq. Again, all cultivars were found to be susceptible to Bayoud disease. Although breeding may be a long-range solution to the problem, it is unlikely to contribute a solution in the near future.

Using biotechnology to develop resistance

Using plant biotechnology may provide solutions to at least some of the problems encountered. Indeed, the use of such techniques as *in vitro* mass propagation, *in vitro* micrografting and insertion of resistance genes into sensitive cultivars may lead to disease resistant cultivars of good agronomic quality. These techniques are briefly discussed below.

Mass propagation

At the present time, *in vitro* propagation of date palm is accomplished through somatic embryogenesis and organogenesis using various explants (auxiliary or terminal buds, bases of very young leaves and young inflorescences). Callus regeneration followed by the induction of somatic embryogenesis, plant regeneration and acclimation to *ex vitro* controlled environments has been achieved for many plant species (Reuveni and Lilien-Kipnis, 1974; Reynolds and Murashige, 1979; Ammar and Bendadis, 1977; Tisserat *et al.*, 1979, 1981). These techniques are being used commercially in many laboratories in Europe, USA and South Africa, and the regenerated plants have been successfully transplanted into field conditions.

Since the use of callus in mass production often results in genetic variation in fruit quality, disease resistance and other morphological and horticultural characters (D'Amato, 1977, 1978), an alternative system of regeneration has been achieved via organogenesis. With this technique, genetic variation is usually much reduced since the callus stage is avoided (Murashige, 1974). The author is presently using such a method on a large scale to mass propagate date palm with hundreds of thousands of plants in the field and tens of thousands already bearing fruits with no genetic variation for all the genotypes produced. These regeneration systems are not only being used for mass propagation of elite cultivars and new selections, but are also crucial to the genetic transformation approach for Bayoud resistance.

Grafting is an obvious solution to the problem, producing a two-part tree made up of a resistant rootstock and a horticulturally superior scion. It has heretofore been excluded from consideration because of the widespread belief that grafting is limited to gymnosperms and dicotyledonous angiosperms, which both contain a continuous cambium layer. However, the presence of a cambium is not an absolute prerequisite for successful graft union. All that is required for grafting is the presence of a meristem. A number of mono-cotyledonous plants have been successfully grafted, including some intergeneric grafts, by taking advantage of the intercalary meristems. These include successful grafts obtained in *Vanilla* (Daniel, 1919a; Muzik, 1958); *Pennisetum purpureum*, var. *merkerii* (merker grass); *Bambasa longispiculata* (bamboo); *Saccharum officinarum* (sugarcane); *Panicum maximum* (guinea grass); and *Panicum purascens* (para grass) (Daniel, 1919b). A successful intergeneric graft has been achieved between para grass as scion and merker grass (Muzik and La Rue, 1954). Reciprocal grafts were also possible between *Zebrina pendula* and *Tradescantia fluminensis* and between *Zebrina pendula* and *Rhoeo discolor* (La Rue, 1944).

Monocotyledonous lianas such as Solomon Islands ivyarum (*Schindapsus aureus*), *Phylodendron nachodomi* and *Nephthytis afzelii* were also successfully grafted, including an intergeneric graft between *Nephthytis afzelii* and *Scindapsus aureus* (Muzik and La Rue, 1952, 1954). Successful unions have also been possible between cultivars of onion (*Allium cepa*), the graft being performed by placing one bulb half onto another (Monakina and Solovey, 1939). Muzik and La Rue (1954) have demonstrated that graft union in grasses follows the same general pattern as observed in dicotyledonous plants, differing only in degree.

Shoot-tip grafting, first developed by Murashige *et al.* (1972), then refined by Navarro *et al.* (1975) and slightly modified by De Lange (1978), has been extremely important for virus elimination in citrus (Roistacher *et al.*, 1976; Navarro *et al.*, 1976; Roistacher and Kitto, 1977; Nauer *et al.*, 1983; Navarro *et al.*, 1980). The technique was also adopted for plums and peaches (Alskief, 1977; Mosella *et al.*, 1979a,b, 1980), apricots (Martinez *et al.*, 1979), apples (Alskief and Villemur, 1978; Huang and Millikan, 1980), grapes (Engelbrecht and Schwerdtfeger, 1979) and avocados (Murashige, 1982, personal communication). Martinez *et al.* (1979, 1981) found that shoot-tip grafting can be used for early detection of graft incompatibility between peach and apricot, between apricot and myrobolan plum and between peach and myrobolan plum.

Ball (1950, 1955) concluded from his experiments on micrografting of apices in date palm that a gradient of graftability existed, with the potential for grafting being higher on the lower side of the apices than at the top. Upon separation of the original apex unit, each part became a new, independent centre of growth and produced a new meristem and shoot. Using shoot tip apices and very young rootstocks of the same (autografts) and of different (heterografts) cultivars, the author has achieved evidence of *in vitro* graft union. In fact, the scion (apex) grew to 10 or 15 cm and produced well-developed leaves. Anatomical studies revealed a good contact between the two partners (Brakez, 1989).

Thus, if high-quality scion cultivars, such as Mejhoul, Bou Feggous, Jihal and Deglet Nour, can be established on Bayoud resistant rootstocks cultivars, like Bou Stammi (noire and blanche), Tadment, Iklane, Sair Layalet, Bou Feggous ou Moussa, Bou Zeggor, Bou Khani, resistant males, chance seedlings or those resulting from controlled pollination, grafting would enable the immediate production of high-quality and disease-resistant planting stock.

Genetic transformation
Although monocotyledonous plants were thought to be more recalcitrant to genetic manipulation, transformation has been successfully achieved with asparagus (Bytebier *et al.*, 1987; Conner *et al.*, 1988), maize (Koziel *et al.*, 1993; Privalle *et al.*, 1994), rice (Shimamoto *et al.*, 1989; Hayakawa *et al.*, 1992; Kimura *et al.*, 1992; Ajisaka and Maruta, 1994; Yahiro *et al.*, 1994), barley and a few others. It is well known that the choice of anti-fungal gene to be used to protect plants depends on the type of strategy the plant employs against fungal attack. Once the strategy has been defined, the approach consists of pinpointing the areas where molecular interference can alter the host–pathogen interactions. Such interference could be to fortify the plant cell wall matrix, thereby making it difficult for the pathogen to enter the plant tissues; or increasing the biosynthesis of antimetabolic proteins such as proteinase inhibitors and amylase inhibitors (Cervone *et al.*, 1990; Huynh *et al*;, 1992a; Laskowski and Kato, 1980; Garcia-Olmedo *et al.*, 1987; Thornburg *et al.*, 1987; Klopfenstein *et al.*, 1991; Richardson *et al.*, 1987; Thornburg, 1990), phytoalexin (Dixon, 1986; Hain *et al.*, 1993) and other metabolites with anti-fungal properties; or cell wall degrading enzymes such as glucanases and chitinases (Broglie *et al.*, 1991; Hyunh *et al.*, 1992b) and lysoenzymes (Trudel *et al.*, 1995), leading to the death of the pathogen.

Even though the interactions between the date palm and the causal agent of Bayoud disease are not yet fully understood, the availability of genes potentially useful in controlling the disease, and technologies to insert such genes and rapidly produce and grow transformed date palms, will have significant technological, economical, sociological and environmental impacts. Indeed, if successful, such approaches will pave the way for the application of such technologies to improve other traits of importance to the date palm industry (fruit quality, faster growth, insect resistance, reduced height, etc.).

Finally, tissue culture derived plants have been successfully established in soil and water with 12 g l^{-1} salinity. If Bayoud resistant genotypes are obtained, a land reclamation process can be seriously considered.

References

Ajisaka, H. and Maruta, Y. (1994) Evaluation of transgenic rice carrying an antisense glutelin gene in an isolated field. In: Jones, D.D. (ed.) *Proceedings of 3rd International Symposium on the Biosafety Results of Field Tests of Genetically Modified Plants and Microorganisms*. Division of Agriculture and Natural Resources, University of California, Oakland, pp. 291–298.

Alskief, J. (1977) Sur le greffage *in vitro* d'apex sur des plantules décapitées de pêcher (*Prunus persica* Batsch). *Académie des Sciences (Paris), Comptes Rendus Hébdomadaires des Séances, serie D.* 284, 2499–2502.

Alskief, J. and Villemur, P. (1978) Greffage in vitro d'apex sur des plantules décapitées de pommier (*Malus pumila* Mill.). *Académie des Sciences (Paris), Comptes Rendus Hébdomadaires des Séances, serie D.* 287, 1115–1118.

Ammar, S. and Benbadis, A. (1977) Multiplication végétative du palmier dattier (*Phoenix dactylifera* L.) par la culture de tissus de jeunes plantes issues de semis. *Académie des Sciences (Paris), Comptes Rendus Hébdomadaires des Séances, serie D.* 284, 1789–1792.

Ball, E. (1950) Isolation, removal and attempted transplants of the central portion of shoot apex of *Lupinus alba* L. *American Journal of Botany* 37, 117–136.

Ball, E. (1955) On certain gradients in the shoot tip of *Lupinus alba* L. *American Journal of Botany* 42, 509–552.

Brakez, Z. (1989) Histological and biochemical aspects of date palm shoot apex micrografts. *Certificat d'Etudes Approfondies en Biologie Générale*, Cadi Ayyad University, Marrakech, Morocco.

Brochard, P. and Dubost, D. (1970) Progression du Bayoud dans la palmeraie d'In Salah (Tidikelt, Algérie). *Alawamia* 35, 143–154.

Broglie, K., Chet, I., Holliday, M., Cressman, R., Biddle, P., Knowlton, S., Mauvais, C.J. and Broglie, R. (1991) Transgenic plants with enhanced resistance to the fungal pathogen *Rhizoctonia solani*. *Science* 254, 1194–1197.

Bulit, J., Louvet, J., Bouhot, D. and Toutain, G. (1967) Recherches sur les fusarioses: I. Travaux sur le Bayoud, fusariose du palmier dattier en Afrique du Nord. *Annales des Epiphyties* 18(2), 213–239.

Bytebier, B., Deboeck, F., De Greve, H., van Montagu, M. and Hernalsteens, J.P. (1987) T-DNA organization in tumor cultures and transgenic plants of the monocotyledon *Asparagus officinalis*. *Proceedings of the National Academy of Sciences, USA* 84, 5345–5349.

Carpenter, J.B. and Elmer H.S. (1978) *Pests and Diseases of the Date Palm*. USDA Agricultural Handbook No. 527, 42 pp.

Cervone, F., DeLorenzo, G., Pressey, R., Darvill, A.G. and Albersheim, P. (1990) Can phaseolus PGIP inhibit enzymes from microbes and plants? *Phytochemistry* 29(2), 447–450.

Conner, A.J., Williams, M.K., Deroles, S.C. and Gardner, R.C. (1988) Agrobacterium-mediated transformation of asparagus. In: McWhirter, K.S., Downes, R.W. and Read, B.J. (eds) *9th Australian Plant Breeding Conference, Proceedings*. Agricultural Research Institute, Wagga Wagga, pp. 131–132.

D'Amato, F. (1977) Cytogenetics of differentiation in tissue and cell cultures. In: Reinert, J. and Bajaj, Y.P.S (eds) *Applied and Fundamental Aspects of Plant Cell, Tissue and Organ Culture.* Springer, Berlin, pp. 343–357.

D'Amato, F. (1978) Chromosome number variation in cultured cells and regenerated plants. In: Thorpe, T.A. (ed.) *Frontiers of Plant Tissue Culture.* University of Calgary Press, Canada, pp. 287–295.

Daniel, L. (1919a) Etude anatomique des autogreffes de selaginelle et de vanille. *Revue Bretonne de Botanique Pure et Appliquée* 14, 13–20.

Daniel, L. (1919b) Nouvelles études sur les greffes herbacées. *Revue Bretonne de Botanique Pure et Appliquée* 14, 29–44.

DeLange, J.H. (1978) Shoot tip grafting: A modified procedure. *Citrus and Subtropical Fruit Journal* 539, 13–15.

Dixon, R.A. (1986) The phytoalexin response: Elicitation, signalling and control of host gene expression. *Biological Reviews* 61, 239–291.

Engelbrecht, D.J. and Schwerdtfeger, U. (1979) *In vitro* grafting of grapevine shoot apices as an aid to the recovery of virus-free clones. *Phytolactica* 11, 183–185.

FAOSTAT (1997) http://apps.fao.org/lim500/nph-wrap.pl?Production.Crops.Primary& Domain=SUA

Garcia-Olmedo, F., Salcedo, G., Sanchez-Monge, R., Gomez, L., Royo, J. and Carbonero, P. (1987) Plant proteinaceous inhibitors of proteinases and α-amylases. *Oxford Surveys of Plant Molecular and Cell Biology* 4, 275–334.

Hain, R., Reif, H.J., Krause, E., Langebartels, R., Kindl, H., Vorman, B., Wiese, W., Schmelzer, E., Schreier, P.H., Stocker, R.H. and Stenzel, K. (1993) Disease resistance results from foreign phytoalexin expression in a novel plant. *Nature* 361, 153–156.

Hayakawa, T., Zhu, Y., Itoh, K., Kimura, Y., Izawa, T., Shimamoto, K. and Toriyama, S. (1992) Genetically engineered rice resistant to rice stripe virus, an insect transmitted virus. *Proceedings of the National Academy of Sciences, USA* 89, 9865–9869.

Huang, Shu-ching and Millikan, D.F. (1980) *In vitro* micrografting of apple shoot tips. *HortScience* 15, 741–743.

Huynh, Q.K., Borgmeyer, J.R. and Zobel, J.F. (1992a) Isolation and characterization of a 22kDa protein with antifungal properties from maize seeds. *Biochemical and Biophysical Research Communications* 182(1), 1–5.

Huynh, Q.K., Hironaka, C.M., Levine, E.B., Smith, C.E., Borgmeyer, J.R. and Shah, D.M. (1992b) Antiqunal protein from plants: purification, molecular cloning and anti-fungal properties of chitinases from maize seeds. *The Journal of Biological Chemistry* 267 (10), 6635–6640.

Kada, A. and Dubost, D. (1975) Le Bayoud à Ghardaia. *Bulletin d'Agronomie Saharienne, Algeria* 1(3), 29–61.

Kellou, R. and Dubost, D. (1975) Organisation de la recherche et de la lutte contre le Bayoud en Algérie. *Bulletin d'Agronomie Saharienne, Algeria* 1(1), 5–13.

Kimura, Y., Hyakawa, T., Zhu, Y. and Toriyama, S. (1992) Environmental risk evluation of transgenic rice expressing coat protein of rice stripe virus. *Japanese Journal of Breeding* 42 (suppl. 1), 124–125.

Klopfenstein, N.B., Shi, N.Q., Keman, A., McNabb, S.E., Hall, R.B., Hart, E.R. and Thornburg, R.W. (1991)Transgenic populus hybrid expresses a wound-inducible potato proteinase inhibitor II-CAT gene fusion. *Canadian Journal of Forestry Research* 21, 1321–1328.

Koziel, M.G., Beland, G.L., Bowman, C., Carozzi, N.B., Crenshaw, R., Crossland, L., Dawson, J., Desai, N., Hill, M., Kadwell, S., Lainis, K., Lewis, K., Maddox, D., McPherson, K., Meghji, M.R., Merlin, E., Rhodes, R., Warren, G.W., Wright, M. and Evola, S.V. (1993) Field performance of elite transgenic maize plants expressing an insecticidal protein derived from *Bacillus thuringiensis* toxins. *Bio/Technology* 11, 194–200.

La Rue, C.D. (1944) Grafts of monocotyledons secured by the use of intercalary meristems. *American Journal of Botany* 31, 36–48.

Laskowski, M. and. Kato, I. (1980) Protein inhibitors of proteinases. *Annual Review of Biochemistry* 49, 593–626.

Louvet, J. and Toutain, G. (1973) Recherches sur les fusarioses: VIII. Nouvelles observations sur la fusariose du palmier dattier et précisions concernant la lutte. *Annales de Phytopathologie* 5, 35–52.

Louvet, J., Bulit, J., Toutain, G. and Rieuf, P. (1970) Le Bayoud, Fusariose vasculaire du palmier dattier – symptômes et nature de la maladie – moyens de lutte. *Alawamia* 35, 161–181.

Malençon, G. (1934a) Nouvelles observations concernant l'étiologie du Bayoud. *Académie des Sciences (Paris), Comptes Rendus Hébdomadaires des Séances, serie D.* 196, 1259–1262.

Malençon, G. (1934b) La question du Bayoud au Maroc. *Annales de Cryptogamie Exotique (Paris)* 7(2), 43–83.

Malençon, G. (1934c) Les palmeraies du Drâa et le Bayoud. *Bulletin de la Sociologie et de l'Histoire Naturelle d'Afrique du Nord* 25, 112–117.

Malençon, G. (1949) Le Bayoud et la reproduction expérimentale des lésions chez le palmier dattier. *Mémoires de la Sociologie et de l'Histoire Naturelle d'Afrique du Nord (Hors série)* 2, 217–228.

Malençon, G. (1956) *Inst. Science.* Morocco, 15 pp.

Martinez, J., Hugard, J. and Jonard R. (1979) Sur les dillerentes combinaisons de greffage des apex réalisés in vitro entre pêcher (*Prunus persica* Batsch.) abricotier (*P. armeniaca* L.) et myrobolan (*P. cerasiféra* Ehrh.) *Académie des Sciences (Paris), Comptes Rendus Hébdomadaires des Séances, serie D* 288, 759–762.

Martinez, J., Possel, J.L., Hugard, J. and Jonard, R. (1981) L'utilisation du microgreffage *in vitro* pour l'étude des greffes incompatibles. *Académie des Sciences (Paris), Comptes Rendus Hébdomadaires des Séances, serie III*, 292, 961–964.

Mercier, S. and Louvet, J. (1973) Recherches sur les fusarioses: X – Une fusariose vasculaire (*Fusarium oxysporum*) du palmier des Canaries (*Phoenix canariensis*). *Annales de Phytopathologie* 5, 203–211.

Monakina, T.A. and Solovey, G.T. (1939) Grafting monocotyledonous bulbous plants. *Horticulture Abstracts* 10, 475.

Mosella, L., Riedel, M. and Jonard, R. (1979a) Sur les améliorations apportées aux techniques de microgreffage des apex *in vitro* chez les arbres fruitiers. Cas du pêcher (*Prunus persica* Batsch). *Académie des Sciences (Paris), Comptes Rendus Hébdomadaires des Séances, serie D* 289, 505–508.

Mosella, L., Riedel, M., Jonard, R. and Signoret, P.A. (1979b) Developpement *in vitro* d'apex de pêcher (*Prunus persina* Batsch): possibilités d'application. *Académie des Sciences (Paris), Comptes Rendus Hébdomadaires des Séances, serie D* 289, 1335–1338.

Mosella, L., Signoret, P.A. and Jonard, R. (1980) Sur la mise au point de techniques de microgreffage d'apex en vue de l'élimination de deux types de particules virales chez le pêcher (*Prunus persina* Batsch). *Académie des Sciences (Paris), Comptes Rendus Hébdomadaires des Séances, serie D* 290, 287–290.

Munier, P. (1955) Le palmier dattier en Mauritanie. *Annales de l'Institut des Fruits et Agrumes Coloniaux* 12, 66 pp.

Murashige, T. (1974) Plant propagation through tissue culture. *Annual Review of Plant Physiology* 25, 135–166.

Murashige, T., Bitters, W.P., Rangan, T.S., Nauer, E.M., Roistacher, C.N. and Holliday, B.P. (1972) A technique of shoot apex grafting and its utilization towards recovering virus-free citrus clones. *HortScience* 7, 118–119.

Muzik, T.J. (1958) Role of the parenchyma cells in graft union of Vanilla orchid. *Science* 127, 82.

Muzik, T.J. and La Rue, C.D. (1952) The grafting of large monocotyledonous plants. *Science* 116, 589–691.

Muzik, T.J. and La Rue, C.D. (1954) Further studies on the grafting of monocotyledonous plants. *American Journal of Botany* 41(6), 448–455.

Navarro, L., Roistacher, C.N. and Murashige, T. (1975) Improvement of shoot tip grafting *in vitro* for virus-free citrus. *Journal of the American Society for Horticultural Science* 100, 471–479.

Navarro, L., Roistacher, C.N. and Murashige, T. (1976) Effect of size and source of shoot tips on psorosis A and exocortis content of navel orange plants obtained by shoot tip grafting *in vitro*. In: Calavan, E.C. (ed.) *Proceedings of 7th Conference of the International Organization of Citrus Virologists*. Riverside, California, pp. 194–197.

Navarro, L., Juarez, J., Ballester, J.F. and Pina, J.A. (1980) Elimination of some citrus pathogens producing psorosis-like leaf symptoms by shoot tip grafting *in vitro*. In: Calavan, E.C. (ed.) *Proceedings of 8th Conference of the International Organization of Citrus Virologists*. Riverside, California, pp. 162–166.

Nauer, E.M., Roistacher, C.N., Carson, T.L. and Murashige, T. (1983) *In vitro* shoot tip grafting to eliminate citrus viruses and virus-like pathogens produces uniform bud-lines. *HortScience* 18(3), 308–309.

Pereau-Leroy, P. (1958) Le palmier dattier au Maroc. *Institut Français de Recherches Fruitières Outre-Mer (I.F.A.C.)* Paris, 142 pp.

Privalle, L.S., Fearing, P.L., Brown, D.L. and Vlachos, D. (1994) Characterization of Bt protein expression in transgenic maize. In: Jones, D.D. (ed.) *Proceedings of 3rd International Symposium on the Biosafety Results of Field Tests of Genetically Modified Plants and Microorganisms*. Division of Agriculture and Natural Resources, University of California, Oakland, pp. 471–473.

Reuveni, O. and Lilien-Kipnis, H. (1974) *Studies of the in vitro culture of the date palm (Phoenix dactylifera L.) tissues and organs*. The Volcani Institute of Agricultural Research, Bet Dagan, Israel. Pamphlet No. 145, 42 pp.

Reynolds, J.F. and Murashige, T. (1979) Asexual embryogenesis in callus cultures of palms. *In Vitro* 15(5), 383–387.

Richardson, M., Valdes-Rodriguez, S. and Blanco-Labra, A. (1987) A possible function for thaumatin and a TMV-induced protein suggested by homology to a maize inhibitor. *Nature* 327 (6121), 432–434.

Roistacher, C.N., Navarro, L. and Murashige, T. (1976) Recovery of citrus selections free of several viruses, exocortis viroid and spiroplasma citri by shoot tip grafting *in vitro*. In: Calavan, E.C. (ed.) *Proceedings of 7th Conference of the International Organization of Citrus Virologists*. University of California, Riverside, pp. 186–193.

Roistacher, C.N. and Kitto, S.L. (1977) Elimination of additional citrus viruses by shoot tip grafting *in vitro*. *Plant Disease Reporter* 61, 594–596.

Shimamoto, K., Terada, R., Izawa, T. and Fujimoto, H. (1989) Fertile transgenic rice plants regenerated from transformed protoplasts. *Nature* 338, 274–276.

Snyder, W.C. and Watson, A.G. (1974) Pathogenicity test for identification of *Fusarium oxysporum* f. sp. *albedinis*. *Bulletin d'Agronomie Saharienne, Algeria* I, 25–30.

Thornburg, R.W. (1990) New approaches to pest resistance in trees. *AgBiotech News and Information* 2, 845–849.

Thornburg, R.W., An, G., Cleveland, T.E., Johnson, R. and Ryan, C.A. (1987) Wound-inducible expression of a potato inhibitor II-chloramphenicol acetyltransferase gene fusion in transgenic tobacco plants. *Proceedings of the National Academy of Sciences, USA* 84, 744–748.

Tisserat, B. (1979a) Propagation of date palm (*Phoenix dactylifera* L.). *Journal of Experimental Botany* 30(119), 1275–1283.

Tisserat, B. (1979b) Tissue culture of date palm. *Journal of Heredity* 70, 221–222.

Tisserat, B., Foster, G. and DeMason, D. (1979) Plantlet production *in vitro* from *Phoenix dactylifera* L. *Annual Date Growers' Institute Report* 54, 19–23.

Tisserat, B., Ulrich, J.M. and Finkle, B.J. (1981) Cryogenic preservation and regeneration of date palm tissue. *HortScience* 16(1), 47–48.

Toutain, G. (1965) Note sur l'épidémiologie du Bayoud en Afrique du Nord. *Alawamia* 15, 37–45.

Toutain, G. (1967) Le palmier dattier: Culture et production. *Alawamia* 25, 83–151.

Toutain, G. (1968) Essais de comparaison de la résistance au Bayoud des variétés de palmier dattier: II Notes sur l'expérimentation en cours concernant les variétés Marocaines et Tunisiennes. *Alawamia* 27, 75–78.

Toutain, G. (1970) Observations sur la progression d'un foyer actif de Bayoud dans une plantation régulière de palmier dattier. *Alawamia* 35, 155–160.

Toutain, G. (1973) Lutte contre le Bayoud: I. Reconstitution de palmeraie Bayoudée au Maroc. *Alawamia* 48, 115–146.

Toutain, G. (1974) Progression du Bayoud en palmeraies établies sur terrains salés. *Alawamia* 53, 65–74.

Toutain, G. and Louvet, J. (1972) Resistance au Bayoud dans les variétés de palmier dattier. In: *Proceedings of the first International Seminar and workshop on the Bayoud Disease*, Algiers, Algeria. 208–210.

Toutain, G. and Louvet, J. (1977) Lutte contre le Bayoud: Orientations de la lutte au Maroc. In: Lhoste, J. and. Besri, M. (eds) *Comptes-Rendus des Cinquiemes Journees de Phytiatrie et de Phytopharmacie circum-Mediterraneennes*, Rabat, Morocco, pp. 140–160.

Trudel, J., Potvin, C. and Asselin, A. (1995) Secreted hen lysozyme in transgenic tobacco: Recovery of bound enzyme and *in vitro* growth inhibition of plant pathogens. *Plant Science* 106(1), 55–62.

Ulrich, J.M. and Finkle, B.J. (1981) *HortScience* 16(1), 47–48.

Ulrich, J.M., Foster, G. and DeMason, D. (1979) *Annual Date Growers Institute* 54, 19–23.

Yahiro, Y., Kimura, Y., Hayakawa, T. and Toriyama, S. (1994) Biosafety results of transgenic rice plants expressing rice stripe virus-coat protein gene. In: Jones, D.D. (ed.) *Proceedings of 3rd International Symposium on the Biosafety Results of Field Tests of Genetically Modified Plants and Microorganisms*. Division of Agriculture and Natural Resources, University California, Oakland, pp. 23–29.

Zohary, D. and Spiegel-Roy, P. (1975) Beginnings of fruit growing in the Old World. *Science* 187, 319–327.

The Use of Coat Protein Technology to Develop Virus-resistant Cucurbits

Hector Quemada

Asgrow Seed Company, 2605 East Kilgore Road, Kalamazoo, MI 49002, USA

One of the earliest examples of a transgenic trait which could be introduced into a crop was the use of viral coat protein genes to confer virus resistance. This approach has found its most advanced commercial use in cucurbit crops. Viruses are a major cause of crop losses in cucurbits. Direct damage can reach 80–100%, but because of fluctuating market prices, even less severe damage can result in a 100% economic loss.

Breeders have for many years recognized that the most effective way to deal with virus problems is to breed resistance into commercial varieties. While resistance does exist in the gene pool of cucurbit crop species, combining resistance with other acceptable horticultural characteristics has met with limited success in the past – although recent breeding efforts are meeting with success in the case of one or two virus resistances. In contrast, the use of viral coat protein genes has enabled breeders to combine rapidly strong virus resistance with good horticultural types to produce a commercially acceptable hybrid. The first commercial variety was a yellow crookneck squash, marketed in 1995 under the name 'Freedom II'.

The same virus resistances have been introduced into cantaloupe, cucumber and watermelon, and commercial varieties are under development in these species as well. Because multiple virus infections are the rule rather than the exception in cucurbit fields, multiple resistance is the ultimate goal. New generations of these crops will include increasing numbers of resistance genes, introduced either through transgenic means or through the incorporation of traditional resistance sources.

Introduction

Diseases and pests are responsible for the loss of a great deal of agricultural productivity. The increased amounts of food which could be made available by preventing losses due to pests and disease can contribute significantly to providing a secure food supply, especially in less developed regions of the world. In addition, the potential for decreased inputs from pesticides, as well as conserving inputs which are wasted when crops are lost (fertilizer, fuel, seed, fungicides, pesticides, water, etc.), makes the development of disease- and pest-resistant crops a continually important goal for conventional breeding as well as biotechnology.

In addition to these benefits to the farmer and consumer, the development of disease-resistant crops increases a seed company's income not only through sales of seed at a higher price (given price increases consistent with a farmer's savings on other inputs) but through capturing market share from a competitor who cannot offer similar resistant varieties. Conversely, a seed company that controls a large share of the market through sales of non-resistant varieties risks the rapid loss of that market to a competitor who develops resistance to important diseases.

The cost of taking the more expensive genetic engineering approach to developing disease or pest resistance has to be weighed against the background of the likely return, as well as the potential for developing the same resistances via conventional breeding (or breeding accelerated by the use of molecular markers). The application of virus coat protein technology – or indeed any genetic engineering approach – provides a potential advantage in that a gene for resistance can be applied to more than one species. Thus, for coat protein, the cost of developing a genetic engineering approach can be recovered by income from cantaloupe, cucumbers, squash, watermelon or pumpkin seed sales. This would not be possible with a resistance derived via conventional breeding, since transfer of resistance genes by sexual means is impossible between these species.

Four viruses – cucumber mosaic virus (CMV), zucchini yellow mosaic virus (ZYMV), watermelon mosaic virus 2 (WMV2) and papaya ringspot virus (PRV) – are responsible for the majority of crop losses in the common cucurbit crops (i.e. cantaloupe, cucumber, squash, watermelon and pumpkin). As a result of multiple infection (Table 12.1) losses can reach 80%, due partly to the reduction in actual yield but primarily due to the reduction in marketable yield. Fruits produced by virus-infected plants are distorted and discoloured, and are therefore of insufficient quality. Because produce prices fluctuate, even minor losses in actual marketable yield can often result in a total economic loss, because the price paid for the produce may reach the point where the loss in marketable yield reduces the projected income below the break-even point. Since storage is a limited option, a farmer often has no choice but to abandon the field, resulting in wasted inputs. For these reasons, the development of resistance to these viruses will provide significant benefits to the farmer and (via its environmental benefits) to the consumer as well.

Table 12.1. Viruses detected by ELISA assay of randomly picked leaves of infected squash from two fields in North Carolina.

Sample number	Cucumber mosaic virus	Papaya ringspot virus	Watermelon mosaic virus 2	Zucchini yellow mosaic virus
1	−	+	+	−
2	−	−	+	−
3	−	+	+	+
4	−	+	−	−
5	−	−	+	−
6	−	−	+	−
7	−	+	−	−
8	−	+	+	−
9	−	+	+	−
10	−	+	+	−
11	−	+	+	−
12	−	+	+	−
13	−	+	+	−
14	−	−	−	−
15	−	+	−	−

Asgrow's Development of Virus-resistant Cucurbits

In 1985, Asgrow chose to pursue the coat protein strategy for engineering virus resistance, after consulting with Dr Roger Beachy, who pioneered the use of these genes in tobacco (Powell-Abel *et al.*, 1986). As a consequence of this decision, a programme was started to clone the coat protein genes of CMV, WMV2, ZYMV and PRV, in collaboration with Dr Dennis Gonsalves at Cornell University. At the same time, efforts were conducted to develop transformation methods for all of the cucurbit species of commercial interest.

Successful transformation and demonstration of virus resistance was first achieved in cantaloupe and squash. However, squash became the lead product, for the following reasons:

1. The virus resistance/susceptibility phenotype was more readily discernible in squash compared with cantaloupe, especially in the field; thus, scoring was more easily done and more rapid progress in variety development could be made;
2. Virus resistance was the overriding priority in our squash-breeding programme, while it had to compete with other equally high priorities in cantaloupe; consequently, more resources could be devoted to the rapid development of commercial squash varieties with transgenic virus resistance;
3. We were able to identify a line of transgenic squash – a yellow crookneck type resistant to ZYMV and WMV2 (Fig. 12.1) – from which we could eliminate the NPT2 marker gene via Mendelian segregation. Because the decision concerning which line to use as a lead product occurred at the time when

(a)

(b)

Fig. 12.1. Virus resistance in transgenic line ZW20. (a) ZW20 plants (right foreground) compared with the corresponding inbred line (left foreground). Both lines were artificially inoculated with ZYMV. Lighter foliage of the inbred is symptomatic of virus infection. (b) Fruit from ZW20 (left) compared with fruit from the corresponding inbred line (right). Plants bearing the fruit were artificially inoculated with WMV2.

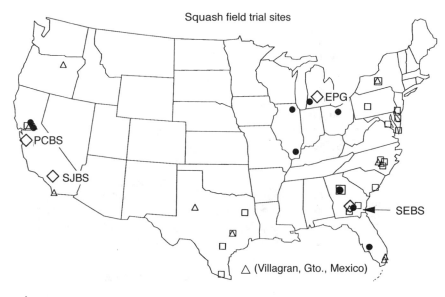

Squash field trial sites

◇ Main site   ● Additional 1992 site   △ Additional 1993 site   □ Additional 1994 site

Fig. 12.2. Locations of field trial sites from 1990 to 1994. Only main sites were planted in 1990 and 1991. Additional sites were planted in the years indicated.

Calgene was involved in submitting a food additive petition to the US Food and Drug Administration (FDA) for the NPT2 protein, Asgrow decided that the opportunity to avoid the questions regarding this protein should be taken in order to reduce the cost of commercial development. The squash line ZW20 provided us with that opportunity. Of course, Asgrow has subsequently advanced lines which contain the NPT2-marker gene, and has obtained regulatory approvals for one such line (Acord, 1996).

The identification of a commercial candidate line prompted its testing (as a hybrid) at a number of locations throughout the United States, and one in Mexico (Fig. 12.2). These tests were conducted by Asgrow or by university collaborators who were able to provide independent assessments of performance under various conditions. One of the lessons we learned from these field trials was that under certain conditions, namely cool temperatures (c. 10°C), virus-like symptoms appeared (Tricoli *et al.*, 1995; Fuchs and Gonsalves, 1995), but soon disappeared when temperatures rose. This phenomenon was also found to be related to the absence of one particular insertion event in the genome (Tricoli *et al.*, 1995). Consequently, we took steps to ensure that the line which was used to generate the commercial hybrid contained the necessary insertion event. The trials conducted by university collaborators allowed us to confirm the benefits that the virus resistance had on marketable yield and also provided additional insight into performance of the hybrid (Fuchs and Gonsalves, 1995; Clough and Hamm 1995; Arce-Ochoa *et al.*, 1995). Finally, trials in farmers' fields under

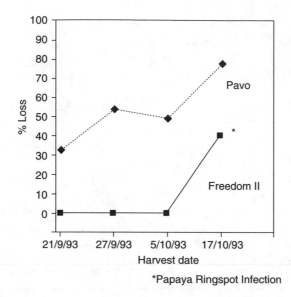

Figure 12.3. Field performance of transgenic hybrid Freedom II compared with the corresponding non-transgenic hybrid (Pavo). Results are given in terms of percentage of yield lost to virus disease. The loss of transgenic yield observed on 17 October 1993 was determined to be the result of papaya ringspot virus infection. Plants were grown in a farmer's field and allowed to become infected with viruses from the surrounding environment.

conditions of natural infections highlighted the fact that these lines could only be claimed to provide benefit when infected by ZYMV and WMV2, but not by other viruses (Fig. 12.3).

Commercialization Issues

The utility of releasing a variety with resistance to only two viruses was debated. However, the decision was made to proceed with commercialization, with the understanding that the introduction of the variety was to be done with very careful claims regarding the limitations of the resistance and that resistance to more viruses was to follow.

Commercialization involved obtaining the appropriate regulatory approvals. The agency with which Asgrow interacted most was the US Department of Agriculture (USDA), from which a determination of non-plant pest status for the transgenic line was needed. Two major issues had to be addressed: (i) weediness of the transgenic crop and of wild relatives with which it might cross; and (ii) the risk of viral transcapsidation by the transgenic coat proteins, and of recombination between infecting viruses and coat protein mRNA. Only part of the information provided to the USDA concerning these issues will be discussed here.

One aspect of the weediness issue was the question of whether or not the transgenic squash itself was likely to be a weed. One way Asgrow addressed this question was to observe the ability of the squash to establish populations when not in cultivated fields. To assess this ability, Asgrow allowed the fruit produced by plants in one experiment to remain in the field and produce volunteers the following spring. The field was left uncultivated, and the ability of these plants to survive in competition with weeds was assessed. Out of approximately 900 volunteer seedlings, only a dozen fruit were collected. These fruit were borne by plants which were severely stunted, yellow and choked by weeds. No evidence of virus infection was found in these plants, despite the presence of an experimentally inoculated field less than 30 m away. Therefore, the stunting and ill health of the plants could not be attributed to viral infection.

Neither could the state of the plants be attributed to toxicity of the soil, since the field – when used as a test plot the year before – was capable of growing healthy plants; furthermore, the adjacent plot of ground was capable of growing healthy plants. The transgene frequency in the seeds collected from the fruit was slightly less than in the previous generation, but the difference was not statistically significant (unpublished; final report to USDA). Therefore, in this single-generation experiment of the fate of plants containing the transgene, no apparent ability to establish wild populations was conferred by the transgene, and therefore the likelihood of these plants becoming weeds appeared to be low.

The more problematic aspect of the weediness issue was the effect that the resistance genes would have on the weediness of wild populations to which they may be transmitted. Asgrow approached this question by asking whether there was any evidence that viruses controlled wild squash populations. In order to assess this, Asgrow requested collaborators from universities in states where wild squash populations occurred to collect samples. These samples were assayed in Dr Dennis Gonsalves' laboratory for the presence of virus. Five different assays were conducted on all the samples in order to ensure detection of virus: (i) visual assessment; (ii) enzyme-linked immunosorbent assay (ELISA) assays; (iii) immunodiffusion assays; (iv) inoculation to indicator hosts; and (v) electron microscope scanning for viral particles. While visual assessment indicated potential disease symptoms in two samples, laboratory testing revealed no viral infection in any samples (Table 12.2). The lack of viral infection did not lend support to the notion that viruses might be controlling wild populations of squash. Consequently, it was difficult to envision how the addition of virus resistance might benefit wild populations which would not be exposed to this selection pressure. These observations are consistent with two other observations: first, virus infection was not observed in volunteer squash growing near inoculated test plots; second, genes for virus resistance are rare in squash or its wild relatives, apparently resistance has not been a trait selected for in these species. These facts taken together suggest that wild squash populations are somehow not subjected to severe virus infections. Therefore, viruses apparently do not impose a biological control on wild squash and virus resistance would not be likely to provide an advantage to these wild populations.

Asgrow approached the viral recombination issue by establishing first that multiple infections were common in the field (Table 12.1), and therefore a significant potential for transcapsidation and recombination already existed. Asgrow then showed that levels of coat protein as well as RNA produced by transgenic plants was much less than the levels in infected non-transgenic plants (Fig. 12.4), and that the transgenic plants were therefore unlikely to present greater levels of risk than was already present in nature. Input from virologists was also obtained by USDA in order to reach a decision.

Eventually the USDA concluded that Asgrow's information, information from experts and the public comments regarding the petition was sufficient to allow USDA to conclude that the transgenic squash did not present a plant pest risk (Medley, 1994). However, recognizing the need to investigate the viral transcapsidation/recombination issue further, USDA and the Biotechnology Industry Organization (BIO) sponsored a workshop to address these questions. This meeting of leading virologists resulted in a report which will prove useful in guiding subsequent decisions regarding crops engineered for resistance to viruses (AIBS, 1995). However, none of the findings of the report presented a reason for the USDA to reverse its decision regarding the squash.

The US Environmental Protection Agency (EPA) does not have a final rule covering transgenic plants. However, Asgrow consulted with the agency during the time it was formulating its proposed rules and was advised that the coat proteins expressed in the squash were pesticidal in nature because they mitigated the effect of a pest, namely a virus. The agency proposed to exempt coat proteins from regulation under the Federal Insecticide, Fungicide, and Rodenticide Act (FIFRA) and the Federal Food, Drug, and Cosmetic Act (FFDCA) (Browner, 1994) and Asgrow was advised that any actions it took with regard to FIFRA were voluntary for the squash. (Because of the status of the rules, and the proposed exemptions, Asgrow concluded that registration was unnecessary.) However, the EPA advised that a petition for exemption from tolerance under the FFDCA needed to be submitted. In order to obtain this exemption, it was important to show that the establishment of a tolerance level was not required to protect the public health. In order to do this, Asgrow sampled cantaloupe and squash for sale at groceries, and measured the amount of viral coat protein found in those samples (Table 12.3). Results showed that levels of viral coat protein in transgenic fruit were within the range found in most of the fruit in the sample, and, in some cases, fruit on the market had many-fold higher levels of coat protein. This result was key to successfully obtaining an exemption from tolerance (Barolo, 1994).

Consultations with the FDA were also part of the commercialization process. Key pieces of information submitted to the FDA included: substantiation of the lack of the NPT2 gene; data regarding levels of viral coat protein in market fruit; and data concerning the nutritional composition of the squash (Quemada, 1996). The market fruit data were useful in this case to support the argument that these proteins presented no toxicity or allergenicity problems. Finally, nutritional analysis of transgenic versus non-transgenic squash helped to answer

Table 12.2. Results of a survey of wild squash populations sampled for evidence of virus infection (modified from data submitted to USDA).

Sample	Location	Visual assessment	Double diffusion serology*	Host indexing*	ELISA*	Electron microscopy*
1-AR	Faulkner County, AR	Slight chlorosis	No virus	No virus	No virus	No virus
2-AR	Washington County, AR	No symptoms	No virus	No virus	No virus	No virus
1-LA	Red River Parish, LA	No symptoms	No virus	No virus	No virus	No virus
2-LA	Red River Parish, LA	No symptoms	No virus	No virus	No virus	No virus
3-LA	Red River Parish, LA	Powdery mildew?	No virus	No virus	No virus	No virus
4-LA	Red River Parish, LA	No symptoms	No virus	No virus	No virus	No virus
1a-MS**	Warren County, MS	No symptoms	No virus	No virus	No virus	No virus
1b-MS**	Warren County, MS	No symptoms	No virus	No virus	No virus	No virus
1c-MS**	Warren County, MS	No symptoms	No virus	No virus	No virus	No virus
1d-MS**	Warren County, MS	No symptoms	No virus	No virus	No virus	No virus
1e-MS**	Warren County, MS	No symptoms	No virus	No virus	No virus	No virus
1f-MS**	Warren County, MS	No symptoms	No virus	No virus	No virus	No virus
2-MS	Issaquena County, MS	No symptoms	No virus	No virus	No virus	No virus
3-MS	Washington County, MS	No symptoms	No virus	No virus	No virus	No virus

* Conducted in the laboratory of Dr Dennis Gonsalves.
** Replicate samples from the same site.

the question of substantial equivalence (Table 12.4). The comparison of the nutritional levels of transgenic squash with those of non-transgenic squash, as well as with published values, were sufficient to support the conclusion of no concerns, in keeping with the FDA guidelines published in 1992 (Kessler, 1992).

Performance and Acceptance of Transgenic Variety

Having obtained the necessary approvals or conducted the necessary consultation, the squash was commercialized in the summer of 1995, under the name 'Freedom II'. Commercial success was mixed. In areas where ZYMV and WMV2 were the prevailing viruses, performance was good and farmers were extremely satisfied with the new variety. Repeat sales were anticipated for 1996. However, as predicted, in areas where other viruses were present in significant amounts, there was no advantage over non-transgenic varieties. Furthermore, these areas tended to be in the more southern states where excessively high temperatures and high levels of whitefly infestation induced morphological problems inherent in the genotype into which the genes were introduced – problems which were recognized features of the hybrid which Freedom II replaced. Thus, instances of 'attached tendril' – a phenomenon in which a fruit developed with a fused tendril or branch – were reported, as were green streaks and doubled fruit. (These all seemed to be different mani-festations of the same developmental problem.) Fruit colour was also reported to be unacceptable in these cases, a consequence of the silverleaf reaction caused by whitefly infection. Although these problems were not the result of the transgenic process and farmers were previously willing to tolerate them when paying the standard seed price, these problems were unacceptable when paying triple the standard price for seed. Consequently, Seminis Vegetable Seeds (SVS), the company resulting from the merger of Asgrow Seed Company's vegetable division and PetoSeed Company, has decided to await the development of resistance to all four main viruses before further commercialization.

At the same time that the transgenic squash was being developed, virus resistance has also been progressing in the cantaloupe, cucumber and water-melon programmes. With the introduction of virus resistance into these species, greater potential for income will be tapped and a wider range of farmers will be benefitted.

Future Developments

Seminis Vegetable Seeds has recently decided to proceed with commercializa-tion of the ZYMV- and WMV2-resistant squash, in the form of a hybrid made from a cross between the ZW20 transgenic line and another parent with improved fruit characteristics. This will be followed by commercialization of a

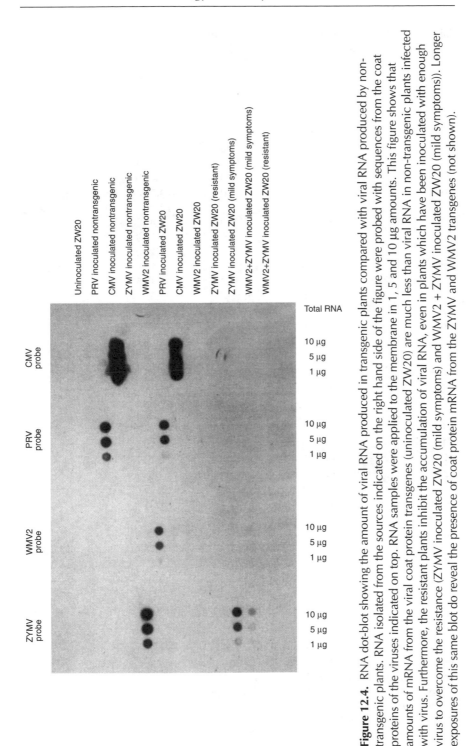

Figure 12.4. RNA dot-blot showing the amount of viral RNA produced in transgenic plants compared with viral RNA produced by non-transgenic plants. RNA isolated from the sources indicated on the right hand side of the figure were probed with sequences from the coat proteins of the viruses indicated on top. RNA samples were applied to the membrane in 1, 5 and 10 μg amounts. This figure shows that amounts of mRNA from the viral coat protein transgenes (uninoculated ZW20) are much less than viral RNA in non-transgenic plants infected with virus. Furthermore, the resistant plants inhibit the accumulation of viral RNA, even in plants which have been inoculated with enough virus to overcome the resistance (ZYMV inoculated ZW20 (mild symptoms) and WMV2 + ZYMV inoculated ZW20 (mild symptoms)). Longer exposures of this same blot do reveal the presence of coat protein mRNA from the ZYMV and WMV2 transgenes (not shown).

Table 12.3. Measurements of viral coat proteins in supermarket fruit, as measured by ELISA assay. (From Quemada, 1996, with permission.)

Fruit Sample	CMV	PRV	ZYMV	WMV2
ZW20	nd	nd	68.4	430.6
C1	355,200	18,000	14,400	10,320
C2	130,464	5,472	10,944	115,488
C3	nd	252,000	28,800	720
C4	nd	nd	864	nd
C5	>2,400,000*	1,200	8,400	nd
C6	>3,216,000*	nd	14,000	nd
C7	>3,216,000	nd	12,864	nd
H1	nd	7,200	9,480	nd
H2	nd	6,840	1,800	nd
H3	nd	nd	2,200	nd
H4	359	4,752	3,888	173
H5	269	3,168	3,168	260
H6	238	nd	2,592	nd
H7	nd	5,928	1,824	137
H8	664	13,272	1,896	190
H9	82	960	24	24
H10	nd	nd	250	nd
H11	nd	nd	1,560	nd
H12	nd	nd	480	nd
H13	nd	nd	2,200	nd
H14	nd	3,120	720	nd
H15	nd	10,080	1,700	nd
H16	nd	nd	3,100	nd
Y1	nd	nd	11,424	nd
Y2	nd	nd	nd	nd
Y3	nd	nd	1,152	nd
Y4	nd	nd	13,056	nd
Z1	nd	nd	140	nd
Z2	nd	nd	nd	nd
Z3	nd	nd	454	nd
Z4	nd	nd	nd	nd
Z5	nd	nd	nd	nd
Z6	nd	nd	576	nd
Z7	43	nd	2,592	nd
Z8	14	nd	2,900	nd

Amounts are µg kg^{-1} fresh weight of fruit. CMV, cucumber mosaic virus; PRV, papaya ringspot virus; ZYMV, zucchini yellow mosaic virus; and WMV2, watermelon mosaic virus 2. Fruit samples: C1–C7 are from cantaloupe; H1–H16 are from honeydew melon; Y1–Y4 are from yellow crookneck squash; and Z1–Z8 are from courgette. nd = not detected. * = exceeds the upper limit of the assay. ZW20 is the transgenic line.

Table 12.4. Compositional analysis of transgenic squash, the corresponding commercial variety and values in the literature (Pennington, 1989). Freedom II and non-transgenic ranges measurements are obtained from three different field sites. (Modified from Quemada, 1996, with permission.)

Component	Freedom II	Non-transgenic	Literature
Protein (g 100 g^{-1})	0.8–1.2	0.8–1.4	0.9
Moisture (g 100 g^{-1})	93.6–94.5	93.7–94.8	94.2
Fat (g 100 g^{-1})	<0.1–0.1	<0.1	0.3
Ash (g 100 g^{-1})	0.4–0.7	0.5–0.8	
Total dietary fibre (g 100 g^{-1})	1.0–1.2	1.0–1.1	1.1
Carbohydrates (g 100 g^{-1})	4.1–4.3	3.9–4.4	4.0
Calories (calories 100 g^{-1})	16.4–18.9	14.4–18.9	18.5
Fructose (g 100 g^{-1})	0.9–1.2	1.1–1.3	
Glucose (g 100 g^{-1})	0.8–1.1	1.0–1.2	
Sucrose (g 100 g^{-1})	<0.2	<0.2–0.2	
Lactose (g 100 g^{-1})	<0.2	<0.2	
Maltose (g 100 g^{-1})	<0.2	<0.2	
Vitamin C (mg 100 g^{-1})	15.1–22.4	14.1–23.2	7.7
Beta carotene (mg 100 g^{-1})	<0.03–0.05	<0.03–0.04	
Vitamin A (IU 100 g^{-1})	<50–80	<50–70	338
Calcium (mg 100 g^{-1})	13.3–29.4	15.7–29.7	21.5
Iron (mg 100 g^{-1})	0.287–0.367	0.372–0.478	0.477
Sodium (mg 100 g^{-1})	<2.50–3.98	<2.50–2.86	

CMV-, ZYMV- and WMV2-resistant hybrid next spring (D.M. Tricoli, SVS, 1997, personal communication). Competitor companies – Rogers and Harris-Moran – have released or are releasing ZYMV- and WMV2-resistant courgette hybrids derived from conventional breeding efforts.

References

Acord, B.R. (1996) Asgrow Seed Co.; Availability of determination of nonregulated status for squash line genetically engineered for virus resistance. *Federal Register* 61, 33484–33485.

AIBS (1995). *Transgenic Virus-resistant Plants and New Plant Viruses*. American Institute of Biological Sciences, Washington, DC, 47 pp.

Arce-Ochoa, J.P., Dainello, F., Pike, L.M. and Drews, D. (1995) Field performance comparison of two transgenic summer squash hybrids to their parental hybrid line. *HortScience* 30, 492–493.

Barolo, D.M. (1994) Watermelon mosaic virus-2 coat protein, zucchini yellow mosaic virus coat protein, and the genetic material necessary for the production of these proteins in transgenic squash plants; tolerance exemption. *Federal Register* 59, 54824–54825.

Browner, C.M. (1994) Plant-pesticides subject to the Federal Insecticide, Fungicide, and Rodenticide Act and the Federal Food, Drug, and Cosmetic Act. *Federal Register* 59, 60496–60518.

Clough, G.H. and Hamm, P.B. (1995) Coat protein transgenic resistance to watermelon mosaic and zucchini yellow mosaic virus in squash and cantaloupe. *Plant Disease* 79, 1107–1109.

Fuchs, M. and Gonsalves, D. (1995) Resistance of transgenic hybrid squash ZW-20 expressing the coat protein genes of zucchini yellow mosaic virus and watermelon mosaic virus 2 to mixed infections by both potyviruses. *Bio/Technology* 13, 1466–1473.

Kessler, D. (1992) Statement of policy: foods derived from new plant varieties; notice. *Federal Register* 57, 22984–23005.

Medley, T. (1994) Availability of determination of nonregulated status for virus resistant squash; notice. *Federal Register* 239, 64187–64189.

Pennington, J.A.T. (1989) *Food Values of Portions Commonly Used*, 15th edn., J.B. Lippincott, Philadelphia, Pennsylvania.

Powell-Abel, P., Nelson, R.S., De, B., Hoffman, N., Rogers, S.G., Fraley, R.T. and Beachy, R.N. (1986) Delay of disease development in transgenic plants that express the tobacco mosaic virus coat protein gene. *Science* 232, 738–743.

Quemada, H. (1996) Food safety evaluation of a transgenic squash. In: *Food Safety Evaluation*. OECD, Paris, France, pp. 71–79.

Tricoli, D.M., Carney, K.J., Russell, P.F., McMaster, J.R., Groff, D.W., Hadden, K.C., Himmel, P.T., Hubbard, J.P., Boeshore, M.L. and Quemada, H.D. (1995) Field evaluation of transgenic squash containing single or multiple virus coat protein gene constructs for resistance to cucumber mosaic virus, watermelon mosaic virus-2, and zucchini yellow mosaic virus. *Bio/Technology* 13, 1458–1465.

The Biotechnology of Oil Palm

Suan-Choo Cheah

Palm Oil Research Institute of Malaysia (PORIM), PO Box 10620, 50720 Kuala Lumpur, Malaysia

It is predicted that palm oil will be the leading oil in the world market for oils and fats for many years to come. By the year 2005, palm oil will account for 20% of total production of oils and fats. In the effort to improve the oil, the Palm Oil Research Institute of Malaysia (PORIM) identified biotechnology as a promising technology. Research has been implemented in two areas, namely genetic engineering and molecular genomics. These studies are aimed at developing the tools and techniques for the creation of value-added palm oil. It is the objective of the genetic engineering programme to produce high oleic acid palms. A multidisciplinary team is isolating genes controlling the synthesis of this fatty acid and developing methods for oil palm transformation. In the molecular genomics programme, the genome of the oil palm is being examined using the techniques of gene linkage mapping, molecular cytogenetic analysis and cDNA sequencing for the generation of expressed sequence tags (ESTs). Although it is the primary objective of the programme to enable marker-assisted selection, it was found that the molecular probes developed were also useful for fingerprinting tissue culture clones. Efforts were thus made to commercialize this process on a small scale.

Introduction

Oil palm and palm oil

In taxonomic terms, the oil palm belongs to the genus *Elaeis* which is a member of the tribe *Cocoineae* in the family *Palmae*. It is now recognized that the genus

has only two species, the African *Elaeis guineensis* and the South American
E. oleifera (Schultes, 1990). The species are closely related and they can be
intercrossed (Hardon and Tan, 1969). The commercially important oil palm is
of the *E. guineensis* species.

The oil palm has a solitary unbranched stem with a crown of large
pinnate fronds. It can grow to a height of over 15 m, producing 24 fronds
year^{-1}. The palm reaches maturity at approximately 2–3 years of age. It can
live for more than 100 years, but its economic life does not extend beyond 35
years.

The palm is monoecious, and male and female inflorescences are produced
on the same palm. However, only one inflorescence matures at any one time on
the same palm, thus ensuring cross-fertilization. In mature palms, each fruit
bunch consists of 1000–3000 fruits.

The fruits accumulate oil in the mesocarp and the kernel. Palm oil refers to
the oil produced in the mesocarp. Its fatty acid composition (FAC) is different
from that of the oil produced in the kernel. Palm oil consists predominantly of
C16:0 (palmitic acid) and C18:1 (oleic acid) fatty acids while palm kernel oil is
rich in the C12:0 (lauric acid) fatty acid (Table 13.1). As a result, these two oils
have different uses. As much as 90% of palm oil is used in food applications. The
presence of very low amounts of linolenic and other highly unsaturated fatty
acids confers good stability to the oil at high temperatures. It is thus an
excellent medium for frying. This property has also enabled it to be used in
margarines and bakery fats with minimal hydrogenation. Palm kernel oil, on
the other hand, is valued as a feedstock for the oleochemical industry as it
contains short- and medium-chain fatty acids. It is widely used in the
manufacture of soap, shampoo and detergent.

Table 13.1. Fatty acid compostion (%) of palm oil and palm kernel oil. Source:
PORIM Technology No. 3 and 6.

Fatty acid	Palm oil	Palm kernel oil
C6	–	0.3
C8	–	4.4
C10	–	3.7
C12:0	0.2	48.3
C14:0	1.1	15.6
C16:0	44.0	7.8
C16:1	0.1	–
C18:0	4.5	2.0
C18:1	39.2	15.1
C18:2	10.1	2.7
Others	0.8	0.1
Iodine value	53.3	17.8
Saponification value	196	245

Palm oil yield

A most remarkable finding that revolutionized oil palm cultivation was the discovery of the single-gene inheritance of the thickness of the kernel shell. It was found that the hybrid (termed *tenera*) of the thick (homozygous *dura*) and thin (homozygous *pisifera*) shelled variety had more mesocarp and thus produced more oil. This discovery, reported by Beirnaert and Vandeweyen in 1941, is unique in that it allowed the manipulation of just one gene to increase yield by as much as 25–30%. Further yield improvements have subsequently been made through the breeding for *duras* and *pisiferas* with good combining abilities (Yong and Chan, 1996). Oil yields as high as $11-12$ t ha^{-1} year^{-1} are possible today, but this is still below the predicted yield potential of 17 t (Corley, 1983).

Oil palm cultivation for sustainable productivity

Of the major oil crops, the oil palm yields the highest amount of oil per unit area (Table 13.2). This combined with the fact that it is a perennial crop has allowed it to be price competitive in the world market. Palm oil's rise to its current position as the second major vegetable oil produced worldwide (after soybean) is attributed to rapid expansion of oil palm cultivation in Southeast Asia in the last 30 years, especially in Malaysia. World production stood at 15.9 million tonnes in 1996, of which Malaysia's production accounted for 53% (Bernama, 1997). It is predicted that palm oil production will continue to rise, and the oil will overtake soybean as the major oil in the first decade of the next century (Oil World, 1994).

Table 13.2. Comparative yields for vegetable oils (1995–1996). Source: Oil World Annual 1996.

Crop	Product	Oil yield (t ha^{-1} year^{-1})
Oil palm*	Palm oil	3.20
Oil palm	Palm kernel	0.46
Soya	Seed	0.37
Groundnut	Seed	0.45
Cotton	Seed	0.19
Rapeseed	Seed	0.60
Sunflower	Seed	0.53
Coconut	Copra	0.36

* Refers to commercial yield of Malaysian hybrid *tenera* variety.

PORIM and Oil Palm Biotechnology

Although palm oil will be the leading oil in the future, Malaysia's ability to retain its current position as the major producer is hampered by an impending shortage of labour and suitable land for cultivation. In order to allow for future growth of the industry, it is realized that there is a necessity to increase value-addedness of the oil and to reduce production costs. The Palm Oil Research Institute of Malaysia (PORIM) has identified biotechnology as one of the expedient means of achieving these objectives. As the techniques of *in vitro* gene manipulation and molecular breeding are deemed to be especially useful, two programmes, one on genetic engineering and the other on molecular genomics, were implemented at the institute.

Genetic engineering for oil quality

Elucidation of the pathways of oil synthesis in plants started in the early 1950s (Stumpf, 1996). Since then, the area has been intensively studied in several laboratories (Ohlrogge and Browse, 1995; Harwood, 1996). This has recently led to much interest in genetically modifying oil crops to change oil quality (Murphy, 1995; Töpfer *et al.*, 1995). The quality of an oil is a reflection of its FAC, and this determines its functionality. It is thus envisaged that changing the FAC of an oil could improve or extend its usage. The first successes in genetically engineering plants for oil quality were in the production of high stearate- and laurate-containing rapeseed oils (Kridl *et al.*, 1993).

The FACs of palm oil and palm kernel oil render these oils applicable to both edible and non-edible uses. However, in order to venture into new markets, a change in FAC is desirable. PORIM's current interest in applying gene technology to the oil palm is thus for the production of oil with a high content of the monounsaturated fatty acid, oleic acid. Such an oil has the potential to open markets for palm oil in the liquid oil sector. In addition, the oleic acid is useful as an industrial feedstock. A multidisciplinary approach has been taken at PORIM for the implementation of a concerted programme to develop the tools and techniques required for genetically engineering the oil palm (Cheah *et al.*, 1995). A strategy was adopted to alter the expression of genes of the enzymes controlling the levels of palmitic and oleic acid in palm oil so that as much of the palmitate as possible is converted to oleate.

Preliminary results obtained have provided some answers to the question of why palm oil contains high levels of palmitic acid. It was shown that the mesocarp contains an active acyl-acyl carrier protein (ACP) thioesterase that very effectively removes palmitic acid from the biosynthetic pathway (Sambanthamurthi and Oo, 1991). On the other hand, the B-ketoacyl ACP synthase II (KASII) enzyme that converts palmitic acid to stearic acid is relatively inactive, thus, palmitic acid accumulates.

The biochemical studies also showed that KASII activity in the mesocarp is positively correlated with the iodine value (IV) of the oil produced (Umi *et al.*, 1995). The enzyme thus has an important role to play in the control of the levels of palmitate and oleate in oil synthesized in the mesocarp. In addition, purification of the thioesterase enzyme clearly demonstrated the presence of two thioesterase proteins, one with a marked preference for oleoyl-ACP and the other for palmitoyl-ACP (Abrizah *et al.*, 1995). This distinction augurs well for the effort to reduce palmitoyl-ACP thioesterase by antisense techniques without coreducing the oleoyl-ACP specific enzyme. Efforts to clone the genes of the thioesterase and KASII enzyme are under way. It is envisaged that introducing a more active promoter to the KASII gene and reducing palmitoyl-ACP thioesterase activity by antisense techniques will drive the pathway towards the conversion of palmitic acid to oleic acid.

In the effort to manipulate the oil palm for improved oil production, one has to ensure that the target genes are expressed in the right tissue and at the right time. Since we aim to genetically engineer mesocarp oil, the manipulated gene has to be expressed in the mesocarp tissue during oil synthesis. Work in this area has identified a mesocarp-specific cDNA clone whose expression correlated with the period of oil accumulation (Abdullah *et al.*, 1995).

The ability to introduce foreign genes into cells and to regenerate these cells into whole plants in the process of transformation is a prerequisite of *in vitro* gene manipulation. We have therefore given much emphasis in our programme to developing a suitable transformation system for oil palm. Our efforts have included the use of a particle gun for bombarding cells with DNA and the uptake of DNA by germinating pollen and cultured tissues. Promoters conferring efficient expression of foreign genes in oil palm have also been identified (Parveez *et al.*, 1995). Various parameters governing the entry of reporter gene constructs into callus cells via particle bombardment have been optimized (Parveez *et al.*, 1997). There is genetic evidence that calli bombarded with the particle gun have the foreign genes integrated into the genome. The bombarded calli are currently undergoing selection and regeneration.

Molecular analysis of the oil palm genome

The oil palm genome has been estimated to contain 2 pg DNA per $2n$ nucleus (Jones *et al.*, 1982) thus giving it a diploid genome size of about 1900 megabase pairs. Our efforts in karyotyping have confirmed that the palm is a diploid with a $2n$ chromosome number of 32 (Madon *et al.*, 1995).

The major impediments to genetic studies in the oil palm are its long life cycle and the requirement for large tracts of land. It is envisaged that molecular markers being neither phase nor tissue specific may help overcome these constraints. For this purpose, we have developed molecular probes for the oil palm employing restriction fragment length polymorphism (RFLP) and random amplified polymorphic DNA (RAPD). Work on applying amplified fragment length polymorphism (AFLP) has been initiated (Singh and Cheah, 1996).

It has been the contention that there is low genetic variability among the oil palm planting materials in Malaysia as they were derived essentially from four mother palms. We therefore attempted to quantify this variability using RFLP (Cheah *et al.*, 1993) and RAPD techniques. Interspecific variability between the *E. guineensis* and *E. oleifera* species was found to be about 30%. When six varieties of *E. guineensis* were examined, there was only 7% variability among them.

Our studies on the application of these probes in tissue culture have shown that the technology can serve as a means of quality control for the process (Cheah and Wooi, 1995). We have lately demonstrated how such a quality control system can be implemented. The probes have been used to monitor the uniformity of tissue culture lines. In this exercise, we were able to identify plants which obviously do not belong to a particular line, perhaps a result of culture mix-up in the laboratory. In the process of recloning palms, the technique has been applied to checking the identity of the ramets (plants derived from tissue culture) to be recloned. The molecular probes were also found to be able to distinguish clones. Such a means of identification will have important implications when intellectual property rights (IPR) can be provided for oil palm clones.

Realizing the commercial potential of this technique, the use of DNA probes for DNA fingerprinting of oil palm clones was launched during PORIM's Transfer of Technology Seminar in May 1996 (Cheah *et al.*, 1996). In response to this launch, a project for the fingerprinting of 12 clones was commissioned by a commercial tissue culture laboratory. The project has been successfully completed and two other projects to fingerprint commercial oil palm clones are under way.

Perhaps a more important application of molecular marker technology is in breeding. The availability of probes for traits of interest will allow breeders to select at the nursery stage, thereby reducing the cost and time scale of breeding programmes. We have initiated a programme to map the oil palm genome using RFLP, AFLP and microsatellites. A partial map consisting of RFLP markers has been obtained. Our target is to map the genes controlling oil quality, kernel shell thickness and tissue culture induced flowering abnormalities. In the future, we plan to locate the genes conferring other characters such as slow height increment, disease resistance and amenity to tissue culture. The availability of probes for these characters should provide an enormous boost to our ability to produce planting materials faster and with greater precision.

As our efforts in mapping make use of cDNA probes as RFLP markers, it will be more meaningful if the identity of these cDNAs is known. For this reason, we initiated a project on single-pass sequencing with the objective of identifying the cDNAs in our collection of libraries by searching for homologies in publicly accessible electronic DNA sequence databases. This has been a fruitful effort as about 20% of the cDNAs sequenced can be identified. The collection of sequences is being compiled in a DNA database (*PalmGenes*) incorporated in PORIM's in-house database which is accessible on the Internet (http://porim.gov.my).

The location of molecular probes on oil palm chromosomes has been achieved by several techniques. Cytogenetic analysis of the ribosomal RNA gene loci in oil palm employing the polymerase chain reaction (PCR)-based PRINS (primed *in situ*) method has shown that there are two nuclear regions in the oil palm genome. This gene has also been located using the technique of fluorescence *in situ* hybridization (FISH). Multiple labelling FISH has located the 5S genes on the proximal arms of the longest chromosome pair. The 18S–5.8S–25S ribosomal RNA genes were located on the telemetric regions of an acrocentric pair of chromosomes and on the satellite DNA of a medium length chromosome pair (Madon *et al.*, 1996).

Several new approaches for studying the oil palm genome have recently been initiated. These include the development of microsatellite markers for future saturation of the molecular map. This is deemed necessary as the RFLP markers currently employed occur in the expressed regions of the genome only. In anticipation of the possibility of cloning genes by map-based methods (positional cloning), we plan to develop large insert DNA libraries of the oil palm genome.

The Road Ahead

It has been about 10 years since PORIM's biotechnology programme was implemented. Initial misgivings about whether the techniques of recombinant DNA were useful in a tree crop such as the oil palm have proven to be unfounded. In fact, there are certain advantages in applying these techniques to a long-lived crop. For example, marker-assisted selection (MAS) has the potential to shorten the time required for the development of improved varieties. As well, instability in the inheritance of transgenes is less of a worry when the same crop is harvested for over 30 years.

Future biotechnology applications in oil palm will take advantage of the palm's inherent high oil productivity. In the palm, most of the acetyl-coenzyme A (CoA) intermediate is presumably channelled into lipid biosynthesis. Using the principles of metabolic engineering (Bailey, 1991), it is possible that some of this metabolic intermediate could be redirected towards the synthesis of other value-added products. For example, transforming the palm with bacterial genes for the synthesis of polyhydroxybutyrate (PHB) (Poirier *et al.*, 1992) could divert the acetyl-CoA from lipid to the synthesis of biodegradable thermoplastics.

In plants, acetyl-CoA is also the precursor of the isoprenoid pathway (Chappell, 1995) which leads to the biosynthesis of several biologically important compounds such as the carotenoids, sterols, ubiquinone and monoterpenes. Several of these compounds are potentially useful as therapeutic drugs. In palm oil, these are present as minor components (Goh *et al.*, 1981).

However, knowledge of their biosynthesis at the molecular level should provide the key to how the palm can be engineered to become a major source of pharmaceuticals.

Presently, the most likely candidate for this *in vitro* approach is the genetic manipulation of carotene and sterol biosyntheses. Knowledge of the molecular control of these pathways is relatively advanced (Bartley and Scolnik, 1994; McGarvey and Croteau, 1995) compared with that of the tocopherols and tocotrienols (Draper, 1980), another class of compounds with therapeutic potential in palm oil. Given the productivity of the palm, these techniques are likely to transform the oil palm from a commodity crop to a pharmaceutical factory early in the next millennium.

Acknowledgements

The author thanks the Director-General of PORIM for permission to publish this chapter. The contributions of my colleagues, Dr S. Ravigadevi, Siti Nor Akmar Abdullah, Abrizah Othman, Umi Salamah Ramli, Mohamad Arif Abdul Manaf, Ahmad Parveez Ghulam Kadir, Rajinder Singh and Maria Madon, are gratefully acknowledged.

References

Abdullah, S.N.A., Shah, F.H. and Cheah, S.C. (1995) Construction of oil palm mesocarp cDNA library and the isolation of mesocarp-specific cDNA clones. *Asia-Pacific Journal of Molecular Biology and Biotechnology* 3, 106–111.

Abrizah, O., Othman, O., Noor Embi, M. and Ravigadevi, S. (1995) Purification and characterization of Acyl-ACP thioesterase from the oil palm (*Elaeis guineensis*) mesocarp. In: Svasti, J., Rimphanitchayakit, V., Soontaros, S., Tassanakajorn, A., Limpaseni, T., Pongsawasoli, P., Wilairat, P., Sonthayanon, B., Boonjawat, J., Packdibamrung, K. and Kamolsiripichaiporn, S. (eds) *Biopolymers and Bioproducts: Structure, Function and Applications, Proceedings of the Eleventh FAOBMB Symposium.* Samakkhisan Public Co. Ltd, Bangkok, Thailand, pp. 585–589.

Bailey, J.E. (1991) Towards a science of metabolic engineering. *Science* 252, 1668–1675.

Bartley, G.E. and Scolnik, P.A. (1994) Molecular biology of carotenoid biosynthesis in plants. *Annual Review of Plant Physiology and Plant Molecular Biology* 45, 287–301.

Beirnaert, A. and Vanderweyen, R. (1941) Contribution a létude génétique et biometrique des varietes d'*Elaeis guineensis* Jacq. *Publis. Institut National pour l'Etude Agronomique du Congo Belge, Series Sci.* No. 27, 101.

Bernama (1997) Malaysia looking abroad to grow oil palm. *Sun,* 19 April 1997.

Chappell, J. (1995) The biochemistry and molecular biology of isoprenoid metabolism. *Plant Physiology* 107, 1–6.

Cheah, S.C. and Wooi, K.C. (1995) Application of molecular marker techniques in oil palm tissue culture. In: *Proceedings of 1993 ISOPB International Symposium on Recent Developments in Oil Palm Tissue Culture and Biotechnology.* Palm Oil Research Institute of Malaysia (PORIM), Kuala Lumpur, Malaysia, pp. 163–170.

Cheah, S.C., Abdullah S.N.A., Ooi, L.C.L., Rahimah, A.R. and Madon, M. (1993) Detection of DNA variability in the oil palm using RFLP probes. In: *1991 Proceedings of the International Palm Oil Conference – Agriculture (Module 1)*. Palm Oil Research Institute of Malaysia (PORIM), Kuala Lumpur, Malaysia, pp. 144–150.

Cheah, S.C., Sambanthamurthi, R., Abdullah, S.N.A., Abrizah, O., Mohamad Arif, A.B., Umi, S.R. and Ahmad Parveez, G.K. (1995) Towards genetic engineering of oil palm (*Elaeis guineensis* Jacq.). In: Kader, J.-C. and Mazliak, P. (eds) *Plant Lipid Metabolism*. Kluwer Academic Publishers, Dordrecht, The Netherlands, pp. 570–572.

Cheah, S.C., Ooi, L.C.L. and Rahimah, A.R. (1996) Quality control process for oil palm tissue culture using DNA probes. *PORIM Information Series, No. 40*, 4 pp.

Corley, R.H.V. (1983) Potential productivity of tropical crops. *Experimental Agriculture* 19, 217–237.

Draper, H.H. (1980) Biogenesis. In: Machlin, L.J. (ed.) *Vitamin E. A comprehensive treatise*. Marcel Dekker Inc., New York, USA, pp. 268–271.

Goh, S.H., Choo, Y.M. and Ong, S.H. (1981) Minor components of palm oil. *Journal of American Oil Chemists Society* 62, 237–240.

Hardon, J.J. and Tan, G.Y. (1969) Interspecific hybrids in the genus *Elaeis* I. Crossability, cytogenetics and fertility of F1 hybrids of *E. guineensis* × *E. oleifera*. *Euphytica* 18, 372–379.

Harwood, J.L. (1996) Recent advances in the biosynthesis of plant fatty acids. *Biochimica et Biophysica Acta* 1301, 7–56.

Jones, L., Barfield, D., Barret, J., Flook, A., Pollock, K. and Robinson, P. (1982) Cytology of oil palm cultures and regenerant plants. In: Fujiwara, A. (ed.) *Plant Tissue Culture 1982*. Maruzen, Tokyo, Japan, pp. 727–728.

Kridl, J.C., Davis, H.M., Lassner, M.W. and Metz, J.G. (1993) New sources of fats, waxes and oils: the application of biotechnology to the modification of temperate oilseeds. *AgBiotech News and Information* 5, 121N–126N.

Madon, M., Clyde, M.M. and Cheah, S.C. (1995) Cytological analysis of oil palm *Elaeis guineensis* (tenera) chromosomes. *Elaeis* 7, 122–134.

Madon, M., Clyde, M.M. and Cheah, S.C. (1996) Fluorescence *in situ* hybridization of rRNA probe to *Elaeis guineensis (Tenera)* chromosomes. *Elaeis* 8, 29–36.

McGarvey, D.J. and Croteau, R. (1995) Terpenoid metabolism. *The Plant Cell* 7, 1015–1026.

Murphy, D.J. (1995) The use of conventional and molecular genetics to produce new diversity in seed oil composition for the use of plant breeders – progress, problems and future prospects. *Euphytica* 85, 433–440.

Ohlrogge, J. and Browse, J. (1995) Lipid biosynthesis. *The Plant Cell* 7, 957–970.

Oil World (1994) Oil World 2012. *ISTA Mielke GmbH*, Hamburg, Germany.

Parveez, G.K.A., Chowdury, M.K.U. and Salleh, N.M. (1995) A preliminary study of the performance of different promoters for oil palm transformation. In: Svasti, J., Rimphanitchayakit, V., Soontaros, S., Tassanakajorn, A., Limpaseni, T., Pongsawasoli, P., Wilairat, P., Sonthayanon, B., Boonjawat, J., Packdibamrung, K. and Kamolsiripichaiporn, S. (eds) *Biopolymers and Bioproducts: Structure, Function and Applications, Proceedings of the Eleventh FAOBMB Symposium*. Samakkhisan Public Co. Ltd, Bangkok, Thailand, pp. 563–568.

Parveez, G.K.A., Chowdury, M.K.U. and Salleh, N.M. (1997) Physical parameters affecting GUS gene expression in oil palm (*Elaeis guineensis* Jacq.) using the biolistic device. *Industrial Crops and Products* 6, 41–50.

Poirier, Y.P., Dennis, D.E., Klomparens, K. and Somerville, C.R. (1992) Production of polyhydroxybutyrate, a biodegradable thermoplastic, in higher plants. *Science* 256, 520–523.

Singh, R. and Cheah, S.C. (1996) Development of AFLP markers for mapping the oil palm genome. In: Osman, M., Clyde, M.M. and Zamrod, Z. (eds) *Proceedings of the Second National Congress on Genetics*. Bangi, Selangor, Malaysia, pp. 402–407.

Sambanthamurthi, R. and Oo, K.C. (1991) Thioesterase activity in the oil palm (*Elaeis guineensis*) mesocarp. In: Quinn, P.J. and Harwood, J.L. (eds), *Plant Lipid Biochemistry, Structure and Utilization, The Proceedings of the Ninth International Symposium on Plant Lipids*. Portland Press Ltd, London, UK, pp. 166–168.

Schultes, R.E. (1990) Taxonomic nomenclatural and ethnobotanic notes on *Elaeis*. *Elaeis* 2, 172–187.

Stumpf, P.K. (1996) Plant lipid biochemistry in the 1940s, 1950s. *Inform* 7, 102–105.

Töpfer, R., Martini, N. and Schell, J. (1995) Modification of plant lipid synthesis. *Science* 268, 681–686.

Umi, R.S., Othman, O., Noor Embi, M. and Ravigadevi, S. (1995) Purification and characterization of β-ketoacyl ACP synthase II from the oil palm (*Elaeis guineensis*) mesocarp. In: Svasti, J., Rimphanitchayakit, V., Soontaros, S., Tassanakajorn, A., Limpaseni, T., Pongsawasoli, P., Wilairat, P., Sonthayanon, B., Boonjawat, J., Packdibamrung, K. and Kamolsiripichaiporn, S. (eds) *Biopolymers and Bioproducts: Structure, Function and Applications, Proceedings of the Eleventh FAOBMB Symposium*. Samakkhisan Public Co. Ltd, Bangkok, Thailand, pp. 558–562.

Yong, Y.Y. and Chan, K.W. (1996) Breeding oil palm for competitiveness and sustainability in the 21st century. In: *Proceedings of the 1996 PORIM International Palm Oil Congress – Agriculture Conference*. Kuala Lumpur, Malaysia, pp. 19–31.

Issues Surrounding the Development, Transfer, Adaptation and Utilization of Agricultural Biotechnology for Emerging Nations

14

Making a Difference: Considering Beneficiaries and Sustainability while Undertaking Research in Biotechnology

Joel I. Cohen

Intermediary Biotechnology Service, International Service for National Agriculture Research (ISNAR), PO Box 93375, 2509 AJ The Hague, The Netherlands

Over the past three decades, experience has been gained regarding implementation, integration and management of biotechnology research. Much of this experience has been gained with temperate crops suitable for use in industrialized countries. However, over the last 10 years, research has also begun addressing agricultural objectives for the application of biotechnology to tropical crops or farming practices in developing countries.

The international community has played an important role in stimulating such research and, over the past 4 years, the International Service for National Agricultural Research (ISNAR) has built and disseminated a knowledge base regarding this research. A meeting was held to better understand emerging needs of developing countries with regards to biotechnology and to assess the potential for international collaboration, leading to a study of international biotechnology research and advisory programmes, and their base of support. A group of 46 international biotechnology programmes is identified which share the objective of developing and transferring products from biotechnology which address developing country needs.

A combined picture of progress for these programmes is seldom presented, as we work much unto ourselves. There are few mechanisms and opportunities to present aggregate analysis, to consider their implications and look strategically at actions expected or needed. Working in isolation means that potential benefits and clients are often misunderstood, sustained funding is questioned and potential impact is minimized. At the individual level, it is difficult to derive lessons learned, and to determine how individuals have made the difference in facing and often overcoming significant hurdles presented by biotechnology, as well as becoming managers responsible for mobilizing public resources and helping to build national capacity and expertise. Finally, whether at the individual or programme level, we minimize the chance to demonstrate responsibility to the public regarding this research and expected products.

These programmes indicate that research opportunities are increasing. Yet, much knowledge and experience is lacking regarding effective management and implementation of technologies to meet the needs of developing countries. In addition, rapid changes have occurred to the global food system which affect agricultural production in the developing countries. This changing environment for agriculture challenges national research leaders as to where best to place emphasis on biotechnology and how to ensure that products address global and national needs.

A number of such challenges were identified during the Agricultural Biotechnology Policy Seminar held in September 1994 in Singapore for six Southeast Asian countries. Consequently, a unique management course is being developed in collaboration between ISNAR and Japan International Research Center for Agricultural Sciences (JIRCAS), through support from the Government of Japan. This course, Managing Biotechnology in a Time of Transition, responds directly to needs identified by countries attending the seminar series. It will provide key agricultural personnel the opportunity to enhance management and leadership skills, with special emphasis on biotechnology.

Introduction and Objectives

Over the past three decades, experience has been gained regarding implementation, integration and management of biotechnology research. Much of this experience has been gained with temperate crops suitable for use in industrialized countries. Over the last 10 years, however, agricultural scientists began applying biotechnology to tropical crops or farming practices in developing countries. The international community has played an important role in stimulating such research, yet the full scope and diversity of this effort has not been readily available, and has seldom been presented in a comprehensive format.

To complement fundamental research undertaken by the international community, the International Service for National Agricultural Research (ISNAR) has initiated innovative approaches to address policy, management and organizational decisions affecting agricultural biotechnology. Three approaches have been developed. First, an accessible base of knowledge for this research was produced. This process began by identifying emerging needs of developing countries regarding biotechnology and assessing potential opportunities for international collaboration. This information provided the foundation for compiling a directory of expertise in biotechnology, based on a study of 46 international biotechnology research and advisory programmes, and their base of support.

The second approach to building and using information entails a series of Agricultural Biotechnology Policy Seminars, held regionally for collaborating countries. These seminars complement technical research by providing opportunities to explore questions of policy, management, needs and priorities posed by the use of biotechnology in developing countries. These seminars have been held for over 25 countries, including countries in Southeast Asia, east and southern Africa, Latin America and west Asia/North Africa.

In its third and most recent approach, ISNAR responded specifically to the management challenges posed to agricultural research by the introduction of new technologies. It has initiated a training programme, New Technologies for Agricultural Research, supported by the Government of Japan. One part of the programme, Managing Biotechnology in a Time of Transition, concentrates on agricultural biotechnology. This course addresses needs identified by our collaborating partner countries. The goal is to provide key agricultural personnel with the opportunity to enhance their management and leadership skills, with special emphasis on biotechnology.

This chapter reviews the first two approaches and lessons derived from international biotechnology research and advisory programmes regarding international collaboration, sustainability and beneficiaries. Then, a more recent approach is introduced which maximizes 'complementarity' through a course on managing biotechnology. This approach uses case studies based on individual perspectives as one tool to increase effectiveness of participants. It addresses management, implementation and policy needs related to agricultural biotechnology in developing countries. The chapter closes by emphasizing the need for continuity of effort and how complementarity can help in this regard.

International Collaboration in Biotechnology: Getting Something Started and Bringing it to Fruition

When describing efforts in international development collaboration, it is not helpful to speak from generalities, yet equally difficult to focus on specific examples and learning by trial and error. A combined picture of progress made and difficulties encountered in this work is seldom presented. While conferences occur for sharing technical advances and information, there are few mechanisms and opportunities to present aggregate analysis, to consider their implications for sustainability and look strategically at actions expected or needed. This means that potential benefits and clients are often misunderstood and sustained funding is questioned, which together minimize the potential impact of internationally sponsored research in agricultural biotechnology.

At the level of the manager, scientist or director, it is difficult to derive lessons learned or to study how individuals have made the difference in facing and often overcoming significant hurdles presented by biotechnology, as well as becoming managers responsible for mobilizing public resources and helping to build national capacity and expertise. Finally, working alone, whether at the individual, programme or international centre level, minimizes the chance to demonstrate both responsibility to the public and contributions towards sustainability regarding research sponsored by the international community and from which products are expected.

Building our Knowledge Base

Over the past 4 years, ISNAR has taken a three-step approach to address the difficulties posed by the issues presented above. First, to better understand emerging needs of developing countries with regards to biotechnology and to assess the potential for international collaboration, a meeting was held at ISNAR in The Netherlands (Cohen and Komen, 1994). This meeting led to a study of international biotechnology research and advisory programmes and their base of support, provided primarily from international donors and foundations. This group of 46 international biotechnology programmes shared a specific objective of developing and transferring products from biotechnology which address developing country needs. These initiatives are categorized as follows:

- research programmes for crops or livestock at national or international public institutes;
- advisory programmes which concentrate on policy and research management issues;
- international or regional biotechnology networks for specific crops or regions;
- bilateral or multilateral donor programmes which support international biotechnology activities (Fig. 14.1; Appendix 14.1).

Focus
Information collected regarding international collaboration is now stored and made available through BioServe, which permits the analysis and dissemination of our knowledge base regarding biotechnology as practised for and with developing countries. Its combined data clearly demonstrate a range of unique opportunities for accessing and developing specific technologies for developing countries. These opportunities are unique because:

- research is undertaken on essential commodities, or foods on which significant numbers of people depend, often with regional significance (Table 14.1; Brenner and Komen, 1994; Cohen and Komen, 1994; IBS, 1994);
- research objectives target a range of new products, including improved crop plants, livestock vaccines and diagnostic probes;
- diseases and pests selected are major problems for enhancing sustainable agricultural productivity in tropical, as opposed to temperate, agricultural systems;
- access to proprietary technologies can be provided.

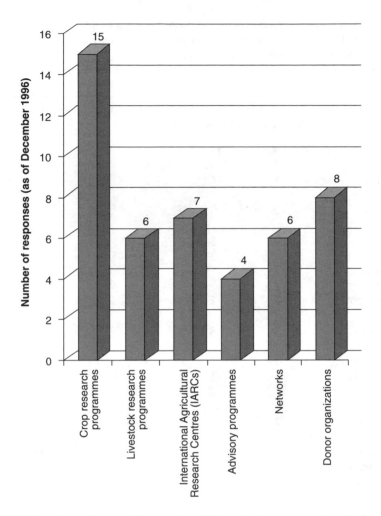

Fig. 14.1. Number and type of international biotechnology research and advisory programmes as identified by IBS.

There are approximately 126 distinct research activities for the 22 international crop research programmes included among the 46 projects identified through the survey. Of these, the majority of work is concentrated on disease and insect resistance, accounting for 60% of the total research effort (Table 14.1). Specific disease and pest examples will be given later. However, as extensive as this list of objectives and activities may seem, when compared with the diversification of biotechnology research in industrialized countries, the list seems far more constrained. The international programmes have a much smaller emphasis on quality traits and no research on herbicide resistance, factors affecting ripening, developing new carbohydrates or proteins, and little research on fungal diseases.

Table 14.1. Five research objectives and related activities undertaken by international biotechnology projects for crops of major importance to developing countries. Source: IBS *BioServe* Database (1994).

Crops	Objectives					
	Disease resistance	Insect resistance	Virus resistance	Quality traits	Micropro-pagation	All
Cereals	**9**	**13**	**8**	**12**		**42**
Rice	5	4	6	6		21
Maize	1	6	2	3		12
Sorghum	1	3		2		6
Other	2			1		3
Root crops	**4**	**5**	**7**	**2**	**1**	**19**
Potato	1	3	2			6
Cassava	1		3	2		6
Yam	2		1		1	4
Sweetpotato		2	1			3
Legumes	**4**	**6**	**4**	**6**		**20**
Bean	1	2	1	2		6
Groundnut	1	1	3	1		6
Chickpea	1	1		2		4
Other	1	2		1		4
Horticulture	**2**		**3**		**1**	**6**
Perennial	**2**	**1**	**2**	**2**	**15**	**22**
Banana/plantain	2		1		5	8
Industrial crops				1	4	5
Coffee		1			4	5
Sugarcane			1	1	1	3
Cocoa					1	1
Forestry species				**2**	**5**	**7**
Miscellaneous	**3**	**3**		**2**	**2**	**10**
All	**24**	**28**	**24**	**26**	**24**	**126**

Figures are based on information gathered from 22 international research programmes that include activities in crop research. For this table, we used those research activities with a specific applied objective, excluding research activities aimed towards general technology development.

Programme elements

An analysis of seven major elements of the international programmes was undertaken, and indications of their respective proportion of effort. Elements[1] of primary importance are: research and development (R&D); human resource development; national programme participation and networking; programme planning, policy, and management; monitoring and evaluation; information and communication; and infrastructure development.

Each programme was asked the proportion of total effort assigned to the above components. Data received reveal that R&D costs account for approximately 50% of total programme costs (Fig. 14.2). In comparison, human resource development totals 18%. This emphasis on research means that the other activities surveyed received comparatively little attention. It also indicates the research-intensive nature not only of biotechnology, but for agriculture in general, and more specifically, for meeting the needs of tropical agriculture.

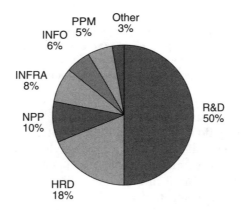

Fig. 14.2. Major programme elements and percentage of total financial effort attributed to them by international biotechnology programmes. R&D = research and development; HRD = human resource development; NPP = national programme participation; INFRA = infrastructure development; INFO = information and communication; PPM = programme planning, policy and management. The chart is based on data on grant funding for the 25 crop and livestock research programmes (totalling US $140 million), in order to keep the information comparable.

---

[1] These elements were defined as follows. First, R&D included all costs for the actual research component of the programmes, whether crop or livestock. Human resource development accounted for training (short and long term), including postdoctoral positions. National programme participation denotes funding reserved to facilitate research and exchanges with national programmes in developing countries. Monitoring and evaluation funding, while limited, gives an important indication of effort planned for the monitoring of biotechnology research. Programme planning included internal management issues and their relation to issues such as biosafety. Information and communication documented expenditures for electronic linkages, newsletters and databases. Finally, infrastructure development included resources for, for example, laboratory and computer equipment.

Policy dialogue

The second approach to building and using information entails a series of Agricultural Biotechnology Policy Seminars, held regionally for collaborating countries. Here, the further integration of biotechnology with agricultural research is an objective, so it may serve broader agricultural or national development objectives and does not become an isolated research endeavour. Attention is given to case study examples where biotechnology research offers solutions to agricultural problems in developing countries. These seminars complement technical research of the international programmes by exploring questions of sustainability, and associated policy, management, needs and priorities posed by developing countries as they consider new technologies and their agricultural research systems.

In doing so, the policy seminars help developing countries address the 2020 Vision statement as set in the section, Strengthening Agricultural Research and Extension in and for Developing Countries. Here, governments are recommended to:

> develop a clear policy on and agenda for biotechnology research; forge partnerships between developing-country research systems, international research institutions, and for the private sector to undertake biotechnology research focused on the problems of developing-country farmers.
>
> (IFPRI, 1995)

The case studies used are explored in this regard by multidisciplinary and representative delegations from developing countries. In facilitating these delegations, the Intermediary Biotechnology Service (IBS) of ISNAR ensures involvement of individuals with responsibility for, or direct interest in, the design, implementation and use of agricultural biotechnology. This range of stakeholder interests enriches each delegation's debate as they identify areas requiring further support, often taking the form of policy dialogues, management recommendations, or responses needed for various international agreements. Each seminar builds on available data and scientific understanding to address the broader needs of stakeholders, including policy makers, managers and researchers, and end users (Komen *et al.*, 1996). By providing this opportunity, IBS complements the heavy emphasis which other international programmes must place on research.

Lessons Derived and Implications for Developing Countries

The previous section summarized sources of information regarding collaborative research and international developments for biotechnology. By reviewing these initiatives, one comes to understand the difficulties these programmes face, gain a more global picture of their objectives and priorities, and consider lessons learned. The picture which emerges indicates that: resources are spread over diverse targets, regions, and time frames; projects address beneficiary needs, as well as concerns for both productivity and sustainability; projects are research-intensive; and new partnerships have been formed. These points are highlighted below.

Resources stretched over diverse targets, regions and time frames

Analysis of the 22 international crop biotechnology research programmes indicates that they address five broad research objectives, containing 126 separate activities (Table 14.1). This work spans over 15 kinds of crop, some of which pose significant research problems and for which a general lack of knowledge exists when compared with crops of commercial interest. There is a paucity of identified genes to address farmer needs, for example, genes for resistance to abiotic stress factors; and technologies for micropropagation, regeneration and transformation often do not exist, or are not yet transferable. These activities have been divided almost evenly among countries in Africa, Latin America and Asia (Cohen and Komen, 1994).

This range of objectives, activities and regional focus also reflects interests of organizations providing financial support. This relationship can create difficulties, as time frames for proposals are constrained to comply with grant requirements. Due to the technical nature of the research, as shown above, longer time frames are needed, with initial periods of at least 5 years. However, beginning in the 1980s and continuing in the 1990s, there was a tendency to minimize time frames for biotechnology, as it was 'sold' as a shortcut to conventional research. Difficulties in implementation, time and funding are experienced as new technologies need testing, for both safety and efficacy, in the collaborating countries. Countries need to consider testing scientific, policy and managerial competency, these areas often have not kept pace with international research.

National programmes in developing countries face this same dilemma, as they confront policies placing emphasis on localized services, which increases the number of crops applicable for research. The technical difficulties are compounded by the growing interest from scientists to apply the tools of biotechnology to the improvement of these crops.

Research-intensive nature of the projects

Major expenditures among the international biotechnology research and advisory programmes are for research. In fact, among the 46 programmes surveyed, all but four are directed towards research. This leaves little funding for socio-economic, policy or managerial issues in order to enhance competency for biotechnology in developing countries. For example, only 10% of expenditure is used for national programme participation, i.e. those costs available for supporting participation of scientists from developing countries in international programmes. These figures indicate that while research may be well financed, adequate support to build developing country capacity and collaboration is funded to a lesser extent.

The research focus and funding constraints of international research programmes were studied by IBS in order to determine the scope of its own activities. It was seen that relatively few programmes could support in-depth analysis and research on the policy, management, needs and priorities of developing countries. This supported the original intention of those advocating a programme such as IBS. Thus, IBS attempts to complement the resources provided by the international research programmes, through its policy seminars and management courses, to enhance competency and capacity among collaborating developing countries.

The research-intensive nature of advanced biotechnology and its related funding implications are major challenges for the national agricultural research systems in developing countries. In fact, these are challenges which the international programmes, including the international centres, have in common with the developing countries. Programmes, whether international or national, are still primarily devoted to fundamental research, with financing available mainly for salaries, equipment and research expenditures. These facts emphasize the capital-intensive nature of this research.

New partnerships possible

Some international programmes can provide access to both public- and proprietary-domain technologies. These new opportunities build on traditional international research which has relied almost exclusively on partnerships with public-sector institutions in developed countries for advances in basic research. Projects, including the Agricultural Biotechnology for Sustainable Productivity (ABSP) project and the sweetpotato research collaboration between Mosanto Company and the Kenya Agricultural Research Institute, illustrate that new partnerships are possible, spanning public–private collaborations. Examples of other commercial technology transfer from international programmes is shown in Table 14.2. Projects with smallholders as beneficiaries are possible, as opportunities are identified which take into account the realities of market economies. In this regard, national and public institutions benefit from collaboration with international biotechnology programmes.

If collaboration with the private sector is an option, then communication with commercial organizations should occur at an early stage. This helps to ensure that products are appropriate for production and geared to the identified clients or users of the research. In such cases, programmes may consider contractual mechanisms for technology transfer, such as collaborative research and development agreements, which delineate the terms of development between public research institutions and private producers.

Table 14.2. Private-sector technology transfer in international research programmes on agricultural biotechnology (Source: IBS, 1994).

International programme	Private-sector collaborator	Technology	Collaborating institute(s)
Agricultural Biotechnology for Sustainable Productivity (ABSP)	ICI Seeds (now Garst Seed Company) (USA)	Maize transformation with *Bacillus thuringiensis* protein genes, for resistance to Asian stem borer	Central Research Institute for Food Crops (CRIFC, Indonesia)
	DNA Plant Technology (USA)	Bioreactor technology for micropropagation of banana, pineapple, coffee and ornamental palms	• Agribiotecnología de Costa Rica (ACR) • Fitotek Unggul (Indonesia)
Feathery Mottle Virus Resistant Sweetpotato for African Farmers	Monsanto (USA)	Transformation technology for the development of virus-resistant sweetpotato	Kenya Agricultural Research Institute (KARI)
International Service for the Acquisition of Agribiotech Applications (ISAAA)	Monsanto (USA)	Transformation technology for the development of potatoes resistant to potato potyvirus X and Y	Center for Advanced Research Studies (CINVESTAV, Mexico)
	Asgrow Seed (USA)	Coat-protein technology for the development of melons resistant to cucumber mosaic virus	• Research Center in Cell and Molecular Biology (CIBCM, Costa Rica) • CINVESTAV
	Pioneer Hi-Bred (USA)	Enzyme-linked immunosorbent assay (ELISA) kits for local maize viruses	National Research Center for Maize and Sorghum (CNPMS, Brazil)
Overseas Development Association (ODA) Plant Sciences Research Programme	Agricultural Genetics Company (UK)	Insect resistance genes for potato and sweetpotato	International Potato Center (CIP, Peru)

Such new partnership capabilities parallel needs of the research organizations in developing countries. There are valuable lessons to learn regarding research relations between public and private sectors as they have evolved and occurred within the international biotechnology programmes. However, the number of such examples are not growing. It is difficult to identify the problems, to clarify appropriate incentive structures, and therefore, to suggest alternative or new approaches for building new partnerships. There are few cases where private–public collaboration, and related incentives, have been taken as a focal point of the programme itself (Cohen, 1993).

Addressing beneficiary needs

Considerations of smallholder farming communities pertain directly to the crops and diseases presented in this chapter. Smallholders or farmers working under resource-limiting conditions are often the 'targeted' beneficiaries of many international biotechnology research projects. The importance of targeting such beneficiaries has recently been emphasized by Dr Ismail Serageldin, who stated:

> The revolutions in molecular biology and in information technology offer us unprecedented opportunities for harnessing new resources on behalf of the poor, but the era of research which produces technological innovations without reference to the needs of the producers is behind us.
>
> (Pearce, 1996)

Data provided in Table 14.1 and Appendix 14.1 indicate that the focus, objectives and activities of the international biotechnology research and advisory programmes targets crops of significance for food security, smallholder farming communities and pest or disease problems which detract from sustainable productivity. Priority setting methods have varied considerably among these programmes (Cohen, 1994). Even so, numerous objectives of important social significance have been identified and beneficiaries well-targeted, as emphasized by Dr Serageldin (Table 14.3).

In IBS policy seminars, beneficiary perspectives are provided by including end-users, usually non-governmental organisations (NGOs), farmer organizations and/or the private sector, as members of the national delegations. The complex and diverse needs they present entail both opportunities and constraints to international research programmes. Discussions emphasize the fact that addressing food security for smallholders means working on a large number of crops, for which technology transfer and delivery of results is often complicated. This raises the question of expectations for biotechnology research, how it will impact and thereby benefit the smallholders, and what may be the consequence if the research never leaves the laboratory.

Table 14.3. Using genetic information to substitute for less effective practices and enhance productivity.

Case study examples	Target for new information	Identified need	Using genetic information	To substitute for prior practices	Beneficiaries
Durable resistance and molecular markers	Rice	Improve resistance to blast fungus and greater understanding of durable resistance mechanisms	Genes from several sources of blast resistance, and use of DNA probes	Spraying fungicides or other chemical controls, as comparable practices are lacking	Smallholder and other farmers in Colombia, as well as agricultural scientists, breeders
Recombinant vaccines	Rinderpest control	Effective vaccine, more capable of withstanding African conditions	Genes cloned which produce surface proteins of virus	Cold chain refrigeration and other technical difficulties for vaccinations	Nomadic herders, smallholders
Viral protection	Sweetpotato	Effective virus resistance for use in household farming situations	Coat protein for feathery mottle virus	Poorly adapted cultural practices	Smallholders, women
Insect resistance	Tropical maize	*Heliothis zea* and other tropical maize pests	Screening of cloned *Bacillus thuringiensis* genes for toxicity; transformation of tropical maize containing *cry-* genes	Poor control methods for tropical insect pests	Wide range of maize producers
Agricultural pests	Pesticides	Locally available, more environmentally-friendly pesticides	Screening of *Bacillus* ssp., for lepidopteran and dipteran pests	Poor control of pests on crops in India, need for availability of local pesticides targeting pests of local importance	Wide range of agricultural growers
Disease-free planting material, biopesticides, and improved maize	Range of crops	Needs determined in four countries (Kenya, Colombia, Zimbabwe and India) setting with Directorate General for International Cooperation, The Government of The Netherlands (DGIS)	Maker-assisted selection of drought tolerant and insect-resistant maize; biopesticide development	Contaminated planting material; lack of adequate pest control measures; enhanced stress performance in maize	Smallholders

Meeting productivity and sustainability needs

Biotechnology and sustainable agricultural systems are often portrayed as antagonistic ends of a continuum. However, this portrayal lacks evidence, especially given that the use of biotechnology-derived agricultural products in agroecosystems is still largely unknown. In fact, there are many applications of biotechnology which seek to minimize the use of chemical inputs as pest, weed or disease control strategies in developing country agriculture. The relation between these applications and broader concerns of sustainability have been recognized (Hauptli *et al.*, 1990).

Developing countries recognize the connection between biotechnology and agricultural sustainability, as identified for Indonesia in the following points of emphasis for biotechnology: improve productivity; breed for resistance; minimize detrimental impacts of pesticides; and increase efficient water use (Moeljopawiro and Manwan, 1995). However, this strategy of substituting genetic information for prior dependency on chemical control measures is being advanced and articulated more directly by the commercial sector (Magretta, 1997). Examples of substitution arising from international research, and their relation to sustainability, are shown in Table 14.3.

A second and related point – how this genetic information can be provided with management practices targeting sustainability – is also being explored, again more directly, by the commercial sector (Hart, 1997). Advancements in technologies alone do not offer complete solutions, and information is needed to manage technology safely, appropriately and sustainably. However, the international programmes and those in developing countries often lack sufficient resources to enhance use of technologies in the context of sustainable agriculture, as the primary focus of their financial resources goes to capital expenditures needed for fundamental research.

A note of caution
The lack of knowledge and unfamiliarity with products coming from this combined research contributes toward heightened concern about safety. This is complicated by the fact that there is a general lack of knowledge regarding how this research deals with management and sustainability requirements. The longer that products take to reach farmers and consumers, the longer misapprehension will dominate the relevant debates. Such feelings have been recently summarized:

> These examples of poor biotechnology acceptability have resulted in a new
> process of overt political assessment rather than the previous system where the
> politicians and the public accepted the scientific judgments of safety and other
> merits. Scientists had to assure us they were beneficial, now they have to assure
> us they are not unsafe.
>
> (Carruthers, 1993)

Complementarity: the Next Step Forward

Learning from one another

The international biotechnology research and advisory programmes identified, whether based at international agricultural research centres or other advanced research institutes, require a decentralized approach to take advantage of centre-specific expertise and mandates. This decentralized approach benefits from 'complementarity', in the sense of completing or addressing gaps which exist because no one programme covers every aspect essential to move research results from laboratory to field to end-users. Complementarity means working together to recognize and address the gaps, where present, between research sponsored through the international arena and needs which exist among developing countries.

IBS works in a complementary manner through its case study methodology and analysis, using cases in policy seminars and in the management course described below. In these approaches, IBS focuses either on the country, institution or programme involved, or instead, on the individuals within these institutions and how they have confronted the challenges which biotechnology has presented. In this way, learning is maximized through one-another's experiences.

Complementarity allows the work of individuals to serve broader objectives and needs by taking into account the diverse priorities and objectives of these programmes, presenting them in a more comprehensive way, and ensuring that each centre or organization maintains control over its work. ISNAR works as one player among many, taking advantage of agricultural research networks, formal and informal, to multiply the efforts of individual programmes. Such complementary efforts in the policy seminars have included many case examples from the international agricultural research centres, as shown in Table 14.4.

Complementarity in Action: Managing Biotechnology in a Time of Transition

To further complement the information and experience provided by the international biotechnology research and advisory programmes, and to follow-up on needs identified from the policy seminars, a new initiative has been undertaken by IBS. This initiative, which moves from the policy arena to that of management, responds to the needs determined by the seminar's national delegations for biotechnology, not in isolation, but in relation to the broader challenges facing developing countries. These include urbanization, decentralization, and the extension of global and regional markets which are affecting the agricultural sector in such a way that new technologies and improved management of agricultural research have become a national priority for most countries around the world.

Table 14.4. Examples of cases and papers used in policy seminars derived from the international agricultural research centres (IARCs).

Policy seminar and region	Participating IARCs	Presenter	Presentation	Case study
Southeast Asia (Thailand, Indonesia, The Philippines, Vietnam, Malaysia and Singapore)	International Rice Research Institute (IRRI)	Klaus Lampe	Agricultural objectives in south Asia: balancing growth, equity and innovation	
East and Southern Africa (South Africa, Kenya, Ethiopia, Tanzania, Uganda and Zimbabwe)	International Livestock Research Institute (ILRI)	A. Lahlou Kassi	Biotechnology for improved livestock productivity in Africa: the challenges ahead	
	International Center for Tropical Agriculture/ Cassava Biotechnology Network (CIAT/CBN)	Ann Marie Thro		Using the results of a case study with cassava farmers in the Tanzanian lake zone
WANA (Morocco, Jordan, Tunisia, Israel and Turkey)	International Center for Agricultural Research in Dry Areas (ICARDA)	Frans Weigand	Potential contribution of biotechnology to agricultural production in the West Asia and North Africa (WANA) region	
Latin America (Peru, Mexico, Costa Rica, Colombia and Chile)	International Potato Center (CIP)	Marc Ghislain		CIP's approach to resource allocations and priorities: potato late blight research
	CIAT	Willy Roca		Developing durable resistance to rice blast; productivity and environmental considerations

A number of challenges regarding biotechnology research programme management and implementation were identified during the first policy seminar held in 1994. This seminar was held in Singapore for six Southeast Asian countries, including Indonesia, Malaysia, The Philippines, Singapore, Thailand and Vietnam (Komen *et al.*, 1995). ISNAR responded to the needs identified by initiating a management training programme, New Technologies for Agricultural Research, supported by the Government of Japan, through its Ministry of Foreign Affairs.

One component of the course concentrated on agricultural biotechnology: Managing Biotechnology in a Time of Transition. Management is defined here as the judicious use of means and resources which together help to achieve a strategy for agricultural biotechnology research. Emphasis is placed on individual effectiveness and the exchange of experiences among people so that management is not merely a series of mechanical tasks but a set of human interactions. It seeks to provide added value to the international research programmes, increasing management capability and exposure.

The goal of this course is to provide key agricultural personnel with the opportunity to enhance their management and leadership skills, with special emphasis on biotechnology. The specific objective of this course is to provide information and tools which equip participants to think strategically about their research programmes, manage biological and human resources for these programmes, enhance collaboration and exchange of experiences among countries, increase awareness of and access to information technologies, strengthen leadership effectiveness and engage policy makers more effectively.

The course also reflects a complementary approach through its faculty – a unique composition of national and international professionals contributing their technical specialties and expertise in the design and management of educational courses. This faculty presents case studies from international biotechnology research programmes of importance to the region, as well as national research leaders presenting case studies from the participating countries. It provides added value to the international research programmes and increases the exposure of management capabilities within the participating countries.

Concluding Remarks: Ensuring Continuity and Complementarity

The examples presented in Table 14.3 illustrate a range of prior practices and external inputs which could be altered by the judicious use of genetic information, as being developed through research supported by the international agricultural research community. Putting the correct information in a plant or vaccine and combining this with appropriate management considerations means that new information can substitute for pesticides, fungicides, costly equipment needs or ineffective cultural practices. These new products contribute to productivity and have the potential to increase factors associated with sustainability requirements and the needs of appropriate beneficiaries.

Will the international agricultural research system realize benefits from biotechnology?

This is a question being asked again and again. As a partial answer to this question, IBS is updating information maintained on international biotechnology research and advisory programmes, including their base of financial support. This update reflects continued commitment, growth and development of the programmes initially surveyed. This updated information, together with the results presented from the examples cited, indicates the continued need for collaboration and support, and the addition of greater complementarity. However, this view contrasts sharply with one recognizing that many organizations supporting such international research and advisory programmes are '... under political pressure from an irresistible combination of budget cutters and greens' seen as responsible for budget declines for such research (Anonymous, 1996). In fact, 'Governments in rich countries, and particularly in Europe, have become reluctant to finance research into genetically-engineered crops for poor countries' (Anonymous, 1996). A loss of support would not only mean that research would be curtailed, but that results would not reach targeted users.

This chapter has reviewed examples of international collaborations in research where expectations regarding biotechnology for agriculture are being met, where research offers products not to be made otherwise and where research contributes directly to a broad range of beneficiaries. The cost of this work is high, with much of the initial investment for research. While numerous activities have been started, only a very few will have the potential and the opportunity to reach completion. While new partnerships have been cited, there are still too few examples to ensure delivery of all research results. Thus, the number of projects adopted for full development must be realistic. If this gap between projects started and those completed grows, it will jeopardize future investments and minimize the knowledge gained from the cases presented.

To increase our chances of reaping benefits in developing countries, biotechnology should be an integral part of the agricultural agenda, a team player which serves broader productivity and social goals, and integrated fully with conventional agricultural research. With regard to the international biotechnology research and advisory programmes, a greater need for complementarity has been suggested and the potential for benefits examined.

References

Anonymous (1996) Will the world starve? *The Economist* November 16, 23–26.

Brenner, C. and Komen, J. (1994) *International Initiatives in Biotechnology for Developing Country Agriculture: Promises and Problems.* Technical Paper 100. OECD Development Centre, Paris.

Carruthers, I. (1993) Going, Going, Gone! Tropical agriculture as we knew it. *TAA Newsletter* 13(3), 1–5.

Cohen, J.I. (1993) An international initiative in biotechnology: priorities, values, and implementation of an A.I.D. project. *Crop Science* 33(5), 913–918.

Cohen, J.I. (1994) *Biotechnology Priorities, Planning, and Policies: A Framework for Decision Making.* International Service for National Agricultural Research, The Hague, The Netherlands.

Cohen, J.I. and Komen, J. (1994) International agricultural biotechnology programmes: providing opportunities for national participation. *AgBiotech News and Information* 6(11), 257N–267N.

Hart, S.L. (1997) Beyond greening: strategies for a sustainable world. *Harvard Business Review* January–February, 66–77.

Hauptli, H., Katz, D., Thomas, B.R. and Goodman, R.M. (1990) Biotechnology and crop breeding for sustainable agriculture. In: Edwards, C.A., Lal, R. and Madden, P. (eds) *Sustainable Agricultural Systems.* Soil and Water Conservation Society, Ankeny, Iowa.

IFPRI (1995) *A 2020 Vision for Food, Agriculture, and the Environment. The Vision, Challenge and Recommended Action.* IFPRI Press, Washington, DC, USA.

IBS (1994) *International Initiatives in Agricultural Biotechnology: A Directory of Expertise.* Intermediary Biotechnology Service, The Hague.

Komen, J., Cohen, J.I. and Lee, S.K. (1995) *Turning Priorities into Feasible Programs: Proceedings of a Regional Seminar on Planning Priorities and Policies for Agricultural Biotechnology in Southeast Asia,* 1994, Intermediary Biotechnology Service, The Hague, The Netherlands, Nanyang Technological University, Singapore.

Komen, J., Cohen, J.I. and Ofir, Z. (1996) *Turning Priorities into Feasible Programs. Proceedings of a Policy Seminar on Agricultural Biotechnology for East and Southern Africa.* Intermediary Biotechnology Service, The Hague, The Netherlands, Foundation for Research Development, Pretoria, South Africa.

Magretta, J. (1997) Growth through global sustainability, an interview with Monsanto's CEO, Robert B. Shapiro. *Harvard Business Review* January–February, 79–88.

Moeljopawiro, S. and Manwan, I. (1995) Agricultural biotechnology in Indonesia: new approach, innovation and challenges. In: Altman, D.W. and Watanabe, K.N. (eds) *Plant Biotechnology Transfer to Developing Countries.* R.G. Landes Company, Austin, Texas, USA.

Appendix 14.1. Summary of international agricultural biotechnology initiatives.

Name (host institution)	Priorities	Agricultural focus (crop/livestock)	Region/country focus
Crop biotechnology programmes Agricultural Biotechnology for Sustainable Productivity, ABSP (Michigan State University, USA)	Genetic engineering of crops for pest/disease resistance Development of micropropagation systems Integration of biotechnology within a general agriculture and business framework	Maize Potato Coffee Sweetpotato Horticultural crops	Indonesia Egypt Kenya
Bean/Cowpea Collaborative Research Support Program B/C CRSP (various US universities)	Control of pests and diseases Increase crop yields Increase nutritional quality	Bean Cowpea	International
Center for the Application of Molecular Biology to International Agriculture, CAMBIA	Novel biotechnologies and methods for agricultural innovation Genetic markers and diagnostics Apomixis	Rice Cassava Bean Agroforestry	International
CATIE – Biotechnology Research Unit (Centro Agronomico Tropical de Investigacion y Ensenanza, Costa Rica)	Enhance regional programme capabilities Genetic improvement of tropical crops	Banana/plantains Coffee Cocoa Roots and tubers	Latin America and the Caribbean
CIAT – Biotechnology Research Unit (International Center for Tropical Agriculture, Colombia)	Increasing the efficiency of CIAT strategic research Institutional development in biotechnology	Cassava Common bean Rice Tropical forages	International
CIMMYT – Applied Biotechnology Center (International Center for Maize and Wheat Improvement, Mexico)	Enhanced resistance to pests and diseases Enhanced stress tolerance	Maize Wheat	International
CIP – Applied Biotechnology Program (International Potato Center, Peru)	Reduce dependence on costly toxic chemical pesticides for potato and sweetpotato production Host plant resistance in potato	Potato	International

Organization	Description	Crops	Region
CIRAD – Plant Breeding Division (Centre de Coopération International en Recherche Agronomique pour le Développement, France)	Develop genetically improved crops	Cotton Rice Sorghum Tropical perennials Tropical fruits Forestry	International
Feathery Mottle Virus Resistant Sweetpotato for African Farmers (Agency for International Development, USA)	Human resource development Production of virus-resistant African varieties of sweetpotato Enhance capacity in biosafety regulation of transgenic crop plants Export of transgenic sweetpotato to Africa for field testing Technology transfer	Sweetpotato	Kenya
ICGEB – Plant Biology Programme (International Center for Genetic Engineering and Biotechnology, Italy/India)	Capacity building Genetically improved rice	Rice	International (ICGEB member countries)
ICRISAT – Molecular and Cellular Biology Program (International Crops Research Institute for the Semi-Arid Tropics, India)	Support and complement conventional crop improvement programmes at ICRISAT	Sorghum Pearl millet Groundnut Chickpea Pigeonpea	International
IIRSDA – Plant Breeding Program (Institut International de Recherche Scientifique pour le Développement an Afrique, Côte d'Ivoire)	Conservation and characterization of yam germplasm Micropropagation and genetic improvement of yam and other crops	Yam African eggplant	Africa

Continued

Appendix 14.1. *continued*

Name (host institution)	Priorities	Agricultural focus (crop/livestock)	Region/ country focus
IITA – Biotechnology Research Unit (International Institute for Tropical Agriculture, Nigeria)	Tackle recalcitrant problems in crop improvement Enhance national research capabilities	Cowpea Yam Cassava Banana/plantain	Africa
International Laboratory for Tropical Agricultural Biotechnology, ILTAB (Scripps Research Institute, USA)	Genetically engineered food crops with virus resistance	Rice Cassava Tomato Sugarcane	International
International Program on Rice Biotechnology (Rockefeller Foundation, USA)	Rice genetic improvement Capacity building	Rice	International
International Service for the Acquisition of Agri-biotech Applications, ISAAA (Cornell University, USA)	Transfer and delivery of appropriate biotechnology applications to developing countries; building of partnerships between institutions in the south and the private sector in the north, strengthening south–south collaboration	Vegetables Fruits Field crops (e.g. cotton) Cereals Forestry	International
Overseas Development Administration – Plant Sciences Research Programme (University of Wales, UK)	Genetically improved crops	Cereals Roots and tubers Legumes Oilseeds Fruit and vegetables Fibres	International
Philippine–German Coconut Tissue Culture Project (Albay Research Center, The Philippines)	Micropropagation of coconut	Coconut	The Philippines

Programme (institution)	Activities/objectives	Diseases/crops	Region
Regional Program of Biotechnology for Latin America and the Caribbean (several UN organizations)	Collaborative research projects; Training	Maize; Potato; Sugarcane	Latin America and the Caribbean
Research on the Date Palm and the Arid Land Farming Systems (Estacion Phoenix, Spain)	*In vitro* propagation; Biological control technology; Date palm farming systems	Date palm	Africa; Asia
Livestock biotechnology programmes CIRAD – Animal Production Division (Centre de Coopération International en Recherche Agronomique pour le Développement, France)	Development of heat-stable vaccines through genetic engineering; Improved diagnostic tests; Determination of genetic resistance to diseases	Cowdriosis; Dermatophilosis; Rinderpest; Livestock diseases of ruminants; Mycoplasmosis; Trypanosomiasis	International
ILRI – Tick-Borne Diseases Program (International Livestock Research Institute, Kenya)	Novel vaccines; Improve current control methods	Theileriosis; Cowdriosis; Anaplasmosis; Babesiosis	International
ILRI – Trypanosomiasis Program (International Livestock Research Institute, Kenya)	Improve diagnosis and parasite characterization; Novel vaccines; Breeding for genetic resistance	Trypanosomiasis	Internatinal
International Laboratory of Molecular Biology for Tropical Disease Agents, ILMB (University of California, USA)	Live recombinant virus vaccines for animal diseases; Technology transfer	Rinderpest; Bovine virus diarrhoea; Equine influenza; Livestock diseases of ruminants; Foot and mouth disease; Vesicular stomatitis virus	International

Continued

Appendix 14.1. *continued*

Name (host institution)	Priorities	Agricultural focus (crop/livestock)	Region/ country focus
International Program on Vectors and Vector-borne Diseases (University of Florida, USA)	Development and commercialization of improved vaccines and diagnostic tests	Heartwater	SADC countries Caribbean
Small Ruminant Collaborative Research Support Program – Animal Health Component (Washington State University, USA)	Improve the efficiency of milk and meat production from small ruminants Virus-vectored vaccines for sheep and goats	Heartwater Contagious caprine pleuropneumonia Nairobi sheep disease	Kenya Indonesia Bolivia
Crop/livestock programmes			
ICIPE – Biotechnology Research Unit (International Centre of Insect Physiology and Ecology, Kenya)	Biological control of pests (plant protection) and vectors Development of anti-tick vaccines Development of diagnostics tools	Maize Sorghum Cowpea Cattle	Africa
Indo-Swiss Collaboration in Biotechnology, ISCB (Federal Institute of Technology, Switzerland)	Research capacity building Human resource development Development, production and commercialization of specific biotechnology products Partnerships between research groups (public and private sector)	Foot and mouth disease Contagious caprine pleuro-pneumonia Plant biopesticides	India
Networks			
African Biosciences Network – Sub-Network for Biotechnology, ABN-BIOTECHNET (University of Nigeria, Nigeria)	Genetically improved crops and farm animals Disease control through new vaccines Capacity building		Africa

Asia Network for Small-Scale Agricultural Biotechnologies, ANSAB	Plant tissue culture Biopesticides Biofertilizers Mushroom technology	Potato Kapok tree Rice Mushroom	Asia
Asian Rice Biotechnology Network, ARBN (International Rice Research Institute, The Philippines)	DNA fingerprinting of pests and pathogens Low-cost marker-aided selection Transgenic rice	Rice	Asia
Phaseolus Bean Advanced Biotechnology Research Network, BARN (International Center for Tropical Agriculture, Colombia)	Constraint identification Technology transfer Information exchange	Beans	International
Cassava Biotechnology Network, CBN (International Center for Tropical Agriculture, Colombia)	Stimulate cassava biotechnology research on priority topics Integrate priorities of small-scale farmers, processors and consumers in cassava biotechnology research planning Information exchange	Cassava	International
Technical Cooperation Network on Plant Biotechnology, REDBIO (Food and Agriculture Organization of the United Nations, Regional Office for Latin America and the Caribbean, Chile)	Generation, transfer and application of plant biotechnology National and regional policies Information exchange	Vegetables Roots and tubers Cereals	Latin America and the Caribbean
Donor agencies			
Australian Centre for International Agricultural Research, ACIAR	Use biotechnology wherever appropriate as a research tool within any of ACIAR's projects		International

Continued

Appendix 14.1. *continued*

Name (host institution)	Priorities	Agricultural focus (crop/livestock)	Region/country focus
DGIS Special Programme Biotechnology and Development Cooperation (Ministry of Foreign Affairs, The Netherlands)	Improve developing country access to biotechnology, with special emphasis on small-scale producers and women Technical cooperation International collaboration and coordination	'Orphan' commodities Cassava	Colombia India Kenya Zimbabwe
FAO/AGP Programme on Plant Biotechnology (Food and Agriculture Organization of the United Nations, Italy)	Information dissemination and cooperation Advisory services Capacity building Promote research, technology transfer and adoption	Rice Roots and tubers Horticulture Industrial crops	International
GTZ Biotechnology in Plant Production (Agency for Technical Collaboration, Germany)	Development of micropropagation systems with diagnostic and pathogen elimination Training and capacity building Integration of biotechnology within BMZ/GTZ supported projects	Potato Cassava Yam Date palm Coconut Banana	Africa Asia
Swedish Agency for Research Collaboration with the Developing Countries, SAREC	Plant and forestry genetics Diagnostics and vaccines in veterinary medicine Environment Biosafety Policy research		Africa Asia
United Nations Development Programme	Productive and sustainable agriculture	Food crops Cash crops Livestock	International
United Nations Educational, Scientific, and Cultural Organization – Biotechnology Action Council	Human resource development		International

World Bank	Invest in biotechnology as a contribution to economic development in World Bank member countries	International
Policy/management programmes		
Biotechnology Advisory Commission, BAC (Stockholm Environment Institute, Sweden)	Provide independent, impartial advice on biosafety development and implementation to developing countries Biosafety capacity building	International
Canada–Latin America Initiative on Biotechnology and Sustainable Development, CamBioTec (Center for Technological Innovation, National Autonomous University, Mexico)	Identify opportunities for biotechnology research and applications by tracking technological trends and carrying out priority-setting exercises Strengthen public policies in biotechnology Promote improved management of innovations Foster partnerships between Canadians and Latin Americans	Latin America
Intermediary Biotechnology Service, IBS (International Service for National Agricultural Research, The Netherlands)	Biotechnology research programme management and policy formulation Country reviews Identify international programme expertise	International
Support to Agricultural Biotechnology Policies (Interamerican Institute for Cooperation in Agriculture, Costa Rica)	Biosafety; intellectual property rights (IPR) Industry development	Latin America and the Caribbean

Rice Biotechnology Capacity Building in Asia

Gary H. Toenniessen

The Rockefeller Foundation, 420 5th Avenue, New York, NY 10018, USA

Biotechnology capacity building in developing countries encompasses the strengthening of human resources and institutional capabilities across a number of disciplines ranging from molecular genetics, cellular biology, breeding and information technology to ecology, economics and management. These capabilities then need to be linked together into research programmes designed to manipulate germplasm in ways that are more powerful, predictive and efficient to produce products that address each country's priority agricultural development objectives. The process often includes the training of national scientists in scientifically advanced foreign countries, encouraging their return to an appropriate position in their home country, providing them with the facilities and supplies necessary to use their training effectively, helping them gain continued access to relevant information and know-how and linking them into larger national and international research networks.

Using the International Program on Rice Biotechnology as an example, this chapter reviews some of the difficulties encountered with biotechnology capacity building in developing countries and comments on alternative strategies for addressing these problems. The strengths and weaknesses of a number of training activities are reviewed and the programme's experience with principal investigator-initiated research proposals and peer review processes discussed.

Biotechnology capacity building in developing countries is similar to general scientific capacity building with the additional requirements that the science and technology be utilized to produce useful biologically based products and that the capacity, once established, be sufficient to keep pace in a rapidly advancing field. While it is not possible to generalize across all forms of biotechnology in all developing countries, this chapter reviews the strategies employed and lessons learned about capacity building within the Rockefeller Foundation's International Program on Rice Biotechnology (IPRB).

The IPRB was launched by the Foundation in 1984. At that time there was essentially no research being conducted in rice molecular biology outside of Japan. The goals of the programme are to:

- assure that the techniques of biotechnology are developed for tropical rice;
- help build sufficient biotechnology capacity in rice-dependent countries to meet current and future challenges to production;
- better understand the consequences of agricultural technical change in Asia, in part to help in setting priorities for biotechnological applications;
- apply this knowledge and capacity to the production and wise deployment of improved rice varieties.

From the beginning it was expected that the IPRB would require a long-term commitment and evolve over time. Initially funds were concentrated on building a rice biotechnology research network involving leading laboratories throughout the world. The first tasks were to generate the enabling technologies which constitute rice biotechnology and to train scientists from developing countries. Selected Asian institutions were invited to nominate candidates for fellowships which provided specialized training in these technologies while fostering collaborative research and technology transfer. As scientific breakthroughs occurred, shorter-term training courses were also offered. Good progress has been made. Rice has become a model plant for cereal biotechnology research and trainees have contributed to most scientific advances. To move the IPRB toward applications, social science research was sponsored to establish regional and national priorities. The priorities took the form of important traits on which subsequent biological research was focused. In recent years greater emphasis has been placed on helping to strengthen the institutional and national capacity in Asia to advance rice science through biotechnology and to apply the results to rice improvement.

Table 15.1 provides a summary of IPRB expenditures to date and the budget for 1997. The funding has been of sufficient magnitude and duration to: assemble a critical mass of researchers capable of producing desired results by sharing and collaborating; build institutional capacities based on the talents of several or more principal investigators, including returned trainees; and in larger countries, build a national rice biotechnology capacity based on the strengths of several or more institutions, including those committed to discovery research as well as applied rice research institutions. The major elements of the programme that contribute to capacity building are shown in Box 15.1.

Fellowships and training courses

In 1984, when the IPRB was initiated, there were a number of scientists in Asia with experience in plant tissue culture but essentially none, outside Japan,

Table 15.1. Rockefeller Foundation expenditure for the International Program on Rice Biotechnology (IPRB) ($ millions).

	1984–1996	1997
Advanced biological research	28.4	1.0
Biological research in developing countries	18.8	2.4
Trainees/fellows from developing countries	16.0	1.6
Biological research at international centres	9.3	1.0
Social science research	3.4	0.2
Networking and information dissemination	1.9	0.2
Totals	77.8	6.4

Box 15.1. Major elements of the IPRB contributing to capacity building.

1. Fellowships and courses offering specialized training, skills maintenance and technology transfer.
2. The enabling environment provided by a network of scientists who are conducting related research and are eager to share ideas and materials.
3. Partnerships with national agencies that assume increasing responsibility for funding and management.
4. Access to relevant information and effective communications systems.
5. A rational process for establishing research priorities.
6. Renewable research grants having application and monitoring processes that place strong public emphasis on the use of rigorous scientific methods and peer review.
7. The emergence of centres of strength capable of playing a leadership role in the future.
8. The work of Foundation field staff scientists located in Asia.

---

conducting significant research in plant molecular biology. The latter was a relatively new field of investigation which had emerged from research at advanced laboratories in the US and Europe. To build capacity in Asia it was necessary to send Asian scientists to these laboratories for training and, following their return home, to help them establish their own research programmes. Since a group of advanced laboratories was beginning biotechnology research focused on rice with IPRB support, the trainees were able to work on rice even in locations such as: Ithaca, New York; Norwich, UK; and Zurich, Switzerland. It was often the Asian trainees who made the scientific advances and breakthroughs that helped to create rice biotechnology and many are lead authors of key scientific publications. As technical advances occurred and relevant knowledge accumulated, shorter-term training programmes were sponsored. And, as trainees began to return home, they were offered fellowships designed to maintain their research productivity and to foster technology transfer. Today, good training in rice biotechnology can be obtained at several locations in Asia as well as in industrialized countries.

As indicated in Table 15.2, over 315 rice biotechnology fellowships of various types have been awarded to individuals selected directly by the Foundation. Most are from a pool of candidates nominated by Asian institutions selected to receive IPRB research grants. The nominating institutions must agree to provide IPRB fellows with an appropriate research position upon their return. Many of the prospective fellows have the opportunity to meet host laboratory leaders at various IPRB meetings and to discuss their training and research options. The Foundation interviews the nominees and, if necessary, helps those selected identify host laboratories where they can develop the type of expertise the home institution needs. After their return home most fellows become the principal investigator on an IPRB research grant and some receive Biotechnology Career Fellowships to help keep their expertise up to date.

Except for PhD fellows from China, most IPRB fellows have returned home following training. Some institutions such as the Philippine National Rice Research Institute, Tamil Nadu Agricultural University in India and the Rural Development Administration in Korea have eight or more PhD scientists trained in rice biotechnology through the IPRB.

Table 15.2. Rice biotechnology training fellowships 1985–1996.

Type	Number	Description
Postdoctoral research	91	Fellow does postdoctoral research at advanced laboratory for 2–3 years.
PhD degree	80	Fellow receives PhD degree training at advanced laboratory for 4–5 years.
Biotechnology career	72	Fellow conducts collaborative research requiring work at advanced laboratory for 3 months per year for minimum of 3 years.
Dissertation research	24	Fellow receives PhD degree in home country with a portion of dissertation research conducted at advanced laboratory for 1–2 years.
Transformation training	22	Six-month training in rice transformation at Scripps Institute (US) or John Innes Centre (UK).
MAS training	11	Six-month training in marker-assisted selection (MAS), Texas A&M University (US).
Visiting scientist	10	Fellow is senior scientist serving as visiting researcher at advanced laboratory for 1–2 years.
Technology transfer	5	Fellow from advanced laboratory conducts collaborative research including work in a rice-dependent country for 3 months per year over 3 years.
Total	315	

Advanced laboratories and international centres receiving IPRB research grants have selected over 100 additional developing-country scientists as research fellows they support from their grants. And more than 200 developing-country scientists have participated in 12 short-term (1–3 week) training courses sponsored by the IPRB. Thus, over 600 scientists, mostly from Asia, have so far received IPRB-sponsored training in rice biotechnology and the programme continues.

Special mention should be made of the Biotechnology Career Fellowships. These fellowships are awarded to highly capable young scientists from developing countries who returned home following a productive research experience abroad. The fellowships allow them to return 3 months a year, for a minimum of 3 years, to an advanced laboratory where they previously worked. They undertake a collaborative research project designed to utilize the strengths of both their home and host institutions. Some 72 Career Fellows have been supported as part of the IPRB, some for 10 years. All have become the principal investigators of an IPRB-supported research project at their home institution, and nearly all collaborate with a host laboratory also supported by the IPRB. The Career Fellows significantly strengthen the linkages between their home and host institutions, making research at both more relevant to the goals of the IPRB. They become effective agents of technology transfer, literally carrying some of the most recent advances in rice biotechnology back to their home country every year. In addition, since the availability of these fellowships was widely announced, they are an effective mechanism for bringing talented new people and new institutions into the IPRB. The Biotechnology Career Fellowships are relatively inexpensive and one of the best investments made within the IPRB.

The network of scientists

Over 700 scientists have participated in the IPRB as researchers, fellows and advisors including many of the world's leading plant molecular biologists and rice scientists. While access to funding is certainly an important attraction, all also welcome the opportunity to be part of a larger mission-oriented programme with the potential to benefit millions of people. In aggregate, the work of these scientists now constitutes a critical mass of effort advancing under its own momentum.

Through the IPRB each participant can interact with a number of scientists throughout the world conducting related research toward a common goal. All are expected to share information and materials with others. Laboratories at the forefront of developing enabling technologies receive funding for broad-scale dissemination of those technologies. Evaluation of advanced laboratories is based on contributions to training and technology transfer as well as the quality of research.

Thus, developing-country scientists have ready access to molecular maps and markers, transformation vectors, potentially useful genes, gene regulatory sequences and rice databases. Many become members of small teams of scientists which draw on the expertise of three or four laboratories to advance particular lines of research. Most importantly, they develop a group of knowledgeable colleagues with whom they can share ideas and discuss results. Over time an increasing number of these colleagues are in their home institution and country. This, combined with IPRB assistance in obtaining supplies and equipment, helps create an environment where significant rice biotechnology research can be done and has been done at Asian institutions.

Partnerships with national agencies

In the larger Asian countries, the IPRB is implemented in collaboration with one or more national agencies that provide leadership, coordination and local currency funding. In 1985, a memorandum of understanding was signed with the China National Center for Biotechnology Development, a funding agency within the State Science and Technology Commission, establishing a Cooperative Program on Biotechnology Development for Rice Improvement in China. In India, coordination and local funding are provided by both the Department of Biotechnology and the Indian Council for Agricultural Research with the Directorate of Rice Research in Hyderabad organizing the national meetings. In Thailand, a smaller but comprehensive programme has been developed in cooperation with the National Center for Genetic Engineering and Biotechnology.

In these countries, as in industrialized countries, capacity for rice biotechnology research was achieved first in public sector institutions staffed with scientists having backgrounds in biochemistry, microbiology and other laboratory sciences, but little or no experience in rice breeding and field research. One of the more difficult tasks facing the IPRB and its national partners is the establishment of integrated national programmes which draw on the full spectrum of national research institutions and facilitate the flow of knowledge and technology from the laboratory to the field. This has worked most effectively when a significant number of laboratory and field scientists (at least 20) first receive funding for their own individual rice biotechnology projects and then are brought together as a national group and encouraged to develop collaborative projects as a means of receiving increased or continued funding. Less successful have been attempts to require cooperation and sharing amongst groups that logically should collaborate, or to designate particular laboratories as service providers that others should utilize.

Information and communications

The IPRB supports various activities facilitating the dissemination, to developing-country scientists and others, of information that can help advance research in rice biotechnology. General meetings of the IPRB involving most participants are held at roughly 18-month intervals. Over 400 scientists attended the 1997 general meeting in Malaysia. National meetings on rice biotechnology are periodically cosponsored with national agencies in India and China, and in 1996 three regional meetings were organized by IPRB scientists themselves and held in east Asia, south Asia and Southeast Asia. All participants learn much during these meetings and developing-country scientists are able to make personal contacts with many scientists they would not otherwise meet. Proposals for collaborative projects are developed during these meetings that attract funding from a variety of sources. In addition to principal investigators, graduate students and postdoctoral scientists working on IPRB projects in Asia attend the meetings and many use this opportunity to arrange for further training in an advanced laboratory. A book of abstracts is prepared for each meeting and widely distributed.

The *Rice Biotechnology Quarterly* is a publication funded by the IPRB to help keep scientists up to date on the scientific literature relevant to rice biotechnology even though they may not have access to numerous journals. The *Quarterly* includes a bibliographic reference for nearly all publications in rice biotechnology shortly after they appear and provides an abstract of most. In many cases additional comments on the significance of the findings are given by one of the authors. News items concerning the IPRB and similar programmes, plus photographs of scientists and scientific teams, add interest and a human dimension. The *Quarterly* is mailed free-of-charge to over 2200 scientists. Funding is included in IPRB research grants for purchasing journals, obtaining reprints and paying page charges. All grantees are encouraged to submit their results to international journals.

Within the IPRB, a special effort has been made to help all participants better understand patenting systems and to more effectively use patents as a source of scientific information. Patent experts give presentations and lead workshops at most IPRB meetings. Abstracts of over 200 patent applications directly relevant to rice biotechnology have been distributed to IPRB participants along with directions on how to obtain full copies at little or no cost.

Theses prepared for PhD dissertations are another excellent but under-utilized source of information, particularly regarding research methodology. Methods that are only briefly mentioned or simply referenced in journal publications are usually presented in detail in dissertations. Information concerning dissertations relevant to rice biotechnology and how to obtain copies is distributed to IPRB participants.

Occasionally, written reports are distributed presenting the results of IPRB workshops and recommending particular research methods. For example, research on rice genetic engineering is prone to false positives. A workshop report was prepared and distributed which presented a set of rigorous assays that should be completed to confirm the integration and expression of an introduced gene. It gives investigators a clear understanding of the need for controls and the types of data required to substantiate a claim. It also helps them prepare their results for publication. Similarly, a workshop report on Rice Biosafety was jointly prepared and widely distributed by the IPRB, The World Bank, and the United States Department of Agriculture/Animal and Plant Health Inspection Service (USDA/APHIS).

The IPRB also aims to facilitate use of modern scientific communication systems by all participants. All Rockefeller Foundation PhD fellowships include funds for the purchase of a personal computer and other IPRB fellows and grantees can include such funding in their equipment budgets. A rice biotechnology e-mail network (RBNET) operates from Ohio State University. It facilitates communication and allows requests for information to be sent across the network. At Cornell University, the RiceGenes Gopher Server allows for Internet connections to the USDA-funded RiceGenes Database. Similar databases in Japan and Korea can also be accessed via the Internet and all are available on CD-ROM. These networking functions, for the most part, are available to anyone working in rice biotechnology.

Establishing research priorities

Support for social science research and its application to priority setting is an important component of the IPRB. Dr Robert Herdt, an economist who spent 10 years at the International Rice Research Institute (IRRI) assessing the constraints to rice production in Asia and developing research priorities, took on the assignment of developing an initial set of research priorities for the Foundation. Using his prior experience, plus input from many rice scientists and biotechnologists throughout the world, he made estimates of the potential value of desired traits and the probability that research using biotechnology would enable successful breeding for such traits. Traits that would especially benefit poorer farmers and/or be resource-conserving were given extra weighting in the priority calculations.

These, however, were priorities for Asia as a whole. To make the results more useful at the national and local level, social scientists in Asian countries are supported to make priority calculations based on more detailed local information and national objectives. Social scientists at IRRI have worked with the national scientists to further refine the methodology and to update priorities as additional and/or more accurate information becomes available. This helps make the work of all national scientists and national institutions more relevant to national needs. In the process, the capacity of national programmes to set agricultural research priorities is strengthened.

Renewable research grants

To help Asian countries develop the scientific capacity to advance rice biotechnology as well as apply it, the IPRB seeks to involve a significant number of Asia's leading biological scientists. Principal investigator-initiated research proposals are welcomed and often solicited from well-qualified scientists, including those at research universities and institutions that are not part of the traditional agricultural research establishment, as well as those that are. Relatively fundamental, strategic and applied research projects are supported and combined to form integrated national programmes. In Asia, IPRB research grants are used primarily to purchase supplies and equipment requiring foreign currency, and usually are more than matched by local currency funding from national agencies.

As part of the process of evaluating proposals and monitoring progress, a concerted effort is made to promote the use of rigorous scientific methods and to encourage collaboration. The prospect of a renewal grant helps encourage compliance with IPRB guidelines and recommendations. The guidelines for new and renewal proposals from developing-country scientists include:

- a summary of the principal investigators' training relevant to rice biotechnology;
- a review of progress to date including reprints;
- a statement of the problems being addressed and the hypotheses being tested;
- the experimental design and methods, including positive and negative controls;
- an estimate of the potential contribution of the research to rice improvement;
- a description of current and/or anticipated collaborations with other scientists.

The IPRB uses and encourages the use of peer review. A scientific advisory committee composed of leading plant molecular biologists and rice scientists makes highly visible contributions to the IPRB. The committee conducts in-depth evaluation of advanced laboratories, monitors overall progress and reports to the full network at the end of each international meeting. This establishes an expectation of rigorous evaluation which carries over to the Asian components of the IPRB.

In China and India, for example, the IPRB helps to sponsor national rice biotechnology meetings which also incorporate peer review. Advisors and consultants participate in these meetings, ask numerous questions, report to both the Foundation and national funding agencies and provide feedback to the principal investigators. All scientists are encouraged to raise questions and much scientific discussion does occur. In China all participants were asked to evaluate each other's presentation based on:

- scientific progress made;
- potential contribution to rice improvement;
- degree of collaboration with others in China;
- clarity of presentation.

As a general policy, the Foundation gives first-time grantees a chance to demonstrate their research capability and productivity and provides continued support only to those that make good progress. Revised proposals are sometimes requested. Overall there is a growing atmosphere of peer review throughout the scientific community in Asia, and the IPRB strongly supports this trend.

Emerging centres of strength

In Asia, several centres of strength in rice biotechnology research, application and training are now coming into existence, in part with IPRB assistance. They are listed in Box 15.2 along with the name of a contact person. At these locations, four or more principal investigators, by sharing facilities and materials and working together, have the capacity to make discoveries and carry them through to application. In the process, excellent training is provided. The IPRB continues to support these centres and is helping to build others that have similar capability. Eventually, these centres should be able to assume significant responsibility for continued advancement of rice biotechnology and its application to rice improvement.

Probably the most advanced centre is at the Korean Rural Development Administration's National Institute of Agricultural Science and Technology and nearby universities. Here, nine PhD scientists are involved in rice genome mapping and map-based cloning, six in molecular breeding, seven in rice transformation and three in database construction and informatics. A large conventional rice breeding programme is located at the same research station. Through effective use of Biotechnology Career Fellowships, this group is at the forefront of advancing the science of rice molecular genetics and expanding its potential for application. The results should benefit all rice-dependent countries.

China and India are large countries which now have a significant number of scientists doing good research in rice biotechnology. In each, there are at least three locations where the talents of several scientists are merging to build a centre of strength. In China, this is occurring at Huazhong Agricultural University and nearby institutes in Wuhan, at the South China Agricultural University in Guangzhou, and in Hangzhou through a combination of Zhejiang Agricultural University and the China National Rice Research Institute. In India, the centres are at Tamil Nadu Agricultural University in Coimbatore, in Hyderabad through a combination of the Directorate of Rice Research, Osmania University and the University of Hyderabad, and in Bangalore through a combination of the University of Agricultural Sciences, the Indian Institute of Science and the Tata Institute for Fundamental Research National Centre for Biological Sciences.

At the National Philippine Rice Research Institute (PhilRice) at Maligaya, several PhD and postdoctoral fellows have recently returned to new facilities and others, still in training, are expected to return in the near future. They are initiating high-quality research programmes and through effective use of Biotechnology Career Fellowships are collaborating with several of the world's leading plant molecular biologists. While a relatively new institution, PhilRice now has the talent to make important contributions to rice improvement in The Philippines and throughout Asia.

Box 15.2. Emerging Asian centres of strength in rice biotechnology.

Korea
National Institute of Agricultural Science and Technology, Suweon
(Dr Tae Young Chung, Fax: 331–290–0307)

China
Biotechnology Center, Huazhong Agricultural University, Wuhan
(Dr Zhang Qifa, Fax: 86–27–739–392)
Genetic Engineering Laboratory, South China Agricultural University,
Guangzhou
(Dr Mei Mantong, Fax: 86–20–875–9069)
Hangzhou Group
　Biotechnology Institute, Zhejiang Agricultural University
　(Dr Li Debao, Fax: 86–571–696–1525)
　Biotechnology Department, China National Rice Research Institute
　(Dr Zheng Kangle, Fax: 86–571–337–1745)

India
Center for Plant Molecular Biology, Tamil Nadu Agricultural University,
Coimbatore
(Dr S. Sadasivam, Fax: 91–422–431–672)
Hyderabad Group
　Department of Biotechnology, Directorate of Rice Research
　(Dr N.P. Sarma, Fax: 91–40–401–5308)
　School of Life Sciences, University of Hyderabad
　(Dr Arjula R. Reddy, Fax: 91–40–301–0145)
　Department of Genetics, Osmania University
　(Dr K.V. Rao, Fax: 91–40–701–9020)
Bangalore Group
　Department of Genetics and Plant Breeding, University of Agricultural
　Sciences
　(Dr Shailaja Hittalmani, Fax: 91–80–333–0277)
　Department of Molecular and Cellular Biology, Indian Institute of Science
　(Dr Usha Vijayraghavan, Fax: 91–80–334–1683)
　Tata Institute for Fundamental Research, National Centre for Biological
　Sciences
　(Dr Villoo Patell, Fax: 91–80–334–3851)

Philippines
National Philippine Rice Research Institute, Maligaya
(Dr Leocadio Sebastian, Fax: 63–2–843–5122)

Field staff in Asia

To implement the capacity-building components of the IPRB in Asia, one and a half full-time equivalent field staff scientists are assigned to the region. Dr John O'Toole, a plant physiologist with 10 years experience at IRRI, was stationed initially in New Delhi and later in Bangkok, and charged with expanding the IPRB into south and Southeast Asia. He also provides advice and guidance to all IPRB participants working on drought tolerance. Dr Toshio Murashige, a well-known tissue culture specialist helps implement the IPRB in China and Korea. He has also assisted numerous IPRB institutions with the utilization of anther culture as a breeding tool. In many cases this was an effective first step toward applying biotechnology in rice breeding programmes.

These field staff scientists are highly accessible to Asian IPRB participants. They spend most of their time visiting grantee institutions, providing advice, facilitating communications, interviewing fellowship candidates, organizing meetings and workshops, reviewing proposal and progress reports, and making recommendations to the Foundation. Over time, they have developed an excellent understanding of the Asian institutions in the IPRB and helped them to build on strengths and overcome weaknesses. Field staff greatly increase the effectiveness of IPRB funds committed to capacity building.

Conclusions

Of the IPRB's primary goals, helping to build rice biotechnology capacity in Asia has been the most difficult to achieve. It has taken several years of significant investments to reach the point where the human resource base is of sufficient magnitude in some countries to permit real building of institutional and national capacity. This is already happening in China, India and Korea and soon should be in The Philippines, Thailand and Vietnam. Most other Asian countries have good scientists making progress on specific components of rice biotechnology but lack the critical mass of talent necessary to establish a comprehensive national programme. They too, however, should be able to produce élite rice varieties through biotechnology by supplementing the talent of their own scientists with collaborators from international centres, neighbouring countries and the rice biotechnology network.

International Biosafety Regulations: Benefits and Costs

16

Robert J. Frederick

US EPA/Office of Research and Development, National Center for Environmental Assessment, Washington, DC 20460, USA

A number of principles on environmental policy were agreed to at the United Nations Conference on Environment and Development in 1992. One of these is the importance of sustainable development so that we manage our environment not only for the health and economic benefit of today, but also for future generations. The second is adherence to a precautionary approach for environmental protection. Fundamental to our efforts to protect the environment is the establishment of appropriate regulatory mechanisms and the use of the best available science to make decisions. In and of themselves, however, regulations do not ensure safe applications of biotechnologies. We must often make decisions with imperfect knowledge. The precautionary approach suggests that when our knowledge is less than perfect, we will be protective of the environment and human health. Consequently, considerable thought must be given not only to the principles of biosafety, but also the practicalities of what constitutes a functioning programme, what the value of having such a programme is and whether that value balances the cost of establishing and maintaining a programme.

The global biosafety regulatory landscape is quickly developing and evolving. The potential of biotechnologies to address national needs and economically attractive possibilities are well known. For practical and philosophical reasons, before countries can take advantage of these technologies, national biosafety mechanisms may be required. Consistent with international acceptance of a moral obligation to protect the environment, the value of having a biosafety programme seems quite high in terms of political capital. This is quite clear in the

The views expressed in this chapter are those of the author and do not necessarily reflect the view or policies of the US Environmental Protection Agency.

CAB INTERNATIONAL 1998. *Agricultural Biotechnology in International Development* (eds C.L. Ives and B.M. Bedford)

treatment of biosafety in Agenda 21 and the Convention on Biological Diversity. For the foreseeable future, technology transfer will necessitate concomitant regulatory oversight.

While many developing countries are forming regulatory frameworks by adapting regulatory guidance already implemented elsewhere, cost sharing still carries a financial burden. The variety and disparity of potential frameworks necessitates normalization of both information/data requirements and relative stringency of regulatory oversight. Other costs include those for biosafety training, data collection and storage, and monitoring programmes. International guidelines and/or a binding protocol under the Convention on Biological Diversity, may help to assist the harmonization process while providing an overall umbrella for biosafety regulation thus reducing the costs of having to establish biosafety regulatory programmes *de novo*. Programmed training approaches that create networks of trained individuals on regional and subregional levels may facilitate the acquisition and dissemination of biosafety expertise at reduced costs. Providing useful, relevant information is a non-trivial task and can be expensive. With unique issues to address, developing countries may find limited use for existing assessments done elsewhere. Acquiring new knowledge may carry a high cost in both time and money.

There are significant benefits to be derived from having biosafety mechanisms in place, but costs associated with the adoption and implementation of regulations may be substantial. Each country will have to weigh the benefits and costs for themselves to determine the best mechanism to achieve national goals for environmental protection and technology development. International cooperation and assistance can be helpful in this regard and, in many instances, will be necessary if all countries are to have national biosafety regulations.

Introduction

In considering an issue as wide ranging as biosafety, it is useful to begin with a definition that sets boundaries on the discussion. For this review of the costs and benefits of biosafety regulations, biosafety is 'the policies and procedures adopted to ensure the environmentally safe applications of biotechnology' (Persley *et al.*, 1992). Developing countries are under growing pressure to develop and implement biosafety regulations (Virgin and Frederick, 1995a). To do so may provide extensive benefit in facilitating technology transfer, provide new opportunities for maximizing the sustainable and productive use of natural biological resources and ensure that these proceed in a safe manner (Brenner, 1995; Brenner and Komen, 1994; Virgin and Frederick, 1995a). There are, of course, costs associated with the development of regulatory instruments and their implementation. Experience has demonstrated that, too often, each stage of regulation development is done without consideration for the succeeding stages. This approach might lead to ineffectual instruments that are not implementable, impede technology transfer, or result in prohibitive burden to government and/or industry.

In attempting to illustrate the potential benefits of having biosafety regulations, a few assumptions have been made. The implementation of biosafety regulations must be financially feasible. They must be scientifically based and amenable to the appropriate use of scientific information and expertise. However, as Tzotzos (1995) points out, risk assessment 'relies on expertise covering a wide range of scientific disciplines' and in 'industrialized countries, regulatory formulation and implementation is being accomplished by national and institutional committees and expert panels, ... the replication of this model in the developing world would require institutions, human and financial resources that are far beyond the means of the great majority of countries'. A more holistic approach that considers the benefits and costs may lead to regulations that are appropriate to the potential risks, yet protective and economically feasible. This chapter intends to present a variety of issues that might enter into such an analysis.

Why Biosafety?

The early development of biotechnology regulation in the mid-1970s is well known. It serves as an interesting contrast *vis-à-vis* current progress in developing countries. Initial uncertainty regarding the risk involved in employing new genetic engineering techniques led scientists to initiate a moratorium on their research and to develop guidelines for experimentation. It was not until nearly a decade later that US federal agencies began the process that would require mandatory review of potential biotechnology products. As the new biotechnology industry emerged, concerns extended beyond academic laboratories and more explicit oversight mechanisms were pursued. While reaction to the increased involvement of the government regulatory arm was mixed, there were many in the industry who saw a need for reasonable regulation if they were to have public acceptance and avoid damaging delays in getting their products to the marketplace.

Biosafety programmes in the United States and the European Community (EC) unfolded slowly. Developing countries have been even slower in the adoption and implementation of biosafety policy and procedure and even today a majority do not have regulations in place (Virgin and Frederick, 1995a). Cohen and Chambers (1991) suggest a list of seven constraints restricting development of national regulations:

- lack of national regulatory framework;
- finances;
- lack of confidence in decision makers;
- multiplicity of actors;
- lack of technical expertise;
- lack of coordination among international organizations;
- apprehension about regulating.

Much of the same sentiment has been expressed at conferences held in developing countries. For example, the conference statement from the African Regional Conference for International Co-operation on Safety in Biotechnology (held in Harare, 1993) included many of the above and added explicitly the need for information collection and exchange, risk assessment research, training, and national and regional collaboration. Such insight suggests that developing countries will take advantage of their late entry in biosafety development by taking into account the experience of their colleagues in industrialized countries.

Potential Benefits

Ensure environmentally safe application

Protection of human health and the environment seems the most obvious and easily agreed upon reason for having some regulatory oversight of new technology. Oversight is used in the sense of 'the application of appropriate laws, regulations, guidelines or accepted standards of practice to control the use of a product based on the degree of risk or uncertainty associated with it' (Medley, 1994). In this case, it is the oversight authority that is ideally charged with carrying out independent, scientifically based reviews of products prior to their use at particular developmental stages or commercialization. The purpose of the reviews is to ensure that, to the extent scientific knowledge and expertise will allow, any foreseeable risk the intended use may present to human health or the environment is identified and evaluated.

Public and political factors also must be considered driving forces for biosafety regulations. These have been extensively discussed elsewhere (see for example Lacy *et al.*, 1991; Leopold, 1995; and Anderson, 1991). Suffice to note that public influence can be substantial and, for whatever reason – moral, ethical or general distrust of new technology – if the public is against the technology it will put pressure on the government to regulate or reject it in the market place. Ensuring public safety through regulatory oversight will help to assuage public concern.

Beyond the national boundaries, biosafety is a high visibility issue. In a recent manifestation of international concern, the Convention on Biological Diversity calls for all countries to have national biosafety regulations. This is part of an international 'push' that is having a significant influence on developing countries (Krattiger and Lesser, 1994; Virgin and Frederick, 1995b) and motivating them to proceed with the development of their regulations while at the same time considering the development of an international biosafety protocol under the Convention. Those with experience of drafting and implementing national regulations will be in a better position to evaluate the implications of language in an international protocol and their ability to abide by it at the national level.

Promote technology development

There are many constraints to the acceptance and inculcation of biotechnology in developing countries (Komen and Persley, 1993). Listed among these is the absence – rather than presence – of biosafety regulations. To some, it may seem counterintuitive that regulatory oversight with its associated costs for compliance and implementation would promote biotechnology development, but there is growing agreement that this is the case. The adoption of biosafety regulations can be seen as an indication of the willingness or desire to have biotechnology. As Medley (1994) points out, 'If structured and administered properly, regulations can facilitate rather that impede expanded development, safe technology transfer, and commercialization of agricultural biotechnology products.' Others have argued that the technology may be excluded altogether from countries that do not have regulation (Virgin and Frederick, 1995a; Persley *et al.*, 1992; Brenner, 1995; Brenner and Komen, 1994). International companies and donor organizations are hesitant to bring genetically modified organisms into countries without local regulatory approval. The resolution from a conference held in Kenya stated participants' thinking quite clearly: '... without biosafety guidelines, countries are being denied the opportunity to access the full potential that the application of biotechnology offers through public and private sector research, development, and implementation' (Zandvoort and Morris, 1996).

Facilitate harmonization

Calls for harmonization of biosafety regulations have been a prevailing theme of the international biosafety scene. As early as 1985, Kuenzi *et al.* recommended that biosafety regulations and guidelines be brought into accord (Kuenzi *et al.*, 1985). In more recent papers, the breadth of the issue has been expanded and reasoning behind it explored at greater length (UNIDO, 1990; Lesser and Maloney, 1993; Virgin and Frederick, 1995a; Frederick, 1995 and 1998). Clearly there are advantages to harmonization:

- A higher level of security or control may be reached with concordant national programmes as opposed to disparate or, worse, contradictory regulatory mechanisms in neighbouring countries.
- A normalization may moderate tendencies for the enthusiasm of acquiring technology from turning into imprudent competition between countries where regulatory mechanisms are minimized at the expense of safety considerations.
- Harmonization may facilitate the formulation, adoption and uniform interpretation of regulatory instruments. This would be of particular advantage to those countries still early in the process of developing their own mechanisms.
- In looking to the future, harmonized regulations may encourage international data collection and information exchange. This is a prominent issue

in the Convention on Biological Diversity and UNEP International Technical Guidelines for Safety in Biotechnology (UNEP, 1995). Note, however, that at the present time, the extent and nature of this information exchange is under considerable debate and individual opinion seems to be based more on intuitive reasoning than empirical evidence. In fact, that information which might be considered necessary to support biosafety decision making is still not well sorted out.

- Finally, harmonized regulations may moderate the burden on biotechnology companies in their compliance with the guidelines and laws of a variety of countries, i.e. what is accepted in one country is accepted in others.

Costs of Biosafety Regulation

Regulatory development

The costs to countries attempting to develop biosafety regulations will vary dependent upon the particular regulatory framework that is chosen and the number and complexity of steps that must be carried out to effectuate the instrument. A variety of approaches are being pursued (Frederick, 1998), but all will have common requirements for successful adoption. Having knowledgeable personnel is first and foremost. Those charged with drafting regulations should have a clear understanding of the regulatory objectives, what alternatives are available, and what the implications of their choices will be on the regulated community. Capacity-building – the strengthening and/or development of human resources and institutional capacities (UNEP, 1995) – is as critical for the preliminary and early stages of regulatory development as it is for the implementation phase. Training and information exchange (explored more fully below) should be included in developmental support budgets as much as standard operating costs for personnel.

Additional items are also worth noting. Outside expertise brought in to advise throughout the development process may have to be supported. If national meetings are necessary to promote debate or garner general support, this will add to the cost. The expenses of vetting draft documents – copying, publication and/or distribution – will have to be added. The latter are not trivial to most developing countries and it should not be assumed that they will be covered under general operating budgets. Without supplemental resources, the pace for establishing regulations has been slowed and, in some cases, stopped altogether.

Regulatory implementation

In and of themselves, regulations or guidelines will not ensure safe applications of biotechnologies. Most important is the way particular biosafety regulatory instruments are implemented. As diverse and disparate as regulations may be, the way they are interpreted and implemented may be even more so (Frederick, 1998; Medley, 1994; UNEP, 1995). This suggests that a wide range in the potential annual cost of implementation programmes is possible. Consequently, considerable attention to the principles and practices of risk assessment and management for biological materials is warranted.

The business of risk assessment has evolved and become more sophisticated over the last 15 years such that an understanding of the art offers many advantages to an assessor. Strauss (1991) defines risk assessment as 'an analytical tool that facilitates the organization of large amounts of diverse data with the goal of estimating the potential risk posed by a process (or event) of interest'. When dealing with biological materials and their potential interactions in the environment, the number of permutations and combinations will easily challenge the most talented and brilliant assessor. Understanding and utilizing risk assessment methodology may be the only acceptable means for making determinations for the safe development and use of biotechnology products. More attention needs to be paid to this critical component of a biosafety programme.

Already limited in the supply of national experts, developing countries with active biotechnology research programmes may find themselves particularly hard pressed to find expert reviewers/assessors without a conflict of interest, e.g. having their own research supported by the same company or one of its competitors. Perhaps this gap may be partially filled through regional cooperation or the use of outside expertise. Alternatively, making experience and information available may help to fill the void. In a practical sense, however, providing useful, relevant information is not a trivial task. With different issues to deal with, e.g. centres of origin or diversity, developing countries may have limited use for assessments made in other countries where these issues have not been dealt with extensively (Frederick and Virgin, 1995).

To satisfy the growing need for national biosafety assessors, there will have to be continued access to training and development programmes. Crude estimates of the scope for such training on a global basis can be made based on certain assumptions of the capacity needed. For example, let us assume that a minimum of 20 technical experts is necessary to fully implement biosafety regulation at the national level. Very probably, the majority of those chosen to be on biosafety panels will be drawn from the ranks of scientists at local universities or research organizations. These, albeit very capable and well trained in their own scientific discipline, may be naive to risk assessment methodology and consequently need additional training. With the further assumption that attending one workshop would satisfy this need, it would require 50 workshops to train 1000 individuals (20 per workshop). This would provide sufficient capacity to support biosafety programmes for only 50 countries. Of course, the intent of such an admittedly simple calculation is only to provide a sense of the extent of effort necessary to achieve a definable goal and, hence, a basis for consideration of options and alternatives.

Harmonization

Harmonization has been broadly defined as the agreement in action, opinion and feeling leading to a common set of biotechnology regulations at the regional or global level (Virgin and Frederick, 1995a). Inherent in this is the establishment of equal or equivalent standards (Medley, 1994), particularly as they might apply to scientific requirements for risk specification (Lesser and Maloney, 1993). Lesser and Maloney (1993) also point out that efforts to achieve harmonization may be affected in the adoption of common language for regulatory instruments or, ultimately, the formation of multinational treaties or a binding protocol as suggested by the Convention on Biological Diversity. It has been argued, however, that beyond the words used in creating them, it is really the interpretation and implementation policies that will determine the level of harmonization (Frederick, 1997). This is well illustrated in the analysis of the European Union's development and implementation of their biosafety regulations under specific directives. Despite the fact that all members of the Union are working under the same directives, the form and implementation of their national regulatory structures could not seem more diverse (Levidow *et al.*, 1996). In the final analysis, it seems most likely that harmonization can only be realized through free and open communication using a common language for biosafety. This will be furthered by making high-calibre training and information transfer available at an affordable cost.

Information Transfer and Cost-effective Training

Information transfer

Marois *et al.* (1991) note that 'information is considered a central limiting factor in the development of policies to assess and manage risks. If information is inadequate, the policies are also inadequate.' For the same reasons, transfer of information is a critical part of capacity-building programmes in developing countries (Frederick, 1995; 1998; Virgin and Frederick, 1995a, 1996). In terms of cost, there immediately comes to mind the expense of obtaining information, maintaining libraries or data systems and, when called for, sorting and dispersing information. Such costs can be adjusted to the financial capacity of the organization maintaining such a system and to some extent what the users deem sufficient to do their job in an acceptable manner. This has the potential to substantially add to the cost (in time as well as money) if it is determined that unavailable data is required and the only way to get it is through new research. Connected to this issue is the capacity for conducting biotechnology research. Technical and scientific capacity has been cited as a limiting factor to biotechnology development, but it must also be considered so for biosafety implementation. If assessments are to be based on scientific considerations, these must be substantively understood by those doing the assessments.

It is worth while to consider this issue in more detail. Actually, knowing what information exists and what does not – something best decided by competent authorities – is only a part of effectively using data in an assessment process. As Tzotzos (1995) rightly points out, 'the management of biological information and availability of tools that permit data interpretation and modelling in risk assessment will ... determine the effectiveness and reliability of methodology.' We have to consider then not only the amount of information that exists, but where it resides, its transferability to the needy recipient and its amenability for translation into local or regional risk assessments. In general, we are really talking about the value of information. Marois *et al.* (1991) presented a very nice graphic to illustrate these thoughts (Fig. 16.1). The greater the degree of accuracy sought, the more information needed to satisfy that need and concomitantly reduce the probability of an error. After a certain point, it might be expected that there are diminishing returns to continued information collection as the likelihood of reducing errors decreases asymptotically. Of most interest to this discussion, however, is the cost/value curves shown in Fig. 16.1b. As we continue to improve accuracy – by increasing the amount of information obtained or the robustness of the analysis performed – the cost of that enterprise increases and the relative value of additional cumulative information decreases. In short, at some point in an information gathering process, the value of additional information may not be sufficient to justify its cost. We cannot leave this discussion without making the observation that someone will have to make a decision on how much information is enough and how the available information will be used to best meet national needs. By no means insignificant, this suggests that the ability to properly interpret and utilize information in a risk assessment context requires an expertise only obtained through education and training.

Cost-effective training

Training has been mentioned at several points in this discussion as necessary for the successful preparation and implementation of biosafety regulations. The potential impact of training programmes cannot be understated. They serve as a primary resource for building biosafety capacity, ideally leading to a sustainable infrastructure of local expertise fully capable of drafting regulations and recommending sound assessments for decision makers.

Collaborative training efforts specifically directed toward biosafety assessment have been organized by a variety of organizations (Cohen, 1996; Frederick, 1995; Krattiger and Wambugu, 1996; Maredia and Dodds, 1996; Virgin *et al.*, 1998) over the last 10 years. These have characteristically taken the form of workshops or colloquia designed, as much as anything, to bring the latest thinking of industrialized countries to the developing countries. Recent analyses of these efforts have led to the question, 'Are we doing this in the most effective way?' (Virgin and Frederick, 1996). In answer, let us look at the *status*

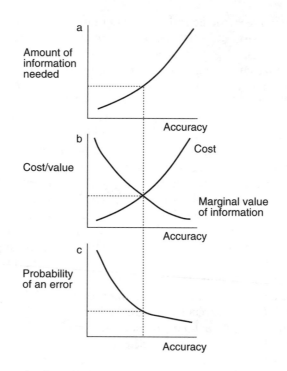

Fig. 16.1. Cost and value of information (adapted from Marois *et al.*, 1991). (Copyright, Purdue Research Foundation, West Lafayette, Indiana 47907. Reproduced with permission.)

quo based on the personal experiences of organizers, instructors, lecturers and/or managers of biosafety workshops. In general, the traditional approach has been much like that depicted in Fig. 16.2. At one workshop a group of developing country participants are instructed by a group of experts. At a second workshop, the same or other experts are lecturing some of the same participants. In a third workshop, many of the same instructors are again giving the same lectures to repeat participants. This approach might be rationalized by accepting that these workshops are often times offered prematurely. That is, much of the potential benefit of the instruction is lost because participants do not have the opportunity to use the information in a risk assessment under the auspices of a national biosafety regulation – the regulations are not in place. Hence there is a need for reinforcement of the instructional material and an appropriate updating based on newly acquired information. On the other hand, having a full complement of industrialized country experts is a significant expense and repeat attendance by participants compromises capacity-building efficiency (fewer 'untrained' persons can attend because the seat is occupied by an already 'trained' individual). A modification in the approach to operating workshops may prove more economically responsible and educationally effective.

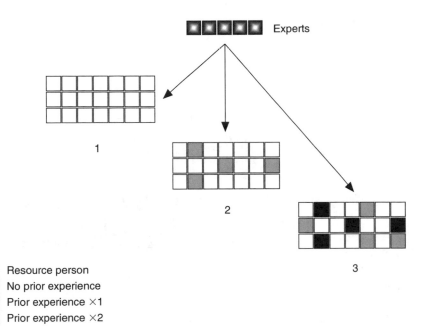

Experts

Resource person
No prior experience
Prior experience ×1
Prior experience ×2

Fig. 16.2. Conceptualization of a traditional approach to biosafety capacity-building efforts through workshops. The numbers indicate a temporal series of workshops where individual experts may or may not be the same. Many participants repeat their attendance even when there is little new in the content of the workshop agenda.

A possible alternative approach for sequential workshops is suggested in Fig. 16.3. Here, a few of the same experts might be used in successive workshops to maintain continuity and direct access to expertise. The full complement of instructors for succeeding workshops, however, is filled out with local experts, e.g. those who have qualifying experience and training. Moving such people to the front of the room involves them in a more substantive way in the instructional process. It reinforces their training and education while better achieving the goals of capacity building. This approach has the added benefits of reducing the cost of workshops and ensuring the articulation of developing country perspectives which is so important to participants. Ultimately, such workshops could be held independent of outside expertise as has been done for local training in Thailand (Attathom *et al.*, 1996).

Who's Going to Pay?

Recognizing there are costs to the development and implementation of biosafety regulations begs the question, 'Who will pay for it?' In an analysis of countries having biosafety regulatory mechanisms in place (Virgin and Frederick,

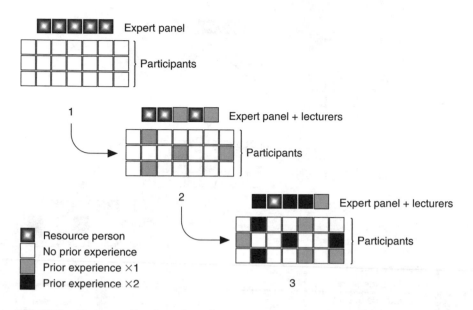

Fig. 16.3. Conceptualization of an alternative approach to biosafety capacity-building efforts through workshops. This suggests a more cost-effective approach that uses local and regional experts as instructors.

1995a), it was clear that those with low economic resources were most likely not to have biosafety regulations. National government support – which for some might be indicated as a line item in the national budget – is important. But the reality is, without outside financial assistance to support capacity building efforts, even minimal access to information may be severely restricted.

As mentioned above, international support has been significant, but how successful has this been in institutionalizing local and regional biosafety expertise? Participants at a recent conference in Sweden looked at this issue critically (Attathom *et al.*, 1996; Bhagavan, 1996; Cohen, 1996; Krattiger and Wambugu, 1996; Lyman and Toenniessen, 1996; Maredia and Dodds, 1996; Thitai and Wafula, 1996; van den Bos, 1996; Wolpers, 1996). Among the needs recognized was that of long-term financial support. Many national overseas development programmes are reducing the amount they spend on such efforts, consistent with an overall trend of diminishing aid (Gaillard, 1990; Thiel, 1996). Other sources may need to be found. If all countries are going to reach the goal of having national biosafety regulatory structures and subsequently being able to implement them effectively, donor commitments of 5–10 years for training programmes may be necessary.

Other assistance efforts in the areas of information exchange, technical expertise and cooperation should be carefully reviewed and modified to take best advantage (i.e. provide the most benefit for the least cost) of what is available. The South African Committee for Genetic Experimentation, for example, charges a fee to applicants (e.g. principal investigators applying for permission to conduct field release experiments) requesting biosafety reviews to support the use of scientists from outside their review committee (Thompson, 1996). It remains to be seen just how much applicants might be willing to pay. While international conglomerates might bear the financial burden for biosafety reviews under the guise of the cost of doing business, those in academia may find this too burdensome and decide to forgo some promising research.

It would be short-sighted to focus solely on financial considerations in biosafety support. To illustrate, consider aid programmes offering basic science support to developing countries. Developing country scientists lack the financial resources available to their industrialized country counterparts. They also lack equipment, vehicles, technicians and scientific documentation. Even more critical, and at the very centre of the scientific enterprise, is insufficient international communication. While some measures are being taken to relieve a sense of isolation felt by developing country scientists, there is a long way to go (Gaillard, 1990). As Gaillard (1990) points out, 'Lack of concertation is detrimental to both the donors, and even more the [developing countries] which would greatly benefit from meeting together to share their attitude to and uses of aid, to assess advantages and drawbacks and to discuss mechanisms that could be used nationally to obtain the most favourable aid conditions.' A resolution will require better cooperation among donors and the managers of assistance programmes (Virgin and Frederick, 1996). To the extent that such advice is applicable to all foreign aid programmes, the need for better cooperation in this era of declining budgets seems not only reasonable but necessary for the successful adoption and implementation of biosafety regulatory programmes in developing countries.

References

Anderson, W.T. (1991) The past and future of agricultural biotechnology. In: Macdonald, J.F. (ed.) *NABC Report 3. Agricultural Biotechnology at the Crossroads: Biological, Social, and Institutional Concerns.* National Agricultural Biotechnology Association. Ithaca, New York, USA, pp. 53–64.

Attathom, S., Wongsasithorn, D. and Sriwatanapongse, S. (1996) Biosafety capacity building in Thailand. In: Virgin, I. and Frederick, R. (eds) *Biosafety Capacity Building: Evaluation Criteria Development.* Stockholm Environment Institute, Stockholm, Sweden, pp. 59–60.

Bhagavan, M.R. (1996) Swedish support to biotechnology research in developing countries. In: Virgin, I. and Frederick, R. (eds) *Biosafety Capacity Building: Evaluation Criteria Development.* Stockholm Environment Institute, Stockholm, Sweden, pp. 23–26.

Brenner, C. (1995) Technology transfer: Public and private sector roles. In: Komen, J., Cohen, J.I. and Lee, S.K. (eds) *Turning Priorities into Feasible Programs: Proceedings of a Regional Seminar on Planning, Priorities, and Policies for Agricultural Biotechnology in Southeast Asia.* Intermediary Biotechnology Service, The Hague, The Netherlands, and Nanyang Technological University, Singapore, pp. 103–108.

Brenner, C. and Komen, J. (1994) *International Initiatives in Biotechnology for Developing Country Agriculture: Promises and Problems.* OECD Development Centre Technical Paper No. 100. Organisation for Economic Cooperation and Development, Paris, France, 60 pp.

Cohen, J.I. (1996) National needs and international opportunities in biosafety, analysis and constraints. In: Virgin, I. and Frederick, R. (eds) *Biosafety Capacity Building: Evaluation Criteria Development.* Stockholm Environment Institute, Stockholm, Sweden, pp. 37–43.

Cohen, J.I. and Chambers, J.A. (1991) Biotechnology and biosafety: Perspective of an international donor agency. In: Levin, M.A. and Strauss, H. (eds) *Risk Assessment in Genetic Engineering.* McGraw-Hill, New York, USA, pp. 379–394.

Frederick, R.J. (1995) International activities in support of safety in biotechnology. In: *Proceedings of the Central and Eastern European Conference for Regional Cooperation on Safety in Biotechnology. 4–6 September, 1995. Keszthely, Hungary.* Netherlands Ministry of Housing, Physical Planning and Environment, The Hague, The Netherlands, pp. 65–74.

Frederick, R.J. (1998) Biosafety regulatory frameworks in developing countries. In: *Proceedings of the International Workshop on Biotechnology for Crop Protection – Its Potential for Developing Countries.* Deutsche Stiftung für Internationale Entwicklung, Feldafing, Germany (in press).

Frederick, R.J. and Virgin, I. (1995) Searching for a balance: Summary of group discussions. In: Frederick, R.J., Virgin, I. and Lindarte, E. (eds) *Environmental Concerns with Transgenic Plants in Centers of Diversity: Potato as a Model. Proceedings from a regional workshop. Parque National Iguazu, Argentina; 2–3 June 1995.* Stockholm Environment Institute, Stockholm, Sweden, pp. 53–57.

Gaillard, J. (1990) Science in the Developing World: Foreign Aid and National Policies at a Crossroad. *AMBIO* 19, 348–353.

Komen, J. and Persley, G.J. (1993) *Agricultural Biotechnology in Developing Countries: A Cross-Country Review. ISNAR Research Report No. 2.* International Service for National Agricultural Research, The Hague, The Netherlands, 45 pp.

Krattiger, A.F. and Lesser, W.H. (1994) Biosafety, an environmental impact assessment tool and the role of the Convention on Biological Diversity. In: Krattiger, A.F., McNeely, J.A., Lesser, W.H., Miller, K.R. St Hill, Y. and Senanayake, R. (eds) *Widening Perspectives on Biodiversity.* IUCN, Gland, Switzerland and the International Academy of the Environment, Geneva, Switzerland, pp. 353–366.

Krattiger, A.F. and Wambugu, F.M. (1996) The ISAAA biosafety initiative. In: Virgin, I. and Frederick, R. (eds) *Biosafety Capacity Building: Evaluation Criteria Development.* Stockholm Environment Institute, Stockholm, Sweden, pp. 51–54.

Kuenzi, M., Assi, F., Chimiel, A., Colins, C.H., Donidian, M., Dominguez, J.B., Fogarty, L.M., Frommer, W., Hasko, F., Hovlans, J., Houwink, E.H., Mahler, J.L., Sandkvist, A., Sargeant, K., Sloover, C. and Tuijnenburg, M. (1985) Safe biotechnology general considerations. *Applied Microbiology Biotechnology* 21, 1–6.

Lacy, W.B., Busch, L. and Lacy, L.R. (1991) Public perceptions of agricultural biotechnology. In: Baumgardt, B.R. and Martin, M.A. (eds) *Agricultural Biotechnology: Issues and Choices.* Purdue University Press, West Lafayette, Indiana, USA, pp. 139–162.

Leopold, M. (1995) Public perception of biotechnology. In: Tzotzos, G.T. (ed.) *Genetically Modified Organisms: A Guide to Biosafety.* CAB International, Wallingford, UK, pp. 8–16.

Lesser, W.H. and Maloney, A.O. (1993) *Biosafety: A Report on Regulatory Approaches for the Deliberate Release of Genetically Engineered Organisms – Issues and Options for Developing Countries.* Cornell International Institute for Food, Agriculture and Development, Ithaca, New York, USA, 65 pp.

Levidow, L., Carr, S., von Shomberg, R. and Wield, D. (1996) Regulating agricultural biotechnology in Europe: Harmonization Difficulties, Opportunities, Dilemmas. *Science and Public Policy* 23, 135–157.

Lyman, J.K. and Toenniessen, G.H. (1996) The Rockefeller Foundation's Crop Biotechnology Program and Support for Biosafety. In: Virgin, I. and Frederick, R. (eds) *Biosafety Capacity Building: Evaluation Criteria Development.* Stockholm Environment Institute, Stockholm, Sweden, pp. 35–36.

Maredia, K.M. and Dodds, J.H. (1996) Capacity building in biosafety: Experience of the USAID Agricultural Biotechnology for Sustainable Productivity (ABSP) Project. In: Virgin, I. and Frederick, R. (eds) *Biosafety Capacity Building: Evaluation Criteria Development.* Stockholm Environment Institute, Stockholm, Sweden, pp. 49–50.

Marois, J.J., Grieshop, J.I. and Butler, L.J. (1991) Environmental risks and benefits of agricultural biotechnology. In: Baumgardt, B.R. and Martin, M.A. (eds) *Agricultural Biotechnology: Issues and Choices.* Purdue University Press, West Lafayette, Indiana, pp. 67–80.

Medley, T. (1994) A regulatory perspective on harmonization of regulations and public perception. In: Krattiger, A.F. and Rosemarin, A. (eds) *Biosafety for Sustainable Agriculture: Sharing Biotechnology Regulatory Experiences for the Western Hemisphere.* ISAAA, Ithaca, New York, USA and Stockholm Environment Institute, Stockholm, Sweden, pp. 71–78.

Persley, G.J., Giddings, L.V. and Juma, C. (1992) *Biosafety: The Safe Application of Biotechnology in Agriculture and the Environment.* International Service for National Agricultural Research, The Hague, The Netherlands, 45 pp.

Strauss, H.S. (1991) Lessons from chemical risk assessment. In: Levin, M. and Strauss, H. (eds) *Risk Assessment in Genetic Engineering: Environmental Release of Organisms,* McGraw-Hill, New York, USA, pp. 297–318.

Theil, R.E. (1996) The helplessness of development policy: An assessment of aid in the modern world. *Development and Cooperation,* No.6. Deutsche Stiftung für Internationale Entwicklung, Feldafing, Germany, pp. 24–26.

Thitai, G.N.W. and Wafula, J.S. (1996) Identified national needs and effects of biosafety workshops. In: Virgin, I. and Frederick, R. (eds) *Biosafety Capacity Building: Evaluation Criteria Development.* Stockholm Environment Institute, Stockholm, Sweden, pp. 67–70.

Thompson, J.A. (1996) Biosafety in South Africa: the role of SAGENE. In: Virgin, I. and Frederick, R. (eds) *Biosafety Capacity Building: Evaluation Criteria Development.* Stockholm Environment Institute, Stockholm, Sweden, pp. 65–66.

Tzotzos, G.T. (1995) Biological risk assessment: an editorial overview of some key policy and implementation issues. In: Tzotzos, G.T. (ed.) *Genetically Modified Organisms: A Guide to Biosafety.* CAB International, Wallingford, UK, pp. 1–7.

UNEP (1995) *UNEP International Technical Guidelines for Safety in Biotechnology.* United Nations Environment Program, Nairobi, Kenya, pp. 31.

UNIDO (1990) *An International Approach to Biotechnology Safety.* UNIDO, Vienna, Austria, pp. 43–47.

van den Bos, R.C. (1996) Safety in biotechnology, reflections from a donors perspective. In: Virgin, I. and Frederick, R. (eds) *Biosafety Capacity Building: Evaluation Criteria Development.* Stockholm Environment Institute, Stockholm, Sweden, pp. 27–30.

Virgin, I. and Frederick, R.J. (1995a) The impact of international harmonisation on adoption of biosafety regulations. *African Crop Science Journal* 3(3), 387–394.

Virgin, I. and Frederick, R.J. (1995b) The push for biosafety regulations. *The Courier*, 153 (September–October), 92–94.

Virgin, I. and Frederick, R.J. (1996) Evaluating biosafety capacity building: a synthesis of views. In: Virgin, I. and Frederick, R. (eds) *Biosafety Capacity Building: Evaluation Criteria Development*. Stockholm Environment Institute, Stockholm, Sweden, pp. 7–11.

Virgin, I., Frederick, R.J. and Ramachandran, S. (1998) Biosafety training programs and their importance in capacity building and technology development. In: *Proceedings of a Workshop on Biotechnology and Biodiversity*. USDA/APHIS, Washington, DC, USA, and Indian Institute of Management, New Delhi, India (in press).

Wolpers, K.H. (1996) Plant biotechnology in German technical cooperation programmes: experiences and perspectives. In: Virgin, I. and Frederick, R. (eds) *Biosafety Capacity Building: Evaluation Criteria Development*. Stockholm Environment Institute, Stockholm, Sweden, pp. 31–34.

Zandvoort, E. and Morris, E.J. (1996) Policy and planning for biosafety: synthesis of an African workshop. Annex 2. Conference resolution: biosafety and biotechnology workshop. In: *Turning Priorities into Feasible Programs: Proceedings of a Seminar on Planning, Priorities and Policies for Agricultural Biotechnology. South Africa, 23–28 April, 1995*. Intermediary Biotechnology Service, The Hague, The Netherlands and Foundation for Research Development, Pretoria, South Africa, 91 pp.

Cassava Biotechnology Research: Beyond the Toolbox

17

Ann Marie Thro

Cassava Biotechnology Network, c/o CIAT, AA 6713 Cali, Colombia

Cassava farmers are hard working and highly concerned for their families and future, yet most cassava farmers are still poor. This is often related to factors such as isolation and poor infrastructure. To the extent that biological and agricultural factors are involved, the Consultative Group on International Agricultural Research (CGIAR or CG) system exists and is mandated to find solutions. These solutions, perhaps more often components of more complex solutions (systems), may be crop varieties that resist diseases or drought; or new processes to add value to agricultural products and reach new markets. It is the CG's job to examine any new research tools that may become available, such as biotechnology, and to capture and focus them on problems of the rural poor in developing countries. There is no other organization with exactly this mandate.

Cassava is underinvested in every region and country, and at every stage of the research and development (R&D) process; human capital is low. A network was required to promote awareness of the importance and needs of cassava and to unite working partners with complementary capacities and resources. Out of a recognition that no one institute alone could accomplish this, the Cassava Biotechnology Network (CBN) was founded in 1988.

From its founding, CBN has shared the goal of the broader cassava R&D community 'To enhance the food security and economic development value of cassava for those who depend on it.' The first phase of CBN (1988–1992) was preoccupied with the urgency of getting started with laboratory work, some of it long-term, that would be needed to develop biotechnology tools for cassava. This initial focus was reflected in CBN's 1988 objectives:

- to make known cassava's potential and its R&D requirements;

© CAB INTERNATIONAL 1998. *Agricultural Biotechnology in International Development*
(eds C.L. Ives and B.M. Bedford)

- to encourage a 'critical mass' and synergy of advanced biotechnology research groups working on cassava;
- to provide a mechanism for voluntary focusing of resources and cost-effective use of existing facilities in cassava biotechnology research.

These objectives remain valid; however, it has become clear that if biotechnology is going to help achieve better food security and a new level of economic development for cassava users, CBN must pay attention to the pathways through which biotechnology tools can, in fact, contribute to the overall goal.

CBN is now equally involved in helping to create the conditions necessary for biotechnology to achieve its intended impact on farms and in rural areas and towns. The network seeks to develop linkages between strategic and applied research and cassava users (farmers, processors, markets and consumers). Ultimately, this function of building action linkages may be more important than the initial one of encouraging the development of the biotechnology tools themselves – although both functions will have been essential to mobilize a new technology to the service of this crop of the poor.

Poverty, Biotechnology and the Consultative Group on International Agricultural Research (CGIAR)

Cassava farmers are hard working, intelligent and concerned for their families and future. Their crop is an important one in a third of the world's low-income, food-deficit countries and especially in Africa, where cassava is the first or second most important locally produced food. Cassava's harvest of starchy roots and high-protein leaves is reliable even in extreme conditions where cereals fail. Cassava becomes even more important in times of social insecurity and war, when other crops are susceptible to theft or destruction. Cassava also provides economic opportunity in poor rural areas. Primary processing for flour and starch (cassava products with growing markets) uses simple technology, and must be done locally because of high transport costs for bulky unprocessed roots. Yet in spite of cassava's advantages and their own efforts, most cassava farmers are poor. Many have little or no access to healthcare, education, decent housing, sanitation or a varied diet. Often the heads of household are women, for whom cassava's dependability is especially critical.

Such persistent rural poverty has many contributing causes. Some are social and economic, such as civil disorder, poor infrastructure or lack of market information. Others are agroecological, including fragile or exhausted soils, irregular rainfall, poor stability or low productivity of local food crops.

The research centres of the Consultative Group on International Agricultural Research (CGIAR, or 'CG') seek to find solutions to agroecological component causes of rural poverty. Among the CG's contributions may be crop varieties that resist diseases or drought, cropping systems that conserve soil, or new ways to add value to local agricultural products such as cassava. When a powerful new research tool – such as biotechnology – becomes available, neither private investment nor national research in prosperous countries provide spill-over benefits for the poor who grow cassava. Cassava-growing countries are for the most part unable to divert resources to investigate new technologies. The CG has the high-risk job of 'capturing' new technologies and focusing them on problems of the rural poor in developing countries.

Realizing the potential contributions of biotechnology for cassava's end-users (smallholder farmers, rural processors, rural and urban consumers) is an enormous and complex job. No single organization has all the capabilities needed. Moreover, cassava is the only major world food crop neither grown in the temperate zone, nor used there in recognizable form. Yet it is in the temperate zone – the 'North' – where most technically advanced countries are located. As a result, the communities who depend on cassava have been separated from research capacity by both space and tradition. Cassava's research needs and opportunities were long neglected.

The founders of the Cassava Biotechnology Network (CBN) saw the need and an opportunity. Although most investment in agricultural biotechnology research capacity is located in the North and focused on temperate crops, there is excess capacity which can be captured for an underinvested tropical subsistence crop such as cassava. A network to encourage globally relevant cassava biotechnology research around a common strategic agenda could enlist the best laboratories in the North, make cost-effective use of existing investment in research and stimulate research links between North and South, as well as stimulating biotechnology development in the South.

The Cassava Biotechnology Network

Origin

CBN originated as an initiative of the Centro Internacional de Agricultura Tropical (CIAT) with the participation of the International Institute of Tropical Agriculture (IITA), the two CG centres that have research expertise on cassava. Other founding members were individual scientists from advanced laboratories who saw the CG centres as an avenue through which their work could benefit the poor in developing countries.

Initially, CBN concentrated on getting started with the medium- and long-term work of adapting existing biotechnology tools, such as genetic transformation and molecular mapping, to cassava. This initial focus was reflected in CBN's original objectives:

- to make known cassava's potential and its research and development (R&D) requirements;
- to encourage a 'critical mass' and synergy of advanced biotechnology research groups working on cassava;

- to provide a mechanism for voluntary focusing of resources and cost-effective use of existing facilities in cassava biotechnology research.

Development of a vision

From its founding, CBN has shared the ambitious goal of cassava research in its broadest sense: to enhance the food security and economic development value of cassava for poor rural areas which depend on this crop. As the Network looked ahead to a future technology transfer stage, it became clear that if biotechnology is to help provide better food security and a new level of economic development for cassava users, CBN must do more than foster the cost-effective development of a powerful new tool kit. It would also be necessary to help create the conditions necessary for generating and passing specific appropriate technologies successfully from concept to adoption.

In 1992, the Government of The Netherlands, through its Special Programme on Biotechnology and Development Cooperation (DGIS/BIOTECH) provided funding for a CBN Coordination office. The objectives of the CBN Coordination project reflect the evolution in the network's concept of its mission:

- integrate farmer priorities into the research agenda for cassava biotechnology;
- foster demand-led cassava biotechnology research;
- promote exchange of information and technologies among farmers, advanced laboratories and national programmes.

Between 1992 and 1996, the first research breakthroughs were made in projects of CBN members (Table 17.1). These included most notably: the genetic transformation of cassava (Li et al., 1996; Raemakers et al., 1996; Sarria et al., 1995; Schöpke et al., 1996); cloning of several agronomically-important genes (Schöpke et al., 1993; Salehuzzaman et al., 1992, 1993; Hughes et al., 1994); and completion of a framework molecular map (Fregene et al., 1997).

In priority setting, and also to plan arrangements for technology transfer, CBN was working closely with intermediate experts in national cassava R&D programmes. At the same time, DGIS/ BIOTECH challenged CBN Coordination to bring cassava's end-users, especially smallholder farmers and rural processors, into the decision-making process for research. DGIS/BIOTECH also helped provide opportunities to explore priorities and experimental solutions interactively with cassava's end-users.

To link biotechnology to end-users, CBN needed a strategy and means of implementation. The framework of CBN's strategy is the cassava R&D cycle (Fig. 17.1). The cycle, in its simplest conceptual form, includes needs assessment and priority setting, basic and strategic research, applied R&D, and technology transfer and impact assessment. The last step feeds back to the first. CBN attempts to encourage integration of appropriate biotechnology tools, or awareness of their potential and limitations, at each point in the cycle, to encourage communication and interaction between all steps of the R&D cycle, and, at each step, to link research with the target end-users in the cassava commodity chain. The cassava commodity chain begins with the producer and passes through the rural (primary) processor, to the trader, additional processors and retailers, and the consumer. In traditional systems, the chain may consist only of producer, processor and consumer. These are often the same individuals, frequently a woman farmer and her children.

Implementation

To implement the strategy, CBN uses a set of services and activities made possible by the coordination grant from DGIS/BIOTECH:

* research priority setting with farmers' perspectives;
* biennial international scientific meetings;
* a small grants programme;
* publications;
* network guidance from steering and scientific advisory committees;
* coordination services (secretariat).

The scientific meetings and the small grants programme have been particularly effective in implementing CBN's strategy, guided by the priority setting exercises.

Research priority setting with farmers' perspectives

CBN has conducted case studies with farmers and small-scale rural processors in Tanzania, China and Colombia. Even more valuable has been collaboration with participatory projects such as PROFISMA (Programa para Fitosanitación de Mandioca) (National Agricultural Research Corporation, Brazil (EMBRAPA) and CIAT), CIAT's Southeast Asia Soil Conservation Project, and a group of Malawian, Tanzanian and Swedish researchers studying farmers' management of cassava biodiversity.

Research priorities are based on case studies and information from interactive research with farmers and rural processors. Priorities are reviewed with disciplinary experts, seeking the most important regional or global needs or opportunities where research can make a contribution. The resulting priorities are discussed in working groups and plenaries at the scientific meeting, and published for further comment and debate until the next cycle (see Table 17.1 for the current CBN research priority areas). Anyone may use these priorities as a planning tool.

Biennial international scientific meetings

CBN meetings are carefully planned technical reviews with a much broader additional function: they are a public forum for input into the international

Table 17.1. Cassava biotechnology research: recent advances and anticipated progress over the next five years.

Number of major projects 1996–1997	Region where objective is important	CBN research priority objective	Recent advances (1992–1997)	Anticipated status at end period 1997–2002 (assumes research funding)
Biotechnology tools: genetic improvement of cassava				
5	Global	Improved plant regeneration systems	Improved method: embryogenic suspension cultures	Additional methods (organogenesis); improved efficiency of embryogenic suspension cultures
5	Global	Improved genetic transformation techniques	First experimental successes with both *Agrobacterium* and direct gene transfer	Efficient methods developed and transferred to national crop improvement programmes
4	Global	Molecular genetic map	Framework map completed	First molecular markers developed and transferred to national crop improvement programmes
6	Global	Useful genes and gene promoters, characterized and cloned	Several genes; two promoters	Additional genes and promoters; genes and promoters tested
3	Global	Molecular/cytogenetic characterization of *Manihot* genomes	Relationships being elucidated	Better understanding of origin, relationships and diversity; and location of diversity
1	Global	Regulation of reproductive biology (flowering, pollen conservation, haploids, apomixis)	Little known (no major projects)	Not yet enough knowledge to predict
Biotechnology tools: conservation/exchange of *Manihot* genetic diversity				
2	Africa, global	Disease diagnostic methods for clean germplasm transfer	First methods being tested	Reliable, low-cost methods
2	Global	Cryopreservation for long-term conservation of genetic diversity	Successful experimental methods with a range of genotypes	Pilot project results on genetic stability; costs
1/±30	Global	Tissue culture for germplasm conservation/micropropagation (advances/application)	Extension to additional national programmes in cassava-growing countries (approximately 30 programmes now use established methods for conservation and exchange)	Research and pilot project information on potential for low cost tissue culture and post-flask methods for faster varietal propagation

Biotechnology applications: realizing opportunities for cassava

3	Asia, LAC*	Starch quantity and quality for diverse end uses	Major cassava starch synthesis genes cloned (proprietary)	First transgenics ready for testing; additional genes; new approaches
0	Asia, LAC	New product development via biochemistry, molecular genetics	First discussions, based on findings in bacteria and a few plant species	Not yet enough knowledge to predict
0	Global	Plant nutrient cycling efficiency	Little known, little or no research	Not yet enough knowledge to predict
1	Global	Range/productivity in stress environments: photosynthesis, drought tolerance	Little known, little research	First molecular tags for drought tolerance established and transferred to crop improvement programmes

Biotechnology applications: solving problems in cassava production and use

5	Africa, global	Resistance to important viral diseases	Cloned genes successfully conferring resistance in model systems	First transgenic cassava ready for testing
4	Africa, LAC	Integrated pest management, including host-pathogen/pest interactions	Improved knowledge of cassava bacterial blight organism (CBB)	Molecular markers for CBB resistance for use in breeding
2	Global	Delayed postharvest deterioration	Increased understanding of biochemistry; gene cloned in wound response pathway (possibly involved); more socioeconomic understanding needed	Molecular markers for plant breeding; better biochemical understanding, possibly cloned genes
6	Africa	Modified cyanogen biochemistry to optimize food security	Major genes in breakdown and synthesis cloned; more socioeconomic and ecological understanding needed	First transgenic prototypes with altered cyanogen metabolism ready for testing
1	Asia	Development of true seed for cassava production	Little knowledge, more socioeconomic understanding needed	Not enough current knowledge to predict

* LAC: Latin America and the Caribbean.

research agenda for cassava biotechnology. CBN meetings are open to anyone with a bona fide interest in CBN's goal, including socio-economic and policy aspects. Plenary sessions and working groups examine research results in light of advances in science and specific development objectives, with discussion and debate. Poster sessions allow all participants to present their work. All presentations and working group reports are published in a proceedings.

From 1992 to 1996, the number of participants from cassava-growing countries tripled. The meeting rotates between major cassava-growing regions of the world in order to distribute opportunities for regional participation, examine cassava's role in each region in turn, and allow visits to cassava farmers in their fields. Field visits are organized in small groups with translators so that all participants can converse with cassava farmers. Although these are only 'snap-shot' visits (too brief for assessing research priorities, for example), they have proven extremely useful in helping laboratory collaborators, who may have little opportunity for field work, to understand fully the information they receive from applied research colleagues and farmer group representatives.

CBN small grants programme

Over 40 CBN small grants in support of cassava biotechnology research have been awarded. Several awards have been to smallholder farmer organizations and processing cooperatives. About half of CBN small grants have been made to national programmes of cassava-growing countries. An additional one-third of CBN small grant projects link national cassava programmes to leading laboratories throughout the world. One-quarter of CBN small grant project leaders are women.

CBN small grant awards have included:

- case studies with farmers and village processors;
- strategic research in developing countries and advanced laboratories;
- operational funds for applied and strategic cassava biotechnology research in national programmes;
- transfer of biotechnology tools to developing countries;
- pilot projects with farmers and processing cooperatives.

Investment of limited additional resources in network coordination through the small grants programme has had an impact on specific advances in the science of cassava biotechnology and has influenced the technical objectives and application of cassava biotechnology research. A few examples follow.

Science
Cassava genetic engineering is a new and long-sought breakthrough. CBN small grants were instrumental at key points: (i) developing some of the techniques that ultimately brought success (in laboratories in China and the UK); (ii) keeping a successful team together during a gap in major funding; and (iii) permitting several teams to share techniques, permitting faster progress.

Research priorities
Biotechnology research on cassava toxicity (cyanogenesis) originally focused
on a single objective – eliminating cyanide precursors completely. CBN's own
case studies with farmers and its contacts with other farmer-oriented groups
revealed the importance of an additional and more complex objective –
preserving benefits of cyanogenesis for cassava production (food security) where
it is important, while at the same time improving cassava food safety.

Application of results
Cassava varietal resistance to bacterial blight (CBB) has been unstable across
environments. Recent work, by L'Institute Française de Recherche Scientifique
pour le Développement en Coopération (ORSTOM) with CIAT, on molecular
genetics of host–pathogen interactions has revealed variation of CBB strains. A
CBN small grant is enabling cassava breeders at the Brazilian Instituto
Agronomico de Campinas to collaborate in a new breeding strategy that will
use the molecular information to produce stably-resistant varieties.

Another CBN small grant is providing start-up funds for a pilot project to
test costs and benefits of using tissue culture for multiplying resistant cassava
varieties to combat a severe epidemic of African cassava mosaic virus in
Uganda. The tissue culture operation will be integrated into the cassava variety
multiplication scheme of NARO (the Ugandan national programme), a scheme
which involves district extension and farmer groups.

Membership

As a result of the activities and services described above, CBN's research
membership has expanded from an original handful of scientists in 1988. By
1997, CBN had about 400 members working on development or application of
cassava biotechnology tools and a similar number of members in national
programmes in developing countries who conduct applied cassava R&D. About
60% of the 'biotechnology' membership are located in 26 cassava-growing
developing countries, about 30% in 13 economically advanced countries and
about 10% in international centres of the CGIAR. They conduct research in three
types of cassava biotechnology: (i) approximately 40% in genetic biotechnologies
(molecular genetics and genetic transformation); (ii) approximately 30% in
tissue culture; and (iii) approximately 30% in fermentation biotechnologies.

Development of collaborative programmes with organizations representing
smallholder cassava farmers and rural processors has been a slow but
rewarding process. Farmer–processor groups, unlike national research
programmes, are usually not yet accessible through the international research
community. Consequently, farmer collaboration requires on-the-ground
exploration and a significant time investment, either by CBN Coordination or
by a collaborating CBN member. By 1997, CBN was collaborating with three
cassava farmer–processor groups in South America, one in Africa, and, through

a collaborating project of national programmes and CIAT, with farmers in several countries in Southeast Asia (see Box 17.1).

The way ahead

In its first phase, CBN has sought impact at two levels. Initially, impact has been developed at the scientific level, by helping to nurture the birth of a new set of technologies for genetic improvement of a major world crop. Second, at the level of the R&D environment, CBN has sought to help establish a tradition for cassava biotechnology research of end-user participation in needs assessment, research agenda formulation, and research development and testing.

Both of these are mid- to long-term goals. Both have had a successful start. Now they must be woven together. There is need for an integrated effort to improve the robustness of the biotechnologies, implement them in cassava-growing countries, and apply them in on-going priority cassava research. Now is the time to proceed purposefully with this task.

In the future, CBN should develop collaborative projects for improving experimental technologies and moving them out of laboratories. It should see more regional initiative and leadership and stronger interregional relationships among research institutes within a region. The regional level should be the integration point for end-user participation. CBN should facilitate opportunities for the growth of South–South collaboration, and for North–South collaboration in forms that build institutions and foster dedication and enthusiasm toward nationally important goals.

Lessons learned

Vision and strategy

Without doubt, the single most important lesson learned by CBN has been that if biotechnology is really to fulfil its promise for a long-neglected crop of small-holder farmers, the network must share downstream responsibility. CBN Coordination has paid increasing attention to the pathways through which biotechnology innovations must pass in order to reach and help cassava farmers.

Cassava as a crop is extremely unique, due to its unusual biology, and cassava systems are usually not supported by any public interventions, and are marginal and highly stressed. Much care and caution are needed in setting priorities accurately, and they should be based on full understanding of the system. Working with farmers to determine their needs and opportunities has proven to be highly demanding, requiring an understanding of, and acceptance by, specific local cultures (Rosling *et al.*, personal communication). It is slow work and requires considerable investment. CBN's own priority-setting exercises with farmers have been invaluable in convincing the network of the truth of this observation, and of the necessity of collaborating with partners who are dedicated to in-depth work with specific cassava-using communities.

Box 17.1. Partial list of CBN collaborators in cassava research and development. Needs assessment, strategic and applied research based on identified needs, and technology transfer.

CBN collaborators in needs assessment

Activity	Collaborators	
Tanzania Village Case Study*	National Root Crops Program	NARS
	Tanz. Home Economics Association	NGO
	Natural Resources Institute (NRI), UK	ARO
	International Institute of Tropical Agriculture (IITA)	IARC
China Village Case Study* Guangdong, Guangxi, Hainan	Centro Internacional de Agricultura Tropical (CIAT)	IARC
	Guangxi Subtropical Crop Research Institute (GSCRI),	NARS
	Chinese Academy of Tropical Agricultural Science (CATAS), Upland Crop Research Institute (UCRI)	NARS
Participatory priority setting, Colombia, Kenya	Government of The Netherlands Special Programme on Biotechnology and Development Cooperation (DGIS/BIOTECH)	Donor
	Comités Campesinos	Farmer committee
	Kenya Agricultural Biotechnology Platform (KABP)	NARS/ Farmer committee
Village case studies, Malawi	Ministry of Agriculture	NARS
	Swedish Agricultural University	ARO
Integrated projects for product/market development Colombia, Ecuador*	CIAT	IARC
	Centre de coopération international en recherche agronomique pour le développement (CIRAD)	ARO
	Colombia National Research Corporation (CorpoICA), Instituto Nacional Autonomo de Investígacion (Ecuador) (INIAP)	NARS
	Union de Asociaciones des Traba jado en Agricultura productores y Procesadores de Yucca (UATAPPY)	Farmer coopera-tive
Participatory cassava breeding, Colombia, Brazil	CIAT	IARC
	CorpoICA, National Center for Cassava and Fruit Crops Research, Brazil/ National Agricultural Research Corporation, Brazil (CNPMF/EMBRAPA)	NARS
	CADETS	Farmer committee

Continued overleaf

Box 17.1 *Continued*

Participatory farm. Systems research – Thailand, China,	CIAT	IARC
	Department of Agriculture (RFCRC), Department of Agricultural Extension, Kasetsart University, Thailand; CATAS, GSCRI, China; Institute of Agricultural Science, Hawaii Agricultural Research Center (HARC), Thai Nguyen University	NARS
	Institute of Soils and Fertilizers, Vietnam	
Participatory IPM research* Brazil, West Africa	CIAT, IITA	IARCs
	CNPMF/EMBRAPA, Institute National de Recherches Agronomiques, Benin	NARS

CBN collaborators in strategic research

Priority: adaptation to marginal environments

Molecular study of genetics of host–pathogen interaction*	L'Institute Française de Recherche Scientifique pour le Développement en Coopération (ORSTOM)	ARO
	CIAT	IARC
	Instituto Agronomico de Campinas	NARS
Biochemistry, isolation of genes/promoters, and genetic engineering for resistance to diseases and pests*	International Laboratory for Tropical Agricultural Biotechnology (ILTAB)/ORSTOM	ARO
	University of Bath	ARO
	Federal Technical Institute (ETH)–Zurich	ARO
	CIAT	IARC
	Centro de Ingenieria Genetica y Biotecnología (CIGB), La Havana	NARS
Development of molecular markers for drought resistance*	CIAT, Assoc. Nac. Productores y Procesadores de Yuca (ANPPY), Colombia	IARC Farmer Cooperative
	CNPMF/EMBRAPA	NARS
	National Center for Genetic Resources and Biotechnology (CENARGEN)/EMBRAPA	NARS

Priority: economic value of cassava

Biochemistry, genes/promoters and genetic engineering for improved or novel processing characteristics:	Agricultural University, Wageningen	ARO
Starch quality, post-harvest perishability*	CIAT	IARC
	University of Bath	ARO
Future: biofactory, biodegradable plastics?	ETH–Zurich (discussion)	ARO

Development of molecular markers for processing qualities	CIAT CorpoICA	IARC NARS
Starch quality, reduced perishability	Comités Campesinos, CADETS	Farmer committee
Biochemistry, genes/promoters, genetic engineering for optimum management of cyanogenesis	IITA University of Newcastle Royal Veterinary and Agricultural University – Copenhagen Ohio State University Uppsala University Central Tuber Crops Research Institute (CTCRI), India	IARC ARO ARO ARO ARO ARO NARS

Priority: conservation and characterization of genetic diversity

Cryopreservation of cassava germplasm	CIAT	IARC
Molecular characterization of cassava genetic diversity*	CIAT CENARGEN/EMBRAPA ORSTOM Escola Superior Agronomica Luiz de Quiroz, University of San Paolo (ESALQ/USP), Piracicaba National Center for Genome Resources (NCGR), New Mexico University of Georgia	IARC NARS ARO NARS ARO ARO
Molecular phylogeny of cassava	Washington University – St Louis CENARGEN/EMBRAPA	ARO NARS

CBN collaborators in applied research

Priority: conservation and use of genetic diversity

Tissue culture for germplasm conservation*	Over 20 countries	NARS
Backstopping:	CIAT, IITA	IARCS
Strategic germplasm exchange*	CIAT IITA Kenya Agricultural Research Institute (KARI)/Kenya Ministry of Agriculture/Malawi	IARC IARC NARS NARS
Backstopping, Kenya & Malawi	IITA Vegetable and Ornamental Plants, Institute (VOPI), South Africa GTZ/DSE Brazil, Colombia	IARC NARS/ ARO Donor/ ARO NARS

Priority: economic value

New product development using lactic acid fermentation*	CIRAD ORSTOM Universidad Estadual de San Paulo (UNESP), Brazil University of Buenos Aires	ARO ARO NARS NARS

Continued overleaf

Box 17.1 *Continued*

Improvement of fungal	IITA	IARC
fermentation for enhanced	ORSTOM	ARO
traditional products and new	NRI	ARO
products, e.g. animal feeds,	University de Parana	NARS
industrial enzymes	Research Institute of Animal	NARS
	Production, Bogor	
	CTCRI, India	NARS
Clean up and valorization of	CIRAD	ARO
cassava processing waste	UniValle, Colombia	NARS
	Central Tuber Crops Research	
	Institute (CTCRI), India	NARS
	UNESP, Brazil	NARS

CBN collaborators in technology transfer**

Priority: planting material		
Pilot project in Uganda*:	National Agricultural Research	NARS
feasibility of tissue culture for	Organization (NARO)	
faster availability of improved	University of Bath	ARO
planting material;	Makerere University	NARS
biotechnology, links,	IITA	IARC
management skills	World Vision, Vision Teruda,	NGO
	Action Aid	
Colombia: pilot project, local	CIAT	IARC
microenterprises	CorpoICA	NARS
	ANPPY	Farmer coopera-tive

*   Activity or research area receiving CBN small grant funds.
** The CBN small grants programme is itself an effective form of technology transfer, involving many more collaborators than those listed here.
Types of collaborating organizations:
ARO, advanced research organization; IARC, international agricultural research centre; NARS, national agricultural research system (member of a ...); NGO, non-governmental organization.

An unresolved question is how to use location-specific results to identify research priorities that are both strongly linked to specific farmer needs, yet general enough to guide long-term strategic research to serve an entire region. Whether it is best to accomplish this through formal analysis or intuitively is not yet resolved. It is particularly challenging because farmer perceptions – even in a single locality – can change from year to year, as farmers react to changes in the biological or socioeconomic environment.

Other lessons
RESOURCES FOR CASSAVA BIOTECHNOLOGY. Reductions in overall funding for agriculture and development have made it difficult to increase the total

resources available for cassava biotechnology. There is evidence, however, that CBN's increasing links to national programmes and farmer–processors have enabled certain donors to maintain funding for cassava biotechnology that otherwise might have been lost. CBN also assists in making more efficient use of existing resources. CBN's small grant projects in Africa, for example, provide complementary funding for specific objectives prioritized by area cassava networks whose primary funding is largely from the United States Agency for International Development (USAID). As another example, South African laboratories have made highly efficient use of CBN to make contacts with national programmes and IITA, and to identify opportunities for relevant cooperation with neighbouring countries.

Modus operandi

ACHIEVING EFFECTIVENESS. CBN is faced with the challenge of how to achieve its goals with neither institutional lines of accountability nor major funding to encourage others to follow its vision. Membership is voluntary, based only on perceived benefit. Consequently, CBN must nurture links and build consensus and conviction. Opportunities are found through CBN services and through collaboration with other cassava-related networks. Personal contact and continuity are very important for national programme and farmer groups working in areas where communication is difficult. Continuity is particularly important at the present time, when staff turnover is very rapid in both national and international programmes.

Informality and flexibility have been assets in helping CBN learn from experience. Informality has allowed growth through genuine examination of ideas and results. In the process, CBN's Coordinator has acted as a facilitator to the network and the scientific community. This modest arrangement allows many strong institutions to collaborate in guiding CBN, making full use of leadership resources within the network. An open membership structure with flexible relationships among members – including many 'networks within networks' - has permitted CBN to become a useful forum for exploring new types of collaborative relationships among a broad range of partners.

CBN's practice of priority setting by consensus has permitted differences of opinion to contribute, with time, to a robust common vision. Success has required a rigorously critical atmosphere plus a strongly-held common goal, a degree of continuity in the scientific community and commitment to the process by the members and donors. Initiation of CBN has been slower than with a voting system, but patiently working to consensus has brought unity and conviction.

Structural flexibility

An open membership structure with flexible relationships among members has permitted CBN to become a useful forum for exploring new types of collaborative relationships among a broad range of partners. Exploring new modes of collaboration has been particularly important in a period of rapid advance in national programme research capacity, especially in South America and Southeast Asia.

Conclusions

CBN has learned two types of lessons: (i) lessons of *modus operandi*, and (ii) strategic lessons. Its lessons of *modus operandi* may be most relevant to a new network with limited resources. These include:

- plan to be flexible and informal;
- expect that consensus and vision may develop in stages;
- take full advantage of leadership, experience and creativity of network members;
- recognize the importance of personal contact and continuity in forming and maintaining links.

CBN's most important strategic lesson concerns farmer links – an essential investment for CBN.

This chapter began with a presentation of CBN's ambitious goal of mobilizing powerful new technologies in the service of a little-understood agricultural sector. CBN has emerged at the end of its first phase with a two-fold strategic lesson: (i) achieving CBN's goal in the long run, will require close links with cassava's end users (the smallholder farmers and rural processors who depend on cassava); and (ii) developing effective farmer links requires significant investment in time and attention.

Cassava systems operate with few, if any, inputs to buffer local conditions and are consequently likely to be in balance with a socioeconomic environment whose customs and needs are unlike those of better understood and more prosperous economies. Identifying useful interventions that may improve such intricate systems requires in-depth familiarity.

Is such familiarity available to CBN, given its limited resources and broad scope of activities? The answer is probably yes. CBN has many contacts with groups who are studying specific cassava-growing communities. In the future, CBN will be active in helping to organize participatory projects that will integrate these communities in interactive R&D projects with the national programme and other institutes.

Acknowledgements

The author wishes to acknowledge the work of past and present members of the CBN Steering Committee and Scientific Advisory Committee, and members of the Ad-Hoc Cassava R&D Group. Without the contributions of these individuals, neither CBN nor its evolution would have been possible. The author also acknowledges the philosophical and financial contribution of the Special Programme on Biotechnology and Development Cooperation of The Netherlands (DGIS/BIOTECH).

References

Fregene, M., Angel, F., Gomez, R., Rodriguez, F., Chavarriaga, P., Roca, W., Tohme, J. and Bonierbale, M. (1997) A molecular genetic map of cassava (*Manihot esculenta* Crantz). *Theoretical and Applied Genetics* 956, 431–441.

Li, H-Q., Sautter, C., Potrykus, I. and Puonti-Kaerlas, J. (1996) Genetic transformation of cassava (*Manihot esculenta* Crantz). *Nature Biotechnology* 14, 736–740.

Raemakers, C.J.J.M, Sofiari, E., Taylor N., Jacobsen, E. and Visser, R.G.F. (1996) Production of transgenic cassava (*Manihot esculenta* Crantz) plants by particle bombardment using luciferase activity as selection marker. *Molecular Breeding* 2, 339–349.

Salehuzzaman, S.N.I.M., Jacobsen, E. and Visser, R.G.F. (1992) Cloning, partial sequencing, and expression of a cDNA coding for branching enzyme in cassava. *Plant Molecular Biology* 20, 809–819.

Sarria, R., Torres, E., Balcazar, N., Destefano-Beltran, L. and Roca, W.M. (1995) Progress in Agrobacterium-mediated transformation of cassava (*Manihot esculenta* Crantz). In: *Proceedings of the II International Scientific Meeting. Cassava Biotechnology Network, Bogor, Indonesia, 25–28 August 1994.* CIAT Working Document No. 150, pp. 241–244.

Schöpke, C., Franche, C., Bogusz, D., Chavarriaga, P., Fauquet, C. and Beachy, R.N. (1993) Transformation in cassava (*Manihot esculenta* Crantz). In: Bajaj, Y.P.S. (ed.) *Biotechnology in Agriculture and Forestry Vol. 23: Plant Protoplasts and Genetic Engineering.* Springer Verlag, Berlin, pp. 273–298.

Schöpke, C., Taylor, N., Cárcamo, R., Konan, N.K., Marmey, P., Henshaw, G.G., Beachy, R. and Fauquet, C. (1996) Regeneration of transgenic cassava plants (*Manihot esculenta* Crantz) from microbombarded embryogenic suspension cultures. *Nature Biotechnology* 14, 731–735.

Fundación Perú: a Path to Capacity Building

Fernando Cillóniz

Fundación Perú, Alcanflores 1245–Miraflores, Lima 18, Peru

In Peru, biotechnology is viewed somewhat like fashion. Government programmes may consider the installation of several biotechnology laboratories for 'organized' small farmers to run them. Often, these programmes fail because there is no genuine interest on the part of the beneficiaries, or because there is a lack of knowledge as to how to operate a biotechnology laboratory.

Very basic things like forming farmers associations and partnerships to establish economically feasible operations with adequate scales; or improving postharvest techniques to avoid the loss of 40% of their crops; or using quality seeds instead of parts of their unsold past crops, have a higher priority than implementing sophisticated biotechnology laboratories.

One obstacle that countries like Peru present to suppliers of biological products is the systematic violation of regulations related to the issues of intellectual properties. Efforts are being made to prevent illegal use and reproduction of software and videotape, but in biological and genetic aspects very little is being done.

Several times the government, probably with good intentions, has undertaken national programmes to donate plants or seeds to alleviate poverty among farmers around the country. In most cases the programmes are sponsored by international institutions. The question is how private suppliers are to compete with competitors who are able to give away their genetic material for free! It may be true that the materials are not as good as the those supplied by the private company, but they are given without charge. Obviously, the private supplier will not survive competition with this unexpected competitor.

Peru does not offer a large market for genetic material. From that perspective, it is not as large a market as Brazil or Argentina. What Peru

can offer is adequate space for foreign investors to participate in the
production of seeds and plants to supply the world market.

   To the benefits of the southern hemisphere's 'counter-season', the
year-round steady season on the coast of Peru should be added. There is
no rain. The water for irrigation comes from rivers that bring it from the
highlands or from wells. There are no hurricanes, droughts, floods or
storms. The valleys on the coast are isolated; natural barriers like the
Pacific Ocean, the Andes and deserts between them facilitate the
phytosanitary aspects of agricultural operations.

   The highlands and particularly the Amazon region have one of the
largest biological reserves on the planet. The wild relatives of all the
commercial varieties of potatoes, tomatoes, groundnuts and other
species are found in Peru. That includes only known species. But how
many unknown genetic materials are in the jungle to be discovered and
developed for the benefit of mankind? Biotechnologists should exploit
these resources for the immense market of food, fibres and drugs.

Fundación Perú

Fundación Perú (Foundation Peru) was officially founded in 1993, with Dr
Richard Sawyer as president and Mr Fernando Cillóniz as the general manager.
Dr Sawyer designed the vision for Foundation Peru, which is today an
institution with a well-defined mission, to assist Peru via economic development
through improved agriculture.

   Foundation Peru is a private nonprofit organization. Its board of directors
represent several important sectors involved in Peruvian agriculture, including:

* growers associations;
* entrepreneurial associations;
* public institutions involved in agriculture;
* banking and commerce associations;
* universities with agricultural programmes.

Foundation Peru has acquired an adequate reputation domestically and
internationally, earning a high level of credibility and uniting people to discuss
important matters. It plays an important role in leading the discussion on
diverse matters affecting Peruvian agriculture, including:

* legislative proposals to stimulate greater participation of qualified institutions
  and private entrepreneurs in Peruvian agriculture – this is an on-going task
  and, as described below, the use and application of biotechnology in Peru will
  depend, largely, on proposed legislation;
* links between national and international institutions specializing in various
  disciplines of agricultural technology;
* information (commercial, financial, technical, legal, statistical, etc.) that
  would interest the various agents involved in Peruvian agriculture; providing

this information has been profitable to Foundation Peru and has provided credibility for the organization; vast amounts of information on Peruvian agriculture can be obtained from Foundation Peru;

• identification of investment opportunities based on strategic advantages in Peru.

Foundation Peru is concerned that all of these efforts provide an adequate balance between production and profitability, economic and social development, and ecological sustainability and conservation of nature.

Foundation Peru and Biotechnology

Anyone who is working in agriculture is working in biotechnology. Many people in Peru may not be aware of this, but it is true. Some of the current activities in agriculture are described below.

In addition to being directly involved in commercial agriculture in products as diverse as rice, sugarcane, cotton, asparagus, grapes, forages, milk and others, Foundation Peru manages an insectary where insects are raised for biological pest control. It also operates, in association with two of the most important universities of Peru, La Molina and San Agustin, a network of soil, water and plant laboratories to serve Peruvian farmers.

In experimental stations along the coast, consisting of seven stations totalling almost 900 ha, Foundation Peru has seed nurseries of rice and cotton. Foundation Peru also has an evaluation plot of more than 20 varieties of asparagus as part of an international programme involving numerous countries. Additionally, Foundation Peru operates a tissue culture laboratory where grapes and strawberries are multiplied.

A view of biotechnology in Peru

The perception of biotechnology in Peru depends on who one asks. There are public officials, researchers and scientists who have conflicting views on biotechnology. The 'general public' may have yet another view.

In Peru, biotechnology may be viewed by the sceptic as something of a fashion to attract government funding for laboratories, despite the lack of trained personnel to operate and maintain them. Money may be spent with little to show for the effort.

Before biotechnology can be used appropriately, basic needs must be met, such as grouping small farmers together to achieve economically feasible scales, avoiding losses of 30–40% of harvests by simply packing and handling produce more effectively, and/or using certified quality seeds instead of planting leftovers of previous harvests. Agriculture in Peru has suffered 30 years of abandonment. One of the worst laws that could have been applied anywhere, the Agrarian Reform Law, destroyed what at the time was considered to be an advanced and competitive agriculture worldwide. Today, Peruvian agriculture needs considerable redevelopment and these tasks are more urgent than implementing a modern biotechnology laboratory. Basically, Peru needs to copy the technologies available in other parts of the world and apply it to its needs.

Specific actions might include the importation of high quality seeds and plants, contracting qualified international consultants, linking Peruvian entities to foreign universities and institutions specialized in the various agricultural disciplines. While there are needs which biotechnology may address in Peru, the technologies that could be employed immediately are at a low level of sophistication.

Intellectual Property Rights

One of the principal obstacles in Peru for the suppliers of biological technology is the systematic violation of the international norms of intellectual property. Although Peru has advanced in the control of illegal uses of software, cassettes and videos, compelled by international commitments in relation to these matters, it is evident that in relation to intellectual property rights (IPR) on biological products there is much to be done. The US is exerting adequate pressure at governmental levels, so that users of biotechnology establish the internal mechanisms of control and sanction violators of intellectual property laws.

Unfair Competition

In Peru, companies or individuals conducting business in agricultural services and supplies sometimes confront unexpected obstacles. The Peruvian government, often with the best intentions, has undertaken massive campaigns of donation of tractors, tools, seeds, plants and even money to alleviate poverty among the 'campesinos' - the indigenous population. In the majority of these cases, the goods and services supplied come as donations from friendly countries, among which the US is the most important donor.

The question is what impact these sporadic campaigns have on the commercial suppliers of those goods and services. A commercial supplier faces bankruptcy when a competitor can afford to give away products for free. Even though the machinery, seeds or plants supplied by the government may be of inferior quality, they will be adopted by the indigenous population because they require no investment. The Peruvian government does not act alone in this peculiar manner. At times the international donor agencies also generate this type of economic disequilibrium and develop unsustainable programmes.

Opportunities

It is evident that Peru does not currently constitute a great market for supplying large quantities of seeds, plants or genetic material in general. Peru is not like Brazil or Argentina, countries that could acquire large quantities of these biological products from exporting countries. Peru is also not like China or India, who will demand increased amounts of certified seeds and improved plants.

What Peru does have is an advantageous position in the production of seeds and plants, not only to supply local markets, which are small, but to supply the world market as well. International seed companies should consider Peru an appropriate site to produce their seeds. At Foundation Peru, the land for such ventures is available.

Peru has an off-season with respect to the northern hemisphere, in addition to a coastal location. Due to the combined effects of the Humboldt cold water current and the Andean highlands that run parallel to the coast, there is very little rain. Water for irrigation comes from rivers that originate in lakes and reservoirs located in the Andes. Additionally, in most of the coastal valleys there is underground water that can be pumped from wells. The coastal areas of Peru do not experience storms like the ones that repeatedly destroy vast areas in other parts of the world, nor does Peru have hurricanes, frosts, floods or droughts.

Changes in temperature are minimal between seasons. For example in Ica, an excellent place to establish seed nurseries of practically any crop, the maximum temperatures vary between 22 and 32°C between winter and summer, and the minimums vary between 11 and 16°C. Compared with the US, Peru has almost no changes between seasons. Finally, valleys on the coast are isolated by sterile deserts, thus facilitating the application of any phytosanitary programmes to be implemented. The Peruvian coast has been compared to a set of immense natural greenhouses.

Biodiversity

While the coast has characteristics that facilitate the production of seeds and plants to service the international market, the Sierra and Tropical forests, especially the latter, are major biological reserves. The wild relatives of most of the commercial varieties of potatoes, tomatoes and groundnuts are found in Peru. While these are well-known species, how much more valuable genetic material is located in Peruvian tropical forests awaiting development? This genetic reserve has stimulated great interest in universities in the US. There is an unexploited opportunity for biotechnologists to isolate, identify and propagate those species that may have commercial value and use in the international food, fibre and pharmaceutical marketplace.

Conclusion

In Peru, biotechnology should be viewed as a useful scientific tool to be used to address important commercial agricultural constraints and meet the demands of a global economy.

Developing and Accessing Agricultural Biotechnologies: International, US and Developing Country Issues, Perspectives and Experiences

Transferring Agricultural Biotechnology: US Public/Private Sector Perspectives

Frederic H. Erbisch

Office of Intellectual Property, 238 Hannah Administration Building, Michigan State University, East Lansing, MI 48824–1046, USA

The handling of agricultural biotechnology intellectual properties by the private and the public sectors is quite different. The private sector looks to new and developing biotechnologies to enhance its position in the field, to move ahead of its competitors and to increase its profit margin. All efforts of the private sector company employees are directed toward these three points. Funding for research and development is primarily from private sector funds under the control of the company and there are no obligations from outside the company on use of the funds or the distribution of biotechnology intellectual property.

Handling of agricultural biotechnologies developed in the public sector, in particular in universities, is very different from that of the private sector. In universities the discovery and development of agricultural biotechnologies is secondary in nature, with education of undergraduate and graduate students being the primary goal. In addition, much of the universities' research is supported by state and federal government funds which may have specific requirements for distribution of technologies developed under their support. For example, the traditional release of biotechnologies from Land Grant universities has been to provide new technologies, including new crop varieties, to the public at no cost. However, this is rapidly changing and free distribution is done infrequently now as universities look more towards the commercial development of their new technologies.

Methods of protection and distribution of new public-supported and public-developed biotechnologies are reviewed in detail in this chapter along with examples of successful transfers of biotechnologies and some of the problems encountered.

In the United States both industry and universities are independently developing new agricultural products through traditional means and biotechnology.

The approach used by each is quite different because of the difference in basic operating philosophy. Industry's motive is profit – the company must make a sufficient profit when selling products in order to continue its operation. The university's motive is education, and research is part of that education. Research support is obtained from numerous sources and is not dependent upon sales of products.

Industry

Industry continues to develop new agricultural products to increase its share of the marketplace. Generally, a company will use its funds to support a development project, use its own research staff and work towards rapid completion of the project so the product can be marketed. If at any time during the development of the new product it appears that the product will not bring a profit, the research/development activities can be brought to a stop and money and researchers are reassigned to a different project. During the development of the new product, the company is studying the market-place and looking for the best means to introduce the new product, so that when the new product is ready for the market the company is fully prepared to take the product forward. The company may market the new product itself, through distributors or sublicense to another company who will market the product and pay a royalty to the developer. If, after introduction, the product fails to make a profit the product is pulled from the market and abandoned.

The scenario just described is typical of industry and applies to all products including agricultural products, whether developed traditionally or through biotechnology. Of course, there are exceptions to this, but they are few because companies cannot operate without a profit.

Universities

Universities operate quite differently and the remainder of this chapter will describe, in general, the steps universities take to develop and market new agricultural products, particularly new seed varieties, followed by a more specific description of how Michigan State University (MSU) handles marketing a new seed variety.

In a university, academic researchers conduct research and are usually assisted by one or more graduate students and an occasional undergraduate. The research conducted by both the faculty member and the graduate student is done to learn basic science. Seldom is university research done for a commercial result; potentially commercially viable products are often a by-product of the research. Often when a researcher does find something which might have a commercial potential, the researcher will choose to learn more about the science of the finding rather than take the idea forward for commercial development.

This research is important to the faculty member because he/she will be able to transfer the most recent developments in a field to the classroom, preparing the student for the future. The undergraduate student usually does not participate in the research laboratory, but the graduate student does actively participate in research. The graduate student, under the direction of a faculty member, learns the research process by conducting library research, designing a research project and then doing the research. During the research period the graduate student will also be taking courses and often will assist with teaching. The actual research project may take several years as the student, working part-time on the research project, learns how to use equipment, repeats tests, perhaps redesigns the project, etc. The faculty member will be heavily involved in the student's research project, partially as an instructor and partially because the student is working on some phase of the faculty member's research.

Research is expensive, especially the acquisition of state of the art equipment. Faculty members are continually looking for research support from a variety of potential sponsors. One of the larger supporters in the US is the Federal Government. Federal Government funds have certain requirements regarding inventions and other intellectual properties developed with this funding. In particular, the university cannot have the exclusive right to the newly developed invention or development – the Government has a non-exclusive, royalty-free, irrevocable licence to use the invention or development for Government purposes.

Sometimes the research supporter is a commodity group, such as a soybean or maize growers group. This group also expects to have certain rights to new inventions or developments in return for its research support. These expectations may cause additional problems for the transfer of new inventions or developments to the commercial arena.

Universities are not-for-profit organizations; they cannot directly market their products, so universities either provide the new products at no cost to the public or license the product to a firm that can sell to the public. The manner in which MSU markets its products will be described in greater detail in a following section.

Industry versus University – a Summary

The approach to developing new technologies or biotechnologies is quite different for industry and universities. Industry is focused on developing marketable products and universities are focused on education. In Table 19.1 a comparison of how these two parties approach research and development (R&D) is provided.

Table 19.1. Comparison of Industry–University development and marketing of a new product.

	Industry	University
Source of funds	Company	Various – Federal Government, state government, commodity groups, industry
Funding requirements	None	Various – each funding party has special requirements
Focus of research	Development of commercially viable product	Education of students, learn more about basic science
Stage of researcher skill	Experienced	Experienced to novice
Rate of project development	As quickly as possible, full time	May take several years, not full time
Recognition of invention or new development	Concentrated effort and continued searching	Unexpected, found by 'accident'
Dedication to development	Only as long as believed profitable	As long as basic science is to be learned
Marketing	Direct	Cannot market directly

Michigan State University – a Case Study

MSU is a Land Grant University which was established in the mid-1800s to assist the farm community in the state of Michigan. The University has provided assistance through a variety of means including crop improvement and extension service. Historically, all of these services were provided at no cost to the farmer. Over time, the farmer came to expect that any service provided by the University was free.

While the University was established to assist the farm community it also developed its educational programme. Today, the University has an enrollment of over 40,000 students and a research budget exceeding US $180 million annually. Most of the more than 2000 faculty at MSU are actively involved in research. This research is important to the faculty member because he/she will be able to transfer the most recent developments in a field to the classroom, preparing the student for the future.

In the Department of Crop and Soil Sciences (CSS) at MSU, a number of researchers are considered crop breeders. These researchers develop new varieties of crop plants, some of which are of economic value. The breeder/ researchers also train students to become plant breeders and require funds to support the training of these students. Marketing of these new varieties can provide supplemental funds for support of students and the training programme. The marketing effort is not what the seed grower or farmer wants. They believe the new variety should be provided at no cost as it had been historically. However, the cost of education and programme maintenance cannot be met by current funding programmes, making more aggressive marketing of new varieties necessary.

Breeders for each crop have a support 'team' of farmers and others who review the progress of new varieties over a period of years. This group, along with the breeder, may recommend that a new variety be considered for commercialization to the CSS Department's Small Grains Commodity Committee (SGCC). The SGCC carefully reviews the new variety characteristics presented by the breeder. If the SGCC decides the variety is ready for distribution it will determine if the release is to be 'public' or 'limited access'. A public release will be handled by the CSS Department through the Michigan Crop Improvement Association (MCIA), an organization outside the University. The MCIA will sell the seed through certified breeders and MCIA or the breeders will pay the CSS Department a research fee for each unit of seed sold. If the CSS Department decides to provide plant variety protection, it will pay all costs.

Before the variety is released from the University, the CSS Department will obtain additional approvals. The Department Chair will inform the Director of the Michigan Agricultural Experiment Station (MAES) (Appendix 19.1)[1] of the desire to release a new crop variety. The MAES Director will either concur with the recommendation or ask for additional information. If the MAES Director concurs, a document will be drafted which requests approval from other administrative officers of the University (Appendix 19.2), in particular, the Vice President for Research and Graduate Studies and the Vice President for Finance and Operations. The Office of Intellectual Property (OIP) obtains these signatures for the MAES Director. Upon receipt of all the signatures, the CSS Department will, for a public release, release the variety to the MCIA for commercialization. If the release is to be a limited access release, OIP will handle this matter for the University and the CSS Department.

Upon receipt of the variety, OIP will prepare an announcement on the availability of the variety. The announcement will briefly describe the variety and the means of obtaining the right to be the distributor of the variety (Appendix 19.3). A proposal is usually requested from a potential licensee of the new variety. If additional information is needed to prepare a proposal, OIP supplies specific information on the variety as prepared by the breeder (Appendix 19.4). Proposals must be received by OIP by a certain date to be considered. A committee, consisting of the Directors of OIP and the MAES, the CSS Department Chair and the breeder, reviews the proposals to select the top two or three. The companies which prepared the selected proposals are invited to meet with the committee. Shortly after meeting with the companies, a

---

[1] Note: permission was granted for the use of any appendix not written by OIP.

decision is made to award the variety to one of the companies. Letters are prepared by OIP – letters of regret to those not selected and a letter of congratulations to the 'winner'. The congratulatory letter (Appendix 19.5) also alerts the company to expect a draft licence agreement and to review it carefully.

OIP drafts a licence agreement which meets the requirements of the MAES Director, the CSS Department Chair and the breeder. Usually OIP will work from a 'standard' licence agreement (Appendix 19.6) and modify it for a particular variety and company. The OIP Director will negotiate the terms of the licence agreement with the company in order to obtain terms which are favourable to both MSU and the company.

All fees and/or royalties required by the licence agreement are to be paid to MSU. After MSU recovers its costs, usually consisting of the fee required for obtaining plant variety protection, remaining fees and/or royalties will be distributed according to MSU policy for royalty distribution. The breeder, Department and University will share the fees with each receiving approximately one-third of the fees.

As can be seen through this particular case, the distribution of university-developed varieties is a long, somewhat complicated process which generally does not result in a profit. Industry could not operate under such a process and remain profitable. Both universities and companies play important, but different roles in the development of agriculture biotechnologies. It is important to remember that the basic differences between universities and companies are inherent in their missions: education for universities and profit for companies.

Appendix 19.1. Memorandum from the Chairman of the Department of Crop and Soil Science to the Director of the Michigan Agricultural Experimental Station requesting permission for the release of a new crop variety.

MICHIGAN STATE UNIVERSITY

DEPARTMENT OF CROP AND SOIL SCIENCES	EAST LANSING Ï MICHIGAN Ï 48824–1325
PLANT AND SOIL SCIENCES BUILDING	Main Office (517) 355–0271
Fax (517) 353–5174	

February 2, 1996

<u>MEMORANDUM</u>

TO: Director, Michigan Agricultural Experiment Station

FROM: Chairman

RE: Release of the Wheat Variety 'Bavaria.'

Based on the recommendation of the CSS Small Grains Commodity Committee and the CSS Variety Review Committee as indicated by the attached copy of the signed variety release form, I recommend the release of the Michigan Agricultural Experiment Station (MAES) soft white winter wheat breeding line D0256 as the variety, 'Bavaria.'

Bavaria was developed by Drs. _____ and _____ and Associates and brought to release by Dr. _____. A complete description and performance record through 1995 is provided in the attached document by _____.

It is also recommended by Dr. _____, the CSS Small Grains Committee and the Commodity Policy and Review Committee that a public announcement be released through the MSU Office of Intellectual Property soliciting proposals for the release mechanism of, and/or the rights to, Bavaria. This would give individuals and agencies throughout Michigan, including MFSA and MCIA, opportunity to propose either a nonexclusive release or an exclusive (limited-access) release, in addition to proposals for such North American production and marketing rights. It is suggested that proposals be reviewed and evaluated by a committee selected by the Directors of the Office of Intellectual Property and the MSU AES, along with the Chair of the Department of Crop and Soil Sciences. I concur with this recommendation for the following reasons:

> The certified seed industry is in a period of transition and change. The competitiveness of the certified seed market has been eroded in Michigan as well as nationally, as privately developed varieties continue to cut into the traditional market for certified seed of publically developed varieties. Furthermore, seed of private varieties (usually uncertified) commands higher market prices because of

Appendix 19.1. *Continued*

professional sales programs of the private seed industry. Thus, it is becoming increasingly clear that marketing practices for certified seed must become more aggressive, more professional, and more controlled if certified seed of publically developed varieties is to be sustained as a viable enterprise in the future. The decreasing market share for certified seed in Michigan and around the country is making it increasingly uncertain whether traditional nonexclusive release policies and marketing practices for certified seed can remain viable in today's competitive environment.

We believe Bavaria should have an opportunity to be released in the way that would permit it to compete with seed of private varieties on the market in its performance category. Conversely, we are sensitive to our traditional certified seed producer clientele, including the Michigan Foundation and the Michigan Crop Improvement Association, as well as their individual grower members.

The call for proposals recommended herein would give all of our clientele an opportunity to propose the kind of release mechanism and possible limited-access (exclusive) release method they believe will result in maximum market penetration over the area of adaptation in the most expeditious manner.

cc:

Recommendation to Release the Soft White Wheat Variety

BAVARIA

February 2, 1996

The Department of Crop and Soil Sciences (CSS) Small Grains Commodity Committee recommends that the soft white wheat variety named Bavaria, tested as D0256 be released.

Background and Pedigree

Bavaria, a soft white winter wheat, was tested as D0256. Bavaria is an F4-derived pure line that originates from the cross number 861823. The parentage of that cross is C0250/B7101// Pioneer Experimental W9021R. C0250 and B7101 are both breeding lines developed at MSU. The parentage of C0250 is E4870//2*ASN/3*GE. The parentage of B7101 is P5517-A1//S92B/5*GE.

Description and Performance Record

A complete pedigree, description and performance record through 1995 is provided in the attached document by _____.

Chair, CSS Small Grains Commodity Committee	Date of Approval
Chair, CSS Commodity Policy and Review Committee	Date of Approval
Chair, Department of Crop and Soil Sciences	Date of Approval
Director, Michigan Agricultural Experiment Sta.	Date of Approval
Vice President for Research and Graduate Studies	Date of Approval
Vice President for Finance and Operations	Date of Approval

DESCRIPTION OF BAVARIA WHEAT

Origin, Background, and Performance of 'Bavaria' Soft White Winter Wheat
Origin

'Bavaria' was developed at Michigan State University. F4:5 head rows derived from cross number 861823 (C0250/B7101// Pioneer Experimental W9021R) were planted in the fall of 1988. F4:6 drill plots were generated by planting families derived from bulk harvesting of selected F4:5 head rows. One of those drill plots was given the stock ID of D0256. D0256 was planted in preliminary yield trials in 1990. A seed purification and increase system independent of yield trials was employed from 1991 until the fall of 1995 when seed was transferred to Michigan Foundation Seed for the production of pre-breeders seed. The seed that was transferred represented an F4:11 pure line.

Performance

Performance data for three years of multi-location testing of Bavaria are tabulated in the accompanying table. In yield tests, Bavaria has exhibited yield and test weight performance comparable or better than both Chelsea and Lowell. Bavaria has excellent resistance to prevailing races of leaf rust, powdery mildew, and wheat spindle streak virus. Bavaria's lodging resistance is moderately good, and it is similar to Chelsea in height. Bavaria has white seeds that are generally several mg smaller than Chelsea's or Lowell's. Bavaria flowers approximately either slightly before or at the same time as Chelsea.

Appendix 19.1. *Continued*

Bavaria (D0256) Performance Data 1992–1995

Year	Trial ID	Locations		Yield BU/Acre	Test weight	Height (inches)	Anthesis D.O.Y.	Milling	Baking	Protein %	Soft. eq.	Flour yield	AWRC	Powder mildew	Leaf rust	WSSV	Lodging
											Quality scores						
1995	95-MP010	7	Bavaria	75.5	57.9	41	161							1.7	1.2	0.9	5.3
			Chelsea	73.1	57.3	39	161							1.6	3.1	1.1	6.1
			Lowell	72.6	55.2	40	159							1.4	5.7	1.1	6
			Trial Mean	71.2	57.7	38	59.3							4.1	2.7	5.4	3.5
			LSD (0.05)	5.2	0.9	3.3	n/a							2.1	n/a	n/a	1.6
			Bavaria	84.6	59.6												
			Chelsea	82.1	58.9												
			Lowell	81.3	57.4												
			Trial Mean	79.1	58.9												
			LSD (0.05)	7	0.8												
1994	94-MP010	7	Bavaria	74.2	57.4	34	160	101.5	94.1	10.9	50.3	71.2	56.8	0.7	2	1.7	3.7
			Chelsea	66.6	57.7	34	162	100.2	95.4	9.6	50.3	70.8	56	0.7	4	1.4	4
			Lowell	71.1	54.5	32	158							1	5	1.2	5.3
			Trial Mean	65.8	57.2	33.5	59.4							3.4	4	3.2	2.6
			LSD (0.05)	7.7	0.9	n/a	n/a							2.3	n/a	2.3	1.8
	94-MP010 (Thumb sites)	3	Bavaria	84.1	59.3												
			Chelsea	73.5	59.5												
			Lowell	72.8	55.5												
			Trial Mean	72.3	58.8												
			LSD (0.05)	9.2	1.5												
1993	93-MP002	7	Bavaria	60.6	55.9	40		98.7	94.5	8.3	58.3	71.4	55.7	2	1.8	2	1
			Chelsea	67.8	56.5	39.7		100	100	7.3	58.7	71.7	56	4.5	5.8	1	2
			Lowell	68	54.9	39.3											

Appendix 19.2. Memorandum from the Director of the Michigan Agricultural Experiment Station to the Office of Intellectual Property requesting signatures to allow the Department of Crop and Soils Sciences to release a new seed variety.

February 23, 1996

MEMORANDUM

TO: Director, Office of Intellectual Property

FROM: Director, Michigan Agricultural Experiment Station

SUBJECT: **Release of the Wheat Variety 'Bavaria'**

I am attaching a copy of a recent memorandum from the Chairperson of the Department of Crop and Soils Sciences concerning the recommended release of the Michigan Agricultural Experiment Station (MAES) soft white winter wheat variety 'Bavaria'.

Two copies of the signed recommendation form to release the variety along with a description of 'Bavaria' including information on its performance, are also attached.

As you will note from the Chairman's memorandum, he and others are recommending that a public announcement be released through the MSU Office of Intellectual Property soliciting proposals for the release mechanism of, and/or rights to, 'Bavaria'.

I concur with the recommendation based on the reasons proposed in the Chairman's memorandum. This recommendation is consistent with discussions you have had with the MAES and plant breeders in the past.

I would appreciate your assistance in getting the additional signatures needed for approval of this release. Also, I would appreciate receiving a signed copy of the approved document for the MAES files.

Your assistance in this matter is greatly appreciated.

Appendix 19.3. An announcement on the availability of a new variety of wheat and request for proposals.

PUBLIC ANNOUNCEMENT
MICHIGAN STATE UNIVERSITY
PROPOSALS ARE INVITED FOR THE
PRODUCTION AND MARKETING RIGHTS FOR CERTIFIED SEED OF
BAVARIA (tested as MSU line D0256)

A SOFT WHITE WINTER WHEAT VARIETY

Type Of Release:

Proposals for both exclusive and non-exclusive will be considered. This includes the traditional 'Public' release and alternative, exclusive 'Limited Access' releases to specific organizations.

Rights will be for the production and marketing of certified seed. Certified seed marketing rights will be limited to the United States.

Foundation seed will be produced by the Michigan Foundation Seed Association in any event.

Proposal Structure:

Proposal should have the following sections:
1. Narrative of organization's structure and history
2. Justification for type of release
3. Details on:
     mechanisms for the production of certified seed
     marketing methods
     royalty collection schemes
4. Contact information
5. Appendices with additional relevant information

Criteria Upon Which Proposals Will Be Judged:

MSU engages in plant breeding to ensure the availability of high quality varieties and to facilitate the training of plant breeders. MSU also recognizes the importance of Michigan-based wheat seed production and marketing to the health of the Michigan wheat system. Specific criteria for proposal evaluation are as follows:
1. Evidence of ability to make and sustain maximum penetration into the appropriate market areas.
2. Evidence of ability to produce high quality seed in adequate quantities for the market area.
3. Evidence of ability to generate competitive income streams for MSU's wheat breeding program.
4. Impact upon Michigan's capabilities in seed production/marketing.

Additional Information:

Bavaria will be protected by the U.S. Plant Variety Protection Act (Title V option) and must be sold by variety name only.

It is projected that 500 bushels of Bavaria will be available for planting breeder and prebreeder seed in the fall of 1996.

Proposals will be accepted through **May 31, 1996** at the following address:

> Director
> Office of Intellectual Property
> 238 Hannah Administration Building
> Michigan State University
> East Lansing, Michigan 48824–1046
> Phone: 517 355–2186
> Fax: 517 432–2186

Six (6) copies of the proposal are to be sent or delivered to the address above by 5 p.m. on Friday May 31, 1996.

Questions or requests for additional information (i.e., characteristics of Bavaria) are to be addressed to the Director at the address or phone numbers listed above.

Specific details for the contractual arrangement are subject to negotiation subsequent to identification of a likely partner. Proposals will be reviewed and a likely partner identified by May 31, 1996.

MSU is an affirmative Action/Equal Opportunity Institution

Appendix 19.4. Description of the new variety and crop production data. This information is provided by the breeder.

DESCRIPTION OF BAVARIA WHEAT

Origin, Background, and Performance of 'Bavaria' Soft White Winter Wheat

Origin

'Bavaria' was developed at Michigan State University. F4:5 head rows derived from cross number 861823 (C0250/B7101// Pioneer Experimental W2021R) were planted in the fall of 1988. F4:6 drill plots were generated by planting families derived from bulk harvesting of selected F4:5 head rows. One of those drill plots was given the stock ID of D0256. D0256 was planted in preliminary yield trials in 1990. A seed purification and increase system independent of yield trials was employed from 1991 until the fall of 1995 when seed was transferred to Michigan Foundation Seed for the production of pre-breeders seed. The seed that was transferred represented an F4:11 pure line.

Performance

Performance data for three years of multi-location testing of Bavaria are tabulated in the accompanying table. In yield tests, Bavaria has exhibited yield and test weight performance comparable or better than both Chelsea and Lowell. Bavaria has excellent resistance to prevailing races of leaf rust, powdery mildew, and wheat spindle streak virus. Bavaria's lodging resistance is moderately good, and is similar to Chelsea in height. Bavaria has white seeds that are generally several mg smaller than Chelsea's or Lowell's. Bavaria flowers approximately either slightly before or at the same time as Chelsea.

Appendix 19.5. Form letter used to notify a successful applicant (potential licensee) for a new variety. The awardee is alerted to review the draft licence agreement which is being prepared for the awardee.

Dear :

   I am happy to inform you that _____ has been selected for the Variety license. Presently I am preparing a license agreement and you should be receiving it in the near future. Please review this draft carefully and contact me when you are ready to negotiate/discuss the agreement.

   Your participation in the Variety licensing program is greatly appreciated. The time and effort needed for the development of your proposal is also acknowledged.

   If you have any questions or need additional information please do not hesitate to contact me. Also, if you have comments or suggestions on how to improve MSU's Variety licensing program, I would appreciate receiving them. Building a program which serves you and MSU is our goal. Your input is important in developing and improving the University's and agricultural community's program.

Sincerely,

Director, Office of Intellectual Property

Appendix 19.6. 'Standard' licence agreement used as a template for the drafting of variety licence agreement.

EXCLUSIVE LICENSE AGREEMENT
For
VARIETY

This Agreement entered into this ____ day of _____, 19__ by and between Michigan State University, a not-for-profit corporation organized under the laws of the State of Michigan, hereinafter referred to as 'Licensor,' having its principal office at East Lansing, Michigan 48824 and _____, a for-profit corporation organized under the laws of the State of _____ located at _____, hereinafter referred to as 'Licensee.'

WHEREAS Licensor is the owner of the right to commercially sell the Michigan State University Variety _____, hereinafter referred to as 'Variety' and is preparing a U.S. Plant Variety Protection Application filed covering Variety, or an alternate name chosen by Michigan State University, and has the right to file corresponding foreign applications.

WHEREAS Licensee is desirous of producing and marketing certified seed of Variety; and

---

WHEREAS Licensor is willing to grant a limited term Exclusive License to Licensee to produce and sell certified seed of Variety covered by such Plant Variety Protection Application.

NOW THEREFORE, the parties agree as follows:

(1) Licensor hereby grants a limited term Exclusive License to Licensee. This License provides Licensee with the exclusive United States production and marketing rights for the certified seed class only as defined in the standards of the Association of Official Seed Certifying Agencies in a particular state and not lower than certified seed as defined by the Michigan Department of Agriculture Regulation 623. Licensee is also granted the right to sublicense Variety to others outside the State of Michigan, within the United States.

(2) (a) The term of the Exclusive License shall be limited to six (6) consecutive years of certified seed production and marketing following the date of this agreement, at which time the License shall terminate. Thereafter, the license shall be subject to renegotiation.
   (b) Licensee may terminate the License granted by this Agreement, provided Licensee shall not be in default hereunder, by giving Licensor ninety (90) days notice of its intention to do so. If such notice shall be given, then upon the expiration of such ninety (90) days the termination shall become effective; but such termination shall not operate to relieve LICENSEE from its obligation to pay royalties or to satisfy any other obligations accrued hereunder prior to the date of such termination. Upon termination of this Agreement, all of the licensed rights shall automatically revert to Licensor.
   (c) The Licensee shall have the right to convert this license, after the six year exclusive term, to a non-exclusive license at the same royalty rate as for the exclusive license, without right to sublicense, and minimum royalties shall not be due thereafter.

(3) In consideration for the License:
   (a) Licensee shall pay to Licensor as an initial agreement fee the sum of _____ U.S. Dollars ($.00), _____ U.S. Dollars ($.00) to be paid upon the signing of this Agreement, and the remainder due within one year of signing,
   (b) the Licensee and it's sublicenses shall pay an earned royalty of _____ ($.00) per pound of Licensee's sales of Variety,
   (c) Licensee shall pay to Licensor an annual research fee equivalent to _____ ($.00) per 60 pound bushel of foundation seed utilized by Licensee and it's sublicensees.
   In the event that the total earned royalties received by Licensor are less than _____ U.S. Dollars ($.00) for the annual period ending December __, 19__; or _____ U.S. Dollars ($.00) for each period thereafter until the termination of this License Licensee shall make further payments to make up the

Appendix 19.6. *Continued*

difference in each of the respective years. These minimums do not apply if Michigan State University (MSU)/Michigan Crop Improvement Association (MCIA) cannot deliver clean, disease-free and pure foundation seed to Licensee in the quantity sufficient to produce certified seed to cover fees.

(4) Licensee shall keep suitable records of operation hereunder and to furnish to Licensor, on the thirty-first day of December, a report giving the pounds of Variety seed sold by Licensee during the current year, and to accompany such report with a payment of the royalties accruing hereunder. Variety seed shall be regarded as sold when delivered or when paid for, whichever occurs first. Royalty checks shall be made payable to 'Michigan State University' and sent to the Office of Intellectual Property, 238 Administration Building, Michigan State University, East Lansing, Michigan 48824–1046. To enable Licensor to verify the accuracy of such reports, Licensee agrees to permit Licensor's accountant to inspect its pertinent records during reasonable business hours. Licensor shall also have the right during normal business hours to inspect the premises and fields where seed of Variety is grown or where it is stored for purposes of verifying the royalties due and the certification of such seed.

(5) Licensor will pay the costs of obtaining a Certificate of Plant Variety Protection in the United States.

(6) (a) Licensee shall use its best efforts to bring Variety to market through a thorough, vigorous and diligent program and to continue active, diligent marketing efforts throughout the life of this Agreement.
   (b) In addition, Licensee shall adhere to the following milestones:
      (i) Licensee shall deliver to Licensor on or before _____, 19__ a business plan acceptable to Licensor for the aggressive marketing of Variety which includes number and kind of personnel involved, time budgeted, examples of advertising, potential advertising sites, means of ensuring uniform pricing, lines of authority over production and advertising decisions and other items as appropriate for the marketing of Variety.
      (ii) Licensee shall provide the name(s) and addresses of Licensee's legal and technical representative(s).
   (c) Licensee's failure to perform in accordance with either Paragraph 6 (a) or (b) above shall be grounds for Licensor to terminate this Agreement.

(7) Licensor believes that Variety seed is ready for planting and sale of the seed; however it makes no representation in this regard. Licensor shall not have any responsibility to Licensee in the event the variety seed does not grow or perform satisfactorily.

(8) Licensee shall maintain the quality of such seed sold under this Agreement as required by the minimum standards for the certified seed class in the state where the seed is produced. Licensee will sell the class

of certified seed only. Licensor reserves the right to inspect the quality of the seed sold under this Agreement to ensure that the quality is as required. The name of Licensor shall not be used in any advertising without the express permission of Licensor. Licensee agrees to hold Licensor harmless from any liability or expense in connection with its marketing of seed.

(9) Foundation seed of Variety will be supplied to Licensee by the MCIA at a cost consistent with the duly established Foundation seed policy of this Association and by procedures mutually agreeable to Licensor, MCIA and Licensee.

(10) In the event that plant variety protection rights to Variety are infringed by a third party, Licensee shall have the first opportunity to sue for infringement and to recover and retain any and all damages after paying Licensor its royalties based upon the infringer's sales. In the event that Licensee does not desire to sue for infringement, the Licensor thereafter will have the right to sue for infringement and retain all damages recovered therefrom. The party bringing the suit shall be responsible for all of the costs of the suit unless there is a further agreement in this regard.

(11) This License shall not be assignable by Licensee without the prior written consent of Licensor, which consent shall not be unreasonably withheld.

(12) Licensor may, at its option, terminate this Agreement by written notice to Licensee, if Licensee shall default in:
   (a) The payment of any royalties required to be paid by Licensee to Licensor hereunder or in the making of any reports required hereunder and such default shall continue for a period of thirty (30) days after Licensor shall have given to Licensee a written notice of such default; or
   (b) The performance of any other agreement contained in this Agreement on the part of Licensee to be performed and such default shall continue for a period of thirty (30) days after Licensor shall have given the Licensee written notice of such default;
   (c) An adjudication that Licensee is bankrupt or insolvent;
   (d) The filing by Licensee of a Petition of Bankruptcy, or a petition seeking reorganization, readjustment or rearrangement of its business or affairs under any law or governmental regulation relating to bankruptcy or insolvency;
   (e) The appointment of a receiver of the business or for all or substantially all of the property of Licensee;
   (f) The making by Licensee of an assignment or an attempted assignment for the benefit of its creditors; or
   (g) The institution by Licensee of any proceedings for the liquidation or winding up of its business or affairs.

(13) Termination of the Agreement shall not in any way operate to impair or destroy any of Licensee's or Licensor's right or remedies either at law or

Appendix 19.6. *Continued*

in equity, or to relieve Licensee of any of its obligations to pay royalties or to comply with any other of the agreements hereunder, accrued prior to the effective date of termination.

(14) Failure or delay by Licensor to exercise its right of termination hereunder by reason of any default by Licensee in carrying out any obligation imposed upon it by this Agreement shall not operate to prejudice Licensor's right of termination for any other subsequent default by Licensee.

(15) Unless sooner terminated in accordance with its terms, this Agreement shall continue through six (6) consecutive production and marketing years; provided, however, that if the Plant Variety Certificate is held to be invalid by a court having jurisdiction in a decision which has become final, unappealable and unreviewable, the Licensee's obligation to future pay royalties shall cease.

(16) Licensee shall provide Licensor with detailed information on the performance of Variety, particularly and information which might reflect adversely on the variety. Licensor shall provide Licensee with the detailed information of Variety which it obtains. Any deliberate failure to report material information in regard to the performance of Variety to Licensor shall be regarded as grounds for the cancellation of this License. It is anticipated that the Licensor will publish the information regarding Variety or at least a summary thereof.

(17) All notices, reports, requests, demands or business communications be given by either party to the other under the provisions of this Agreement shall be forwarded to the respective parties as follows:

LICENSOR: MICHIGAN STATE UNIVERSITY

LICENSEE:

(18) Licensor retains the right to freely distribute seed of Variety for research purposes. Seed for testing purposes may be distributed by either Licensor or by Licensee with mutual consent of both parties.

(19) This Agreement is executed and delivered in Michigan and shall be construed in accordance with the laws of Michigan.

(20) This License shall be effective as of the date first appearing.

MICHIGAN STATE UNIVERSITY	LICENSEE
By: _____	By: _____
Title: _____	Title: _____
Date: _____	Date: _____

International Intellectual Property and Genetic Resource Issues Affecting Agricultural Biotechnology

John H. Barton

*Stanford University, Crown Quadrangle, Stanford, CA
94305–8610, USA*

This chapter will review the emerging proprietary and intellectual property protection regime shaping international trade in advanced agricultural materials and products. One group of issues involves access to germplasm and technology. This involves important questions of plant variety protection, of rights under the Convention on Biological Diversity, and of international licensing and technology transfer. A second group of issues involves access to markets. Even when varieties are available or developed locally, there is a strong possibility that intellectual property provisions, including both plant variety protection and regular patent coverage of biotechnological innovation, may be used by competitors to protect their position in market nations. The chapter will consider appropriate responses, both in the planning of research programmes and in the design of intellectual property strategies.

Introduction

This chapter reviews the emerging proprietary and intellectual property protection regime shaping international trade in advanced agricultural materials and products. The first part considers questions of access to germplasm and technology, including issues of rights under the Convention on Biological Diversity (CBD), of plant variety protection (PVP) and other intellectual property rights (IPR), and of international licensing and technology transfer. The second part explores access to markets. It notes the ways in which IPR, including both PVP and regular patent rights, may be used by competitors to protect their position in market nations and it considers appropriate responses, both in the planning of research programmes and in the design of intellectual property strategies.

Issues of Access to Technology

There are many rights that may be held in breeding materials and products; it is crucial for the breeder to be aware of all of these. They include source nation rights, rights under the PVP statutes, rights under the regular patent laws and rights under trade secrecy laws.

Genetic resources

During the 1970s, there arose concerns in developing countries and in the non-governmental organization (NGO) community that it was unfair for source material contributed by developing countries to be transferred freely, while breeding activities contributed by the developed countries were being rewarded with IPR. These concerns became one of the major factors shaping the United Nations CBD, signed at Rio de Janeiro in 1992. The CBD includes carefully nego-tiated provisions governing genetic resources, as part of a much broader pack-age oriented towards conservation of biological diversity in its natural habitats and in collections. In particular, Article 19 of the CBD affirms the sovereign rights of countries over their genetic materials, but the CBD leaves it clear that those genetic materials that were earlier transferred out of their country of ori-gin have entered the public domain and can be used freely for any purpose. There is pressure to change this arrangement.

The clear implication is that, in general, it is essential for any breeder to enter negotiations for a 'material transfer agreement' (MTA) before undertaking collection efforts in any country. The MTA will be agreed between the collector and appropriate national authorities, and will govern the arrangements under which the material is transferred. These may include an allocation of profits or a provision that the material cannot be used commercially without a further agree-ment allocating profits. There may also be provisions that restrict the acquisition of IPRs on the material or prohibit transfer of the material without building a chain of responsibility. Not all countries have yet adopted the legislation needed to enforce this right that they hold under the CBD. Moreover, some countries, looking to the costs of preparing and implementing these agreements and looking to the benefits of free exchange of genetic material, may choose not to require restrictive MTAs. But the current trend is toward restriction of free flow, and it will sometimes be necessary for a breeder to work with source countries in order to ensure good title to the material used in a breeding programme.

Plant variety protection

Most developed countries enacted PVP legislation in the period since the Second World War, and the European Union (EU) created a Europe-wide PVP system in 1994. A number of developing countries are now adopting parallel legislation in order to comply with the requirements of the Uruguay Round of the General Agreement on Tariffs and Trade (GATT) to protect plant varieties. Such legislation, also known as 'plant breeders' rights', provides an intellectual property system adapted to the needs of traditional plant breeders and is designed to give these breeders an increased incentive to develop new varieties while respecting the traditions of breeding.

These laws typically grant protection to varieties that are novel, distinct, uniform and stable. Novelty requires that the variety had not been sold previously, although there is typically a grace period of one to several years, depending on the country and the species. Distinctness requires that the variety be clearly distinguishable from previous varieties – this is not as severe an inventive step requirement as is typical of patent law. Uniformity and stability require that the plant be uniform and that it breed true to type, but is typically defined in such a way as to allow for the protection of hybrids.

The protection is by means of a certificate granted, most typically, by an office of the Ministry of Agriculture (rather than by part of the national patent office), upon receipt of a relatively simple and inexpensive application. The certificate entitles its holder to be the exclusive marketer of the relevant variety, and also of the product of the variety. This right may, of course, be licensed to others. The certificate does not, however, prevent others from using the variety in efforts to breed further varieties.

The PVP laws are generally adopted in accordance with an international treaty (UPOV 1978 or 1991), after the French language acronym for the International Union for the Protection of New Plant Varieties. Under the older version of this treaty (1978), nations were required both to allow use of protected materials for breeding of additional new varieties, and to allow farmers to reuse their harvest for seed purposes. Article 15 of the new (1991) version, which is likely to come into force in 1997, permits nations to allow farmers to reuse seed, but does not require nations to grant this permission; most nations, however, are likely to opt to allow such reuse. In addition, Article 14 of this new version adopts a concept of 'essentially derived variety', designed in part because of the rise of biotechnology. A breeder remains free to use a protected variety and to make any change in such a variety, but is subject to the rights of the owner of the initial variety if that change is so small as to leave the new variety 'essentially derived'. Examples listed in Article 14 are varieties made 'by the selection of a natural or induced mutant, or of a somoclonal variant, the selection of a variant individual from plants of the initial variety, backcrossing, or transformation by genetic engineering'. Thus, if a biotechnologist uses an existing protected variety as material within which to insert an agronomically important gene and markets the resulting material, the biotechnologist would infringe the rights of the owner of the protected variety. Normally, this problem would be solved through a licence under which the biotechnologist and the breeder would share the profits available from giving the farmer the benefit of both the high-quality background material and the new gene.

For international trade purposes, a particularly important aspect of the new version of UPOV is strong protection for harvested material from plants. Article 14 of UPOV (1991) gives the certificate holder rights over 'harvested material, including entire plants and parts of plants, obtained through the unauthorized use of propagating material of the protected variety ... unless the breeder has had reasonable opportunity to exercise his right in relation to the said propagating material'. It also permits member nations to make similar arrangements with respect to 'products made directly from harvested material'. Article 14 thus protects the holder of a certificate in a market country against the import of material grown elsewhere, perhaps quite legitimately, as if the certificate holder had not obtained protection in the country of origin. And the certificate holder's rights extend to any part of the plant, whether useful for propagation or not, and, depending on the national law, extend to products from the plant.

The likely implication, in an era of high-technology agriculture in high-value crops, is that breeders will seek to develop consumer recognition of their brands and to use PVP protection in market countries to seek to limit competition and imports, and command and protect a premium price. Licensing rights can be used to protect against imports of materials, even though the materials may have been legally grown in a country where they are not protected.

The regular patent system

After initial hesitation, surmounted in the US by *Diamond v. Chakrabarty*, 447 U.S. 303 (1980), and in Europe by interpretations of the European Patent Convention, e.g. *Propagating Material/CIBA-GEIGY*, Case T 49/83, OJ EPO 1984, 112, patent offices began to issue many different types of regular patents protecting biotechnological methods of breeding and biotechnologically produced plants.

Under regular patent systems, an invention or discovery must be, using US terms, novel, nonobvious, useful and enabled in order to be patentable. 'Novelty' means that the invention has not been anticipated by publication or use in the market. (Unlike most nations, the US allows a one-year grace period between the time of publication and the time at which a patent can be filed.) 'Nonobviousness' or 'inventive step' (as in the European Patent Convention) means that the invention is an actual advance in the state of the art. The US definition is that a patent shall be denied if 'the subject matter as a whole would have been obvious at the time the invention was made to a person having ordinary skill in the art to which said subject matter pertains.' The interpretation of this standard varies from system to system. Likewise, the standard of 'utility' or 'industrial application' varies from country to country, but is intended as one way to distinguish basic scientific advances from patentable inventions. 'Enablement' means that the patent describes a way actually to carry out the invention, typically through a description in the patent.

Sometimes enablement may also require deposit of actual genetic material, e.g. a cell line, when this line cannot be reliably produced on the basis of a written description. Such deposit can be made at any of a number of institutions; there is an international treaty allowing each country to recognize deposits in other countries.

The precise character of plant-oriented regular patents depends on the details of the national system and on the limitations it places on patents in the biological area. The possibilities in the US are among the most broad. Here, it is possible to obtain a patent on a gene and its application in a plant, on basic processes and inventions, and, in a number of ways, on a plant itself.

Among the most important forms of patent, in the sense of value to researchers, is the patent covering a gene and transformed plants utilizing the gene. This type of patent can frequently be written with a number of claims covering, for example, an isolated or purified protein, the isolated nucleic acids having a sequence that codes for the protein, plasmids and transformation vectors containing the gene sequence, plants (or seeds for such plants) transformed with such vectors and containing the gene sequence, and the progeny (or seeds) of such plants. The structure of the claims protects the patent holder against use of the gene by another biotechnologist, but leaves anyone free to use and breed with organisms containing the gene naturally. Although some countries exclude 'parts of natural organisms' from patent coverage, and thus arguably exclude natural genes, it appears very likely that such an exclusion would violate the Uruguay Round's Agreement on Trade-related Aspects of Intellectual Property Rights (TRIPS), which excludes only whole organisms from patentability.

Another category of patents involves basic processes and inventions. Here, there are many extremely important patents, covering, for example, transformation processes, specific promoters used in agricultural applications, the use of virus coat proteins to confer resistance, and antisense technology. The variety and scope of these are so broad that it is likely to be very difficult to develop new transgenic plants without infringing one or another of these patents. Many of these patents are likely to be available in almost any legal system, so the coverage in any specific country will depend primarily on where the inventor has chosen to file. The ability of a patent holder to prevent import of plants produced with these technologies will depend on the precise claims of the patent in the importing country and on the scope of the country's rules for excluding products of patented processes.

The final patent considered here is for finished plants. It is possible in some legal systems to obtain a patent claim reaching a plant transformed with a particular gene. It has also been possible to obtain claims covering broad groups of transgenic plants, as exemplified by the Agracetus patents on *all* transgenic cotton and *all* transgenic soybean plants. The breadth of these patents is extremely significant and has been the subject of severe criticism. The underlying legal question is enablement; the claims are supposed to reach as far as the disclosure enables a person of ordinary skill in the art to do the claimed action without 'undue' experimentation. When a person applies for a patent after transforming several strains of a species with several different genes, there is an obvious question as to whether that person has actually enabled transformation of all strains with all genes. Although it is likely that no one knows the answer to this question at the time of patent application, the burden of proof in the US on this issue is on the patent office to show that a claim was not enabled.

In the US, it is possible to obtain not only broad patents but also claims on a specific variety identified by a description or a deposit, e.g. on a specific hybrid maize plant. The obviousness standards for granting these patents can be relatively low, as in *In re Sigco Research*, 36 U.S.P.Q.2d 1380 (Fed. Cir. 1995), which held that it was not obvious to apply conventional plant breeding techniques to obtain true breeding sunflower plants whose oil had an oleic acid level of 'approximately 80% or greater'. Claims for such lines are designed to make it impossible for another to breed with the material – for the claims can reach the use of the material as a parent – and thus provide a way to protect an important line, such as the inbred lines used as parents of a hybrid. This use of the regular patent system, almost certainly correct under *Chakrabarty* – and unlikely to be accepted elsewhere in the world – provides a way to avoid the limitations of the PVP system.

There is controversy over such claims on transformed plants and their progeny. From the business perspective, such claims are essential to obtain effective control of agricultural biotechnology using the gene, and to keep a third party from crossing the inserted gene into a different variety and marketing that variety. With these claims a breeder is much more protected than under the PVP system. Nevertheless, in many countries the legal system prohibits the patentability of living organisms themselves, and such an exclusion is consistent with TRIPS if there is provision for *sui generis* (PVP) protection. Thus, these claims will not always be available. Note, however, that in some countries such as Brazil, it is possible to reach the same result effectively by claims on the process of transformation, which can be enforced against the unpatentable – but infringing – products of the process. Even so, here and in other countries, there may be questions as to the ability to control indirect progeny. And there are separate difficulties in systems like Europe, that prohibit the patenting of plant varieties but not necessarily of plants. Here, the issue is whether a claim on a line of plants containing a particular gene amounts, in fact, to a claim on a variety. According to the 1984 *Propagating Material/CIBA-GEIGY* case mentioned above, propagating material treated with a certain oxime derivative did not amount to a variety for purposes of the European Patent Convention, but the scope of this decision was called into some question by a later case, *Glutamine synthetase inhibitors/PLANT GENETIC SYSTEMS*, T356/93 [1995] EPOR 357.

Trade secrecy

One of the most important forms of intellectual property protection is the trade secret system, a combination of legal principles of contract law and legal principles against misappropriation of another's information. The contractual component recognizes and encourages private enforcement of contracts designed to protect information, e.g. confidentiality agreements between a firm and its employees. The misappropriation component protects the holder of a trade secret against, for example, one who comes into the laboratory and secretly copies laboratory notebooks. To benefit from trade secret protection, a bit of information (which can include genetic material) must have economic value deriving from its secrecy and must be protected by means that are reasonable under the circumstances. The effective term of the protection is as long as the secret is valuable and secret, rather than being limited to a fixed term as with the patent and PVP systems.

This body of law provides a technique for control of inbred lines used as parents of a hybrid. These lines need not be released publicly in order for the hybrid to be marketed. They can be protected through a combination of physical protection of the materials themselves and of contracts with employees and those involved in producing seed. This does not, however, keep a third party from attempting to reconstruct the parental lines from the marketed hybrid.

Firms are therefore attempting to supplement PVP and patent protection by using contractual provisions to prohibit 'reverse engineering' of the material they sell to farmers. When one buys the seeds, the label or the reverse of the sale bill contains a restrictive provision, whose key relevant language is that the material will not be used for breeding. The legal effectiveness of this approach is subject to debate. First, there is a question whether this mechanism of achieving contract agreement is effective and there are cases on both sides of the issue in such contexts as warranty disclaimers on herbicides. Moreover, there has been a tradition in US law that one has a right to 'reverse engineer' products that are commercially marketed, reflecting a sense that maintaining this right permits more rapid scientific advance. Hence, it is possible that, even if they would otherwise be enforceable under contract law principles, these agreements are unenforceable because they are preempted by federal standards on intellectual property protection (or, in other legal systems, by a competition law provision). The leading recent Supreme Court case, *Bonito Boats, Inc. v. Thunder Craft Boats, Inc.*, 489 U.S. 141 (1989), struck down a state statute prohibiting the use of direct moulding processes to copy boat hulls, on the theory that the state statute conflicts with 'strong federal policy favoring free competition in ideas which do not merit patent protection'. Nevertheless, in 1996, a federal judge in the US midwestern area upheld a somewhat parallel agreement governing use of a CD-ROM containing an uncopyrightable database, *ProCD, Inc. v. Zeidenberg*, 86 F.3d 1447 (7th Cir. 1996). Much of the new case's logic could be applied by analogy to the seed labels – but will not necessarily be followed in seed contexts.

International issues

Intellectual property regimes are national rather than international (except in the special cases of European regional regimes) and are territorial. It is the international understanding that a plant variety protection (PVP) certificate or patent confers exclusivity within a country's territory and nowhere else. Thus, one cannot infringe a US patent by one's activities within Canada. This is an extremely important point to remember when learning of a patent issued in the US or Europe; the patent may simply not be relevant to the foreign researcher.

The most important point tempering this territoriality is that countries generally treat it as infringement of the national intellectual property system to import a product that infringes the relevant intellectual property right (IPR) or one that is made by means of a process that would infringe the national right if carried out within the country. The exact scope of this process-oriented protection varies from country to country; in most countries, it is only the direct product of a patented process that cannot be imported. The US seeks to reach further and in its statute governing certain customs actions has no such restriction, while in its statute governing judicial actions, it replaces the limitation of directness by limitations based on whether the product is 'materially changed by subsequent processes', or has become 'a trivial and nonessential component of another product'. For the person seeking to use a patented product or process, this pattern of territoriality and import limitations means that the patent search that needs to be done is that (and only that) of *all* the countries in which the product is produced or marketed.

The various international treaties for plant variety protection, and the Paris Convention (1979) (www.cerebalaw.com/paristm.txt) for regular patents, assume and accept this territoriality. Their primary objective is to create reciprocal rights: each party gives the nationals of other parties the same rights as its own nationals to obtain intellectual property protection in its system and to enforce these rights in its own country, in return for a reciprocal commitment from other parties. Most countries are party to these agreements. The result is that a breeder or inventor can obtain essentially global coverage, but, in order to do so, must file for protection within the systems of each country. Normally, because of the expense, firms choose the market countries that will be most important to them, and file in only those countries. It should be noted that, as a special case, Europe offers a European patent under the European Patent Convention, in which, by one filing, one can obtain what is effectively a bundle of rights, e.g. those of a British patent in the UK, of a French patent in France, etc. And, under an important international agreement, the Patent Cooperation Treaty (PCT) of 19 June, 1970, it is possible to make a single initial filing to begin the protection process in nearly all countries at once. Although it is ultimately necessary to make the more expensive filings in each significant market country, this system can delay the point at which these decisions must be taken, thus permitting the decisions to be taken on a more informed basis.

The Uruguay Round brought a new international standard to intellectual property protection. The treaties described above incorporated some, but relatively weak, minimum standards of protection that had to be satisfied by a member country's intellectual property system. The Uruguay Round's Agreement on TRIPS, Including Trade in Counterfeit Goods establishes more severe standards, particularly in the biological area. Its Article 27 states that 'patents shall be available for any inventions, whether products or processes, in all fields of technology', but permits members to 'exclude from patentability' inventions whose exploitation must be prevented 'to protect *ordre public* or morality, including to protect human, animal or plant life or health or to avoid serious prejudice to the environment.' The Article permits members to exclude from patentability 'plants and animals other than microorganisms, and essentially biological processes for the production of plants or animals other than nonbiological and microbiological processes. However, members shall provide for the protection of plant varieties either by patents or by an effective *sui generis* system or by any combination thereof.' Again, most countries are parties, but the agreement gives countries a number of years (up to 10 years for the least developed countries) to bring their systems into compliance. Countries are now moving to comply with this requirement; if they do not comply, they can be subject to trade sanctions under the World Trade Organization (WTO)'s dispute settlement system. There is a procedure for review of the obligations governing plants and animals 4 years after TRIPS enters into force; thus the review will be in 1999.

Implications

In today's world, the definition of a development and marketing strategy is likely to require consideration of legal questions, for IPRs can be used to achieve competitive advantage and market exclusivity. Patents have enormous implications for the strategy of a breeder seeking to protect new innovations and ensure access to its market.

Sometimes, a breeder may be able to devise a combination of technology and patent coverage that leads to an exclusive ability to fill at least a niche market and to protect a proprietary position within that market. This possibility, and the ability to build consumer recognition for specific varieties, suggest that maintenance of proprietary position may become nearly as important in agricultural biotechnology as in medical biotechnology. From this perspective, the acquisition of a patent on a gene and plants transformed with it offers a very valuable incentive for research and the investment needed in product approval; this is the context in which the patent system is likely to work to benefit innovation.

Nevertheless, the sheer bulk of very broad and very basic patents creates significant barriers to the development and marketing of biotechnology products. The samples described above of patents covering fundamental research and transformation methods and covering broad species ranges show how difficult it is likely to be to operate without risking infringement litigation. Probably some of these patents would be struck down during litigation, but the expense of litigation is itself a significant problem for many firms. Just as there is a concern in ensuring that one has a strong proprietary position, there is thus a concern in ensuring that one will not be sued. It is likely to be necessary to obtain a careful legal opinion examining the possibility of infringing a variety of possible patents before it is safe to invest large sums in developing a new biotechnology-derived variety. Consider, for example, that there are hundreds of US patents mentioning or covering various aspects of *Bacillus thuringiensis* (*Bt*) technology, including patents on a variety of specific toxins, some identified through sequences and some through deposit accession numbers. Needless to say, it is possible that some of the patented sequences are the same as certain of the deposited materials. For this and many other reasons, a firm may not be confident that it has full rights to use a specific technology and may be unable to promise a licensee that it holds valid title to a specific strain. There is substantial ongoing litigation over *Bt* rights (and over other technologies as well); it is likely that firms that have licences will be affected, even though they are not participating in the litigation.

In this context, a smaller, developing world firm or research group has three clear options:

- Avoid the developed-world market: one (which will work for a time) is to avoid the developed-world market and to market in those parts of the world still of little interest to the multinational biotechnology industry. A research group in the developing world is legally restricted in its research only to the extent that developed-world firms have obtained intellectual property protection in the developing countries themselves. Very often, this coverage has not been obtained, or it is unavailable under national laws that have not yet been harmonized to WTO rules. Nevertheless, the developing country firm must still be concerned about two issues: first, it may be developing products for export, and must therefore take into account IPR in market countries (including those over the products of patented processes); and second, it may have to maintain good long-term relationships with important technology suppliers, who may wish it to respect intellectual property concerns beyond those formally applicable.
- Find an ally: a second approach is to ally with one of the major multinational companies, and license technology from it, under arrangements that ensure that the multinational will support it in the event of major litigation. It will have to consider a variety of rights, e.g. over transformation methods, specific promoters, specific genes and specific plant categories, as well as the holders of PVP over the variety that is genetically transformed, in order to be able to legally participate in important agricultural markets. These rights may prove expensive.
- Create bargaining chips: a third approach is to patent an innovation that is important enough that it can be a useful bargaining chip in cross-licence negotiations. In this context – and the point is extremely important for a firm's intellectual property strategy – it may be as important for the firm to obtain patents that others are likely to infringe as to obtain patents to protect its own proprietary position (and patents in developed-country markets are

most important.) A typical commercial response in such a situation is for firms to enter cross-licences, under which each firm permits its cross-licensee to use certain of its own basic technologies in return for the right to use the cross-licensee's basic technology. Patents that everyone in the industry needs provide a basis for entering into a cross-licence structure or for use as a counter threat to protect oneself from suit. Such cross-licences and mergers – the ultimate cross-licences – have become a central feature of the international agriculture biotechnology industry.

From a public policy perspective, developing countries would be wise to enact patent laws with broad experimental use exemptions, to issue relatively narrow patents and to work for antitrust/competition law responses to the concentration in the industry.

Acknowledgements

In preparing this presentation, I have drawn liberally from materials prepared for an Appendix to *Transgenic Plants*, E. Galun and A. Breiman (eds), Imperial College Press, London, and for chapter 2 in Intellectual Property Rights in Agricultural Biotechnology, F.H. Erbisch and K.M. Maredia (eds), CAB International.

References

UPOV (International Union for the Protection of New Varieties of Plant) (1978) International Convention for the Protection of New Varieties of Plants of 2 December 1961, as revised at Geneva on 10 November 1972 and on 23 October 1978.
UPOV (International Union for the Protection of New Varieties of Plant) (1991) International Convention for the Protection of New Varieties of Plants of 2 December 1961, as revised at Geneva on 10 November 1972, on 23 October 1978, and on 19 March 1991.

21

Developing Capacity and Accessing Biotechnology Research and Development (R&D) for Sustainable Agriculture and Industrial Development in Zimbabwe

Joseph Muchabaiwa Gopo

Biotechnology Research Institute, SIRDC, PO Box 6640, Harare, Zimbabwe

Biotechnology is the integration of biological and engineering sciences enabling the use of organisms, cell enzymes and other derivatives, including molecular analogues, in industrial or commercial activities. It may be called bioengineering, genetic engineering or recombinant DNA technology. The process may be classical fermentation or may involve the use of *in vitro* genetic manipulation or other techniques dependent on molecular biology or nucleic acids, enzymes, microorganisms, cells or tissue culture. Biotechnology processes have practical application in areas of agriculture, human health, animal health, fine chemicals, equipment and reagents, household and over-the-counter goods, food and beverages, animal feed, bulk chemicals and fuels, fibres and polymers, waste treatment and pollution control, materials recycling, mineral extraction and recovery, diagnostics and analytical processing, environment, conservation of genetic resources and human genome research. Biotechnology provides the potential for significant advancement in Science and Technology (S&T), leading to greater development in the industrial bases essential for strategic and economic advancement and economic planning.

In Zimbabwe, we need to address the issues of a national policy on biotechnology research and its administration. The government of Zimbabwe has invested in the Scientific and Industrial Research and Development Centre (SIRDC) which has the mission to provide a centre of excellence in research for the provision of S&T leadership in research and industrial sectors. The Biotechnology Research Institute (BRI) must provide scientific and technical know-how and research leadership in

all the areas and industrial sectors which are impacted by biotechnology. BRI must provide leadership in policy formulation, administration of the national policy, setting of national priorities in biotechnology research, supporting sectoral interests and new areas of biotechnologies. It must take a lead in the conservation of genetic resources and provide educational initiatives for human resources development to produce a critical mass of scientists. BRI must also provide leadership in developing legislation on safety in biotechnology, development of intellectual property rights, commercialization of inventions and help create a positive public perception of biotechnology. The institute must also have an atmosphere which promotes international relations and cooperation in biotechnology research and biosafety issues. Above all BRI must provide for future growth in national programmes of biotechnology research and strengthen educational and manpower needs and funding. The government of the Republic of Zimbabwe has demonstrated its commitment to biotechnology research by the establishment of SIRDC and the BRI.

Introduction

Developing countries in Africa, Asia and Latin America need to invest very heavily in biotechnology research and development (R&D) for the development of a science and technology (S&T)-based industrial economy. Developing countries like Zimbabwe need to divert from the current practice of supplying developed countries with raw materials for processing into value-added products and actually process their own raw materials into processed goods. The conversion of raw materials through industrial processing will enable developing countries like Zimbabwe to improve their competitiveness in international commodity markets. Examples from some Southeast Asian countries like Malaysia have amply demonstrated that industrial development has been supported by investment in S&T. As for Zimbabwe and other developing countries in the southern African region, biotechnology R&D, as part of a strong S&T programme will help spearhead industrial and economic development.

Biotechnology is the integration of biological and engineering sciences in order to use organisms, cells, enzymes and other derivatives in industrial or commercial activities. It may also be called bioengineering, genetic engineering or recombinant DNA technology. The processes may be classical fermentation or may involve the use of *in vitro* genetic manipulation or other techniques dependent on molecular biology or nucleic acids, enzymes, microorganisms, cells or tissue culture applied to processes of potential interests. Biotechnology processes have practical application in areas of agriculture, human health, animal health, household and over-the-counter goods, food and beverages, animal feed, bulk chemicals, fuels, waste treatment and pollution control, mineral recycling and extraction, and conservation of genetic resources. Biotechnology offers the potential for significant advancement of S&T, leading to greater development of the industrial base essential for strategic economic advancement.

Zimbabwe has developed the National Policy on Biotechnology Research and its Administration. The Government of Zimbabwe (GOZ) has invested in the Scientific and Industrial Research and Development Center (SIRDC), its mission is to provide a centre of excellence in S&T leadership in research for the industrial sectors. The Biotechnology Research Institute (BRI), one of the institutes within SIRDC, provides S&T know-how and research leadership in all research areas and industrial sectors which are impacted by biotechnology. BRI provides leadership in policy formulation, administration of national policy, setting of national priorities in biotechnology research, supporting sectoral interests in new areas of biotechnology, conservation of genetic resources and educational initiatives for human resources development to produce a critical mass of scientists. BRI also provides leadership in developing legislation on biosafety and development of intellectual property rights, commercialization of inventions and developing positive public perceptions of biotechnology. BRI also promotes international relations and international cooperation in biotechnology research and biosafety issues. Above all, BRI must provide for the future growth of national programmes in biotechnology research, and strengthen educational and manpower needs and funding. The GOZ has demonstrated its commitment to biotechnology research in Zimbabwe by the establishment of SIRDC and BRI.

SIRDC

SIRDC is a national technology centre established by the GOZ whose operating principle is that technology should be targeted to enhance production. This, in turn, requires that the technology transfer agenda be finely tuned to the needs of industry.

Functions of SIRDC

The functions of SIRDC are: to carry out the scientific and industrial R&D required by Zimbabwe industry; adapt foreign technology to suit local needs in agriculture and manufacturing; promote the development and application of clean technologies; and promote excellence, partnerships and total quality assurance.

Administrative structure

The SIRDC Board is responsible for policy determination and control over SIRDC functions. The majority of the board members are from the industrial sector. SIRDC is headed by a director general while each research institute is headed by a director. The technical work is done by engineers and scientists in each institute.

Human resources

The development of human resources is regarded as absolutely crucial to the overall functioning and performance of SIRDC. A programme of continuous professional development for all staff has been instituted. Scientists and engineers with outstanding expertise are to engage in its R&D activities. It is the considered view of SIRDC management that a well-trained workforce will assure product quality and accurate execution of commissioned projects. Management also believes that a highly motivated staff will guarantee efficient use of resources and prompt delivery of results.

Infrastructure

There is a concerted effort to provide SIRDC with adequate infrastructure for all its needs. Laboratories in SIRDC institutes are to be outfitted with state-of-the-art equipment; international links are being established to facilitate accessing information for global technology utilization; and a documentation centre is being set up to serve as a repository for S&T information. The SIRDC library and documentation centre is being set up to be a referral facility for industrial information. Conference and training rooms with adequate facilities are planned.

Targeted groups

The focus of SIRDC service activities will be small- to medium-scale enterprises. The targeted clientele falls into a number of categories, including industrial sector enterprises, public sector enterprises, small-scale farmers in need of biotechnology services, resource-poor entrepreneurs needing technological assistance, universities for collaborative research projects and other research institutions such as the National Agricultural Research Centres in Zimbabwe and the southern African region.

Services offered

The operational units of SIRDC have been equipped to provide a wide range of services to relevant industries. SIRDC will offer R&D services, consultancy services, analytical services and technical advice to both small- to medium-scale enterprises and large-scale industrial enterprises. SIRDC will also have a venture capital company that will assist with technology transfer from the research laboratory to industrial application.

The constituent institutes of SIRDC

There are eight constituent institutes of SIRDC, each with a specific programme of activities. They are:

- Building Technology Institute (BTI)
- Energy Technology Institute (ETI)
- Environment and Remote Sensing Institute (ERSI)
- Electronics Research Institute (ERI)
- National Metrology Institute (NMI)
- Production Engineering Institute (PEI)
- Division of Industrial Management (DIM)
- Biotechnology Research Institute (BRI)

Core programmes of institutes of SIRDC

Building Technology Institute (BTI): Dr M.T. Musarurwa, Director
The core programme areas of BTI are:

- Structural and Geotechnical Engineering
- Housing and Infrastructure Development
- Construction Technologies/Materials
- Design Systems (Architecture, Industry Design, Urban Design)

Energy Technology Institute (ETI): Dr M. Muyambo, Coordinator
The core programme areas of the ETI are:

- Hydro and Thermal Energy
- Biomass–Biogas
- Solar Energy
- Coal Energy
- Power Energy
- Energy Efficiency

Environment and Remote Sensing Institute (ERSI): Dr C. Matarira, Director
The core programme areas of ERSI are:

- Remote Sensing
- Geographic Information Systems
- Natural Resources Management
- Environment Monitoring and Management
- Hydrology/Water Resource Inventory

- Cleaner Production Technologies
- Water Quality Engineering

Electronics Research Institute (ERI): Mr O. Chinyanga, Coordinator
The core programme areas of ERI are:

- Microelectronics
- Information Technology
- Software Engineering
- Telecommunications

Production Engineering Institute (PEI): Mr M.F. Mutyambizi, Coordinator
The core programme areas of PEI are:

- Material Science
- Design Engineering
- Manufacturing Processes

National Metrology Institute (NMI): Dr M.T. Musarurwa, Coordinator
The core programme areas of NMI are:

- Mechanical Department
- Electrical Department
- Heat Department
- Optics Department

Division of Industrial Management (DIM): Mr L. Riyano, Coordinator
The core programme areas of DIM are:

- Project Management
- Industrial Analysis
- Industrial Techno-economic Analysis
- Technology Transfer
- Technology Auditing
- Technology Scanning
- Innovation Management
- Application of Management Science Techniques to Business Decisions

Biotechnology Research Institute (BRI): Dr J.M. Gopo, Director
The core programme areas of BRI are:

- Basic Molecular Biotechnology
- Agricultural Biotechnology (Plant and Animal)
- Medical Biotechnology (Human Health and Animal Health)
- Environmental Biotechnology
- Industrial Biotechnology

SIRDC is poised to play a crucial role in developing and accessing S&T, and transferring accessed and developed technology to Zimbabwe for industrial application. Sustainable S&T policy can be assured through the presence and programmes of the SIRDC.

BRI

The BRI philosophy

Nationally, BRI is viewed as a centre of excellence in the transfer of technology in all aspects of biological, biochemical, biophysical and bioengineering sciences. Technology must be transferred from the laboratory to industrial application because recent advances in molecular biology, molecular genetics and biochemical engineering, using microorganisms, cell tissue culture and monoclonal antibodies, have potentially important industrial applications for the agricultural, industrial and economic development of Zimbabwe.

Status and mission of BRI

BRI was established through a memorandum from the Research Council of Zimbabwe to the President and Cabinet on 30 January 1992. The mission and objectives of BRI are:

- to research animal and plant genetics with a view to producing plant/crop species that are adaptable to the Zimbabwean environment;
- to develop crop varieties suited for growth in marginal zones, including those crops that will benefit the small-scale farmer;
- to provide technical expertise in the collection of indigenous germplasm;
- to develop expertise and provide services in food processing to enable community enterprises and the food industry to improve Zimbabwean agricultural commodities;
- to develop new diagnostic tools and vaccines for animal and human health;
- to adapt biotechnology to local needs and applications;
- to develop technologies and processes that will promote the growth of biotechnology industries in the country;
- to contribute to an adequate national and international legal framework on issues of patents and intellectual property rights;
- to assist industry in producing value-added products in agriculture, food manufacturing, mineral production and the production of new generation drugs;
- to provide leadership in the formulation of national policy on biotechnology and provide a national priority list of biotechnology research.

In order for BRI to fulfil its stated mission, it is important to recognize that research in biotechnology is likely to have a major impact on specific sectors of

Zimbabwe's agriculturally based economy. The four major sectors or areas which are most directly impacted by biotechnology research, or the lack of it, are agriculture, the food industry, medical or health related industries and the environment. If BRI is to conduct sound research in any of these sectors or areas, there must be sound and strong knowledge in the basic sciences. The research programmes at BRI must therefore be supported by a strong programme in basic biotechnology.

Structural organization of BRI

As mentioned earlier, biotechnology research at BRI is organized into the following core research areas:

* Basic Biotechnology Programme
* Agricultural Biotechnology Research Programme (Plant and Animal)
* Medical Biotechnology Research Programme (Human and Animal)
* Industrial Biotechnology Research Programme
* Environment Biotechnology Research Programme

Services offered by BRI

Basic biotechnology
BRI is developing training programmes in the area of basic biotechnology to enable the institute to have expertise in molecular biology, biochemistry, molecular genetics and molecular biotechnology. An emphasis has been placed on DNA fingerprinting for analysis of evidence in criminal cases, for solving paternity disputes, identifying human and animal remains, and collecting genetic profiles by employers and insurance firms. BRI will also focus on the production of new generation antibiotics, vaccines, drugs and single-cell proteins. Additionally, BRI will employ restriction fragment length polymorphism (RFLP), random amplified polymorphic DNA (RAPD), amplified fragment length polymorphism (AFLP) and marker-assisted plant and animal breeding technology and support the development of tissue culture technology for plants and animal cells.

Agricultural biotechnology

PLANT BIOTECHNOLOGY. BRI will have expertise in plant tissue culture, micropropagation, marker-assisted plant breeding, development of transgenic plants for pest and viral resistance, plant viral diagnostics for the provision of virus-free planting material, improvement of crops for drought tolerance and resistance to pests through gene cloning, transformation and plant regeneration in addition to the production of biofertilizers and rhizobium innoculants. BRI will actively apply biotechnology in forestry and horticulture.

ANIMAL BIOTECHNOLOGY. BRI will offer better animal husbandry practices through artificial insemination, embryo transfers, sire evaluation, marker-assisted animal breeding for the improvement of national beef and dairy herds and small ruminants (goats, sheep, etc.), and through vaccine and antigen production.

Medical biotechnology: human and animal health
BRI will have expertise in the production of new generations of vaccines, antibiotics, monoclonal antibodies, allergens and blood derivatives, diagnostic reagents and probes (DNA and RNA) and gene therapy. BRI expects to participate in human genome research activities.

Industrial biotechnology
BRI will develop expertise in food technology, food processing, food quality and nutrition. BRI will offer services in food analysis, fermentation technology and in the production of wines, beers, dairy products and beverages. BRI will produce microbial starter cultures and provide quality control assessments.

In addition, BRI will offer services to industry in such areas as industrial bio-processing, bioconversions, biotransformation, mineral processing, production of industrial enzymes, water purification, and the management and treatment of sewage and urban waste water for recycling in the main potable water systems.

Environmental biotechnology
BRI will offer special services to Zimbabwe and to the southern African region in biosafety issues, drafting and implementing biosafety guidelines/regulations, providing technical assistance in the release and testing of genetically modified organisms (GMOs), and conducting environmental risk assessments. BRI will also offer services in the area of genetic resources and biodiversity conservation and in the prevention of genetic resources erosion. BRI will protect Zimbabwe against biopiracy and intellectual property loss at the national level and will assist individual scientists with information relating to intellectual property.

Current Activities at BRI

Capacity building

Infrastructure
The GOZ has committed to construct state-of-the-art laboratories and provide laboratory equipment and consumable chemicals and enzymes to scientists for research activities.

Human resources development
BRI has developed a programme to train scientific and technical staff in various areas. A staff development training programme was started and 18 scientists

are currently training at The John Innes Institute (UK), Cape Town University (South Africa), Centro Internacional de Mejoramiento de Maiz y Trigo (CIMMYT) (Mexico), Cranfield University (UK), Wageningen University (The Netherlands) and others.

Research activities

Maize Improvement Research Project (MIRP)
The MIRP project is funded by Dutch Ministry of Foreign Affairs (DGIS). The aim is to assist small-scale farmers in producing more maize through the use of improved maize cultivars resistant to insects, pests and viruses, and tolerant of drought. Two scientists are training at CIMMYT in quantitative trait loci (QTL), transformation, regeneration and marker-assisted plant breeding.

Cotton Improvement Research Project
The main objective of this project is to develop insect resistant cotton for Zimbabwe through the use of *Bacillus thuringiensis* (*Bt*) crystal-protein-based biotechnology. *Bt* cotton will be developed through a collaborative programme between BRI, Cotton Growers Association (Zimbabwe) and Dekalb Company. BRI will contract all the research.

Sweetpotato Micropropagation Research Project
This project is funded by DGIS to provide transgenic sweetpotatoes to communal farmers. BRI is developing a tissue culture laboratory. To date, equipment has been bought and two scientists have been trained in tissue culture techniques.

Sweet Sorghum Utilization Project
This project explores the use of sweet sorghum for energy production. The project will study cocultivation of sweet sorghum with sugarcane at Chiredzi, Zimbabwe, and the use of sweet sorghum for the production of ethanol.

Fermentation technology for indigenous foods
This project explores traditional fermentation methods and characterizes all microflora involved in traditional fermentation. A special mutwiwa (traditional African fermented maize product) from Chipinge is being explored.

Biosafety Project
BRI is a regional focal point for biosafety. This project was initially funded by DGIS, but now funding is being sought from the United Nations Environment Programme – Global Environment Fund project. Some 14 African countries are involved in the project and the project officer is Mrs Joy Chigogora, a BRI scientist.

Training centre for industrial technical staff

BRI provides a training centre for technicians from local industries, particularly food manufacturers, in equipment repair and maintenance, and environmental monitoring.

Other project proposals

A number of other research project proposals have been developed and funding is sought from the international donor community.

The Vaccine Production Project
This project proposal aims to locally produce biologicals (vaccines and antigens) such as anthrax spore vaccine, Blackley vaccine, Newcastle disease vaccine and the African horse sickness vaccine. It was jointly prepared by Dr Gopo (BRI) and Professor Mohan (University of Zimbabwe Veterinary Sciences) and submitted to the European Commission.

The Maize Transformation Project
A project proposal was developed with the goal of producing transgenic maize cultivars resistant to maize streak virus (MSV) using coat-protein genes from the MSV genome. This project will mostly benefit the small-scale farmers who cannot afford the cost of an integrated pest management programme. A project proposal has been submitted to a prospective donor.

The Soya-cow Project
BRI, in collaboration with Pro-Soya Incorporated, Ottawa, Canada, is developing a proposal to establish a pilot project to produce soya milk in Zimbabwe at the SIRDC/BRI complex. The project proposal is being prepared for the Canadian International Development Agency (CIDA).

The Forestry Biotechnology Research Project
The African Academy of Science and International Service for the Aquisition of Agri-Biotech Applications (ISAAA) have identified donors who wish to fund research in forestry biotechnology. BRI is preparing and developing a project proposal in this important area. The Provincial Governor of Matabeleland North Province has asked BRI to develop tissue culture methodologies to produce commercial hardwood plant species.

The National Cassava Research Project
BRI, in collaboration with Dr Visser and Mr Munyikwa, Wageningen University, The Netherlands, is developing a research project proposal for cassava transformation for viral and insect pest resistance, and the transformation of cassava starch genes. The Natural Resources Institute (UK) (NRI) has asked BRI to

prepare a project proposal on the development and utilization of non-grain starches from cassava and sweetpotatoes in collaboration with NRI.

Seed-Co: Maize Regeneration and Cultivar Purity Identification Project
BRI has been requested by Seed-Co, Zimbabwe, to develop a cultivar purity identification research programme. Seed-Co is developing a new line of maize cultivar and has requested that BRI develop a method to test seed purity before the bulking of the seed.

Seed-Co is also interested in a simple, cost-effective plant virus diagnostic indexing protocol. BRI staff are in the process of developing a gel electrophoresis procedure which will be used for cost-effective viral diagnosis in both crop plants and horticultural plants (flowers, fruits and vegetables, and others).

Future activities

BRI wishes to develop closer and more effective links with Zimbabwean industry, and with decision makers in the government, private sector, diplomatic and donor communities. BRI believes that it has made a good beginning in this direction, and soon hopes to procure a greater number of contract research projects from the industrial sector, on a national, subregional, regional, international and donor community basis. This will enable BRI to fulfil its mission.

The Technology Transfer System in Thailand

22

Lerson Tanasugarn

Office of Intellectual Property Policy Research, Intellectual Property Institutue of Chulalongkorn University, Phya Thai Road, Bangkok 10330, Thailand

This chapter is an outline of technology transfer and technology diffusion systems in Thailand, with special emphasis on agriculture. Technology transfer and technology diffusion are defined. Classical modes of technology transfer and technology diffusion are covered. Infrastructures supporting technology transfer and technology diffusion systems are briefly reviewed. The legal foundation is further explained, including: (i) traditional intellectual property system including the following legal regimes – copyright, patent and trademark; (ii) current intellectual property regime developments, including plant variety protection; (iii) unfair competition prevention system, including the protection of undisclosed information and the compulsory registration of technology transfer agreements; (iv) currency transfer and foreign exchange regulations; and (v) special laws, regulations and guidelines related to science, technology and commerce. An additional emphasis is then placed on technology transfer and technology diffusion in agriculture, including the mode, extent, critical factors and future.

Introduction

This overview of the technology transfer system in Thailand is organized into seven areas. Definitions of relevant technical terms and conceptual frameworks are presented in order to give the readers some idea of what technology transfer means to Thais. A brief review of the present status and problems of technology transfer in Thailand is followed by outlines of the national technology transfer policy and legislative measures. Finally, the future outlook for technology transfer in Thailand is presented.

© CAB INTERNATIONAL 1998. *Agricultural Biotechnology in International Development*
(eds C.L. Ives and B.M. Bedford)

Definitions

It is the author's observation that agricultural scientists in developed and developing countries often attach different meanings to the same technical term. Common working definitions are therefore needed to eliminate this type of misunderstanding.

Technology refers to the practical knowledge of how something is made or done. In other words, technology is the practical application of science. According to the Asia and Pacific Center for Technology Transfer (APCTT), technology may reside (be embodied) in machine (technoware), knowledge (infoware), people (humanware) and organization (orgaware). These four embodiments of technology interact dynamically (Ramanathan, 1996). Control of *Phytophthora* by *Trichoderma*, absorption of ethylene by chemical or biochemical agents and transgenic cotton are a few examples of agricultural biotechnology (MOSTE, 1996d).

Technological capability, coined by economists in the 1980s, refers to the ability to acquire (search, select, negotiate, procure, etc.), operate, adapt, design and innovate a technology (Kritayakirana, 1986).

In developed countries, people tend to think of *technology transfer* in the sense of 'putting research results (from developed countries) to work in the field (of developing countries)' without any notion of technological capability changes on the recipient side. In this chapter, however, the term 'technology transfer' refers to the movement of technology across national boundaries, whether inward or outward (ESCAP, 1992a; MOSTE, 1996e). Technology is transferred from technology provider (source, donor) to technology recipient (acceptor). The transfer takes place if and only if there is an increase in technological capability of the technology recipient. Emphasis is placed on adaptive, design and innovative capabilities as primary targets of technology transfer on the recipient side.

Technology diffusion refers to the spread of technology inside a country (ESCAP, 1992a). Diffusion mechanisms as practised in Thailand include setting up incentives for local manufacturing, securing grants for technology development and diffusion in local universities, holding provincial techno-fairs and subsequently establishing a university or a technology transfer centre at the site of each techno-fair, and establishing a folk-inventor (indigenous/local inventors) programme especially designed for lay people.

Any technology developed within the country, e.g. as a result of a research and development (R&D) programme, is called *indigenous technology*.

Conceptual Framework

Mechanisms of technology transfer

As seen in Table 22.1, various mechanisms exist to aid technology transfer. Advantages and disadvantages with respect to developing countries are also listed in the table. Several mechanisms exist for gathering industrial information. These include education, literature reading, World Wide Web browsing, engagement in cooperative research and participation in trade or professional meetings.

One type of technology transfer involves reverse engineering. Once the technology is known, improvements can be made to obtain better products or processes. Care must be taken, however, that intellectual property rights (IPR) (mostly patent rights) are not infringed.

It is well-known that a majority of firms in developing countries rely on turnkey operations, purchase of technological services or purchase of embodied technologies. These modes of transfer, however, offer little or sometimes highly variable degrees of technology transfer in terms of an increase in the technological capability of the recipient.

Several government agencies are responsible for controlling and promoting technology transfer. Thailand's Technology Transfer Center, established under the Ministry of Science, Technology and Environment, plays a key, albeit limited, role in promoting technology transfer and licensing.

Constraints and impact of technology transfer and diffusion

During the past couple of decades, several models of technology transfer, diffusion and development have been tested and used in Thailand. Built-in constraints (macro/micro) are capital, technology and trade barriers, and local conditions such as manpower, absorptive capacity, social values, market and price structures, management experience, risk perception, a firm's characteristics and technical infrastructure. The impacts of technology transfer and diffusion may be divided into business impacts, impacts on further development of technology and industry, and direct social benefits through the availability and utilization of technology or products of such technology.

Technology transfer from analysis of patent documents

During the past decade, intellectual property protection has been recognized by Thailand as forming a foundation for industrial development. A new wave of modern intellectual property laws were drafted starting in 1988. In 1992, the Department of Intellectual Property was established in the Ministry of Commerce. This Department assumes the role of Thailand's Patent and Trademark Office, as well as handling copyright matters. During the same period, scientists and engineers started to appreciate the value of patent documents. Various organizations offer patent search services for US and European patents, as well as Japanese Patent Abstracts.

Recently, it has been recognized that patent documents should be analysed in groups of related patents (in addition to analysing each individual patent) as

Table 22.1. Mechanisms of technology transfer. Adapted from Frame (1983).

Mechanism	Advantage	Disadvantage
Education, research, reading and seminar participation		
Education abroad	Very important means of technology transfer. Can be very cost-effective.	Costly: direct cost + opportunity cost. Long gestation period
Site visit and on-the-job training abroad		Lcts of prior competence and purpose to absorb technology. Beware of cheap-labour fraud.
International cooperative research efforts		Usually focused more on transferring scientific information than technology.
Published literature	Relatively cheap transfer of scientific knowledge. Analysis of patent documents can produce invaluable insight.	May be out-of-date. Some may lack detail. Proprietary technologies may not be published at all.
Meetings, seminars and colloquia	More up-to-date than publication. Good for transfer of scientific knowledge.	Travel funds needed. Proprietary technology may not be transferred.
Browsing the World Wide Web (www)	Extremely convenient to use. Can also be used to browse patent information.	Proprietary technology may not be transferred.
Reverse engineering and emulation		
Simple emulation of products or process	Reverse engineering can stimulate innovative capability.	May involve illegal activities.
Primitive acquisition of bundled technology		
Turnkey operation	High quality, state-of-the-art technology, no need for skilled workers.	Costly, minimal acquisition of know-how.
Purchase of technological services	Enormous transfer through education.	Transfer highly variable. Poor transfer if technology personnel are hired and the locals cannot absorb their know-how.
Purchase of embodied technology		Pay more for technology in terms of cost per unit bought

Advanced modes of technology acquisition

Direct purchase of naked technology	Cut cost and efforts in R&D of complex, state-of-the-art.	No incentives to develop in-house technology.
Licensing	Good for recipient with high absorbing capacity. Help reduce duplications of efforts.	Watch for terms in the licence agreement.
Joint ventures	Share cost, technical marketing, production and managerial skill.	Party of stronger technological capabilities may want to avoid transfer of proprietary technology. Recipient needs high absorptive capability.
Patent purchase	Help existing products, move into new lines, avoid patent infringement suits.	
In-house transfer to foreign subsidiaries	Transnational corporations are major technology transfer agent. No concerns for national border and proprietary rights.	May be used to transfer profits out of developing countries.

practised by private companies in industrialized countries like Japan and the US. Currently, the Intellectual Property Institute of Chulalongkorn University is performing a trial run using such methodology in four fields of industry: food; agricultural machinery; rubber; and space communication. Results of such analyses will help in business and R&D planning, i.e. in monitoring competitors' activities, finding trends and niches in technology development, and competitively positioning R&D and commercialization of each local firm. This will help, directly or indirectly, to improve technology transfer.

Present Problems of Technology Transfer in Thailand

Economic problems

For the past decade, Thailand has enjoyed a booming economy with fantastic growth. Two recognized problems were an increasing unskilled and semi-skilled labour cost and the gross lack of a science and technology manpower base (TDRI, 1990). To be internationally competitive in the manufacturing sector, the country relied too heavily on cheap labour as opposed to the accumulation of technology through technology transfer and R&D.

Recently, partly as a result of organized foreign capital speculation, and partly from gross negligence on the part of the Bank of Thailand, the bubble of that fantastic growth has collapsed. The government, in compliance with the guidelines set up by the International Monetary Fund (IMF), has begun a very tight spending policy, which inevitably limits direct and indirect assistance to the private sector in technology transfer.

Technology transfer problems

Being a net technology consumer
Although importing technology is not necessarily bad for the economy, being a major net technology consumer is a problem common to many developing countries of which Thailand is no exception. Starting in 1985, the Technology Transfer Center under the Ministry of Science, Technology and Environment published the Annual Technology Transfer Status of Thailand, which documented economic status, science and technology manpower status, foreign technology purchases, status on R&D, roles of organizations involved in technology transfer, technology transfer policy, investment and licensing recorded, etc. (MOSTE, 1985). Unfortunately, since the new Foreign Exchange Regulations (established c. 1995) do not keep track of technology payment balance, it is difficult to determine the present situation with respect to technology imports, exports and balance between the two.

Some imported technologies obviously are used to produce value-added goods, services or technologies for export. Nevertheless, the majority of technologies, especially those embedded in consumer goods, are consumed without generating inventions or even adaptation. As for production technology, many factories still prefer to use a turnkey technology transfer mode in order to obtain short-term profits.

Lack of capabilities to select, absorb, diffuse and develop

Simply put, there is a shortage of both quantity and quality manpower to support effective technology transfer in various sectors. The lack of selective capability leads to non-optimal technology choices. This is related to a lack of information about competing technologies and their impacts. The lack of absorptive capability comes from a lack of awareness by the technology recipient that effective technology transfer will take place only with active participation by the donor and especially the recipient. Lack of adaptive, design and innovative capabilities means local capabilities in industries seem to stop at operative capability.

Actually, adaptive, design and innovative capabilities do exist in universities and research institutes but a lot of links will be needed to extend these capabilities to actual users in industries. Theoretical treatment of the subject as well as survey data in the fields of biotechnology, material technology and electronics and computer technology have been documented by the Thailand Development Research Institute (TDRI, 1992).

The overall situation in agriculture and agricultural industries is not much different from that in the manufacturing industry. Fertilizers, growth promoters and pesticides are either imported or manufactured using foreign technology with little transfer beyond operative capability. Nevertheless, the Ministry of Agriculture has done a good job in the development of new varieties of plants such as rice and maize.

Lack of information

Information includes supply-source information, information on spare parts, technical information on the subject being studied, etc. For the past several years, a few government and semi-government agencies have specialized as information sources. These include the Department of Intellectual Property (Ministry of Commerce), the Department of Science Service (Ministry of Science, Technology and Environment), and Technology Information Access Center (TIAC, which is an office established under the National Science and Technology Development Agency (NSTDA), a semi-government agency affiliated with the Ministry of Science, Technology and Environment).

Lack of research funding

In Thailand, most research work is done in universities and government research laboratories with support from the public sector. Traditionally, most of the government research funding is in agricultural technology. In response to requests for more funding in other areas of technology during the past decade, the government significantly increased research funding through the NSTDA and the Thailand Technology Fund (TTF) established under the Prime Minister's Office. Annual budgets of each of these agencies have been in the 100 million baht range over the past several years. Now that the Thai Government is reducing overall spending, the government-sponsored funding on research, development and engineering (RD&E) will be inevitably reduced.

Present problems in agricultural technology transfer

Major problems in the plant sector
These problems include inappropriate variety selection and development, misuse of fertilizers and pest controls, inefficiency in planting and harvesting, and poor packaging.

Major problems in the animal sector
Major problems include the use of imported breeding stocks, inadequate disease control and special problems related to harvest and postharvest technologies.

National Technology Transfer Policy

Thailand's technology transfer policy is reflected in the comprehensive National Science and Technology Development Plan, 1997–2006 (MOSTE, 1996e), which covers: (i) a science and technology manpower plan, (ii) a science and technology transfer plan, (iii) a research and development plan, and (iv) a development of science and technology infrastructure plan. The objectives of the present National Technology Transfer Policy are: to accelerate the transfer of technology in manufacturing and service sectors in order to foster international competitiveness; to establish mechanisms necessary for sustained technology transfer and diffusion for improved quality of life; and to encourage investments in production and R&D in strategic sectors.

Goals

To develop absorptive, transfer and innovative capabilities in the manufacturing sector, using clean technology in compliance with ISO 9000 and 14000
According to the Plan, both agricultural and industrial sectors must retain competitiveness with respect to cost and quality. Technology transfer must be linked to 'megaprojects' (large, government-sponsored development projects) in the production and public-service sectors. Technology transfer in a few selected key industries will be promoted. Radiation technology will be used to preserve agricultural goods as well as in high value-added export industries. These activities will need manpower with expertise in searching, assessing and receiving technology that will be transferred from abroad.

To establish agencies and mechanisms to transfer and develop new technologies in the capital and provincial areas
Professional societies and technical consultants in each industry will be utilized as executing agencies and/or mechanisms. Infrastructure in remote areas, including technology news feed (dissemination of technology information via electronic means (tv, radio, etc.)) and intellectual property information, will be established during the next decade.

To establish fiscal and monetary policies that promote the manufacturing and service sectors to acquire, transfer and develop technologies
Adequate funding for science and technology development is crucial. Technology development banks and tax incentives will be used to encourage technology transfer.

Strategies and measures

Priority setting
A specialized organization will be responsible for technology information surveys, determination of short-, medium- and long-term target industries, as well as conducting monitoring and follow-up activities. Methodology for technology assessment will be developed for use in the planning of target industries. A white-paper on technology transfer will be assembled with the help of the public and private sectors. The government will take advantage of information technology to disseminate technological information to the people.

Awareness in government
The plan includes campaign funding for seminars for high-ranking government executives and Members of Parliament in order to develop an appreciation of the necessity for technology transfer.

Manpower preparation
Emphasis will be placed on curriculum redesign, with multilateral cooperation on technology and technology management. This will naturally demand more technical foreign language education. Both foreign and Thai consulting firms will be promoted.

Attract multinational companies to Thailand
Measures that will be used to attract private industries include elimination of double-taxation for multinational companies and their employees, reduction or elimination of corporate income tax as well as taxes on royalty payments and investments in promotional incentives for transnational companies that establish research and training centres in Thailand.

Utilize megaprojects
Government's megaprojects, e.g. public infrastructure, will be required to promote technology transfer to Thai personnel. Thai consulting firms will be utilized as well.

Incentives for Thai and foreign experts
Experts from foreign countries, as well as Thai experts living in a foreign country, will be eligible for incentives such as special treatment for visa application and reduction or elimination in personal income tax in order to access technology.

Incentives to create training institutes
Government funding will be provided for institutes with technical training. Corporate income tax for these institutes will be exempted. Donations to these institutes will be tax-deductible.

Diffusion mechanisms
Promotion will be given to industries which use local raw materials. Provincial universities will be supported to become technology transfer centres. Science and technology organizations will be encouraged to work with local folk-inventors to develop and diffuse indigenous technologies.

Utilize intellectual property system
Responsible agencies for intellectual property management will be promoted to a more professional level. Services must seek users and not the other way around. Private firms will be encouraged to take advantage of the new intellectual property system.

Promote science and technology culture
Science and technology-based culture will be promoted in a society where most people still believe in superstition.

Legislative instruments
In order for the National Technology Transfer Policy to be realized in practical terms, laws and regulations are needed to implement these policies. Some of the necessary laws and regulations are now in place and others are in the drafting process. A few, unfortunately, are still on a wish-list.

Screening measures for modern technology

Certain laws have the effect of screening for modern technologies (S. Vongkiatkachorn, 1985 personal communication).

Alien Business Law (National Executive Council Announcement No. 281 of December 24, 1972)
According to this law, before foreign entrepreneurs are issued permits to engage in a business on the 'Control List', they need to convince the authorities that their proposed business requires technology that Thailand does not possess.

Investment Promotion Act of 1977
Promoted projects have to be technologically sound as well as economically sound. This law provides the government with screening controls for both technology and payment fees.

Immigration Act of 1979
Applicants for immigrant visas must have academic and professional abilities, among other requirements, while applicants for non-immigrant visas may come to Thailand, among other things, for the purpose of scientific research in a research institution in the Kingdom.

Working of Alien Act of 1978
Applicants for work in Thailand must show the authorities that their expertise is necessary and not available in Thailand.

Industrial Product Standards Act of 1968
This law sets out a framework for industrial product standards.

Applications of technology

Some technology, particularly information technology, may require special laws or amendments in order to allow practical applications. An example is the paperless system of Electronic Data Interchange (EDI), which requires amendments in the law of evidence.

Promotion of technology transfer

Measures to promote effective technology transfer should be built into laws and regulations related to government procurement, especially in megaprojects. These laws are still on the author's wish list.

Protection of intellectual property rights

At present, Thailand has enacted most of the intellectual property laws as required by the World Trade Organization (WTO). On the literary and artistic side, The Copyright Act of 1978 was amended in 1994 to include computer programs. On the industrial property side, Thailand has in place the Patent Act of 1979, amended in 1993, and the Trademark Act. In the works are laws on the protection of integrated circuit layout design, trade secret, geographical indications and plant varieties.

Gathering of information and statistics

This set of laws serves to gather information on investments, royalty transfers, licensing agreements, etc. Examples are the Exchange Control Act (administered by the Bank of Thailand), the Investment Promotion Act (administered by the Board of Investment) and the Alien Business Act (administered by the Department of Commercial Registration, Ministry of Commerce).

Future Outlook

Given the economic turmoil that Thailand is now facing, there is an increasing need to strengthen the technology transfer system as a platform for the economic rebound that is expected to take place a couple of years from now. In the author's opinion, the most difficult problem facing the Thai Government is how to encourage private industries to invest in active technology transfer practices in order to increase technological capabilities for the long-term while financially surviving in the short-term. Meanwhile, laws and regulations will need to be established in order to implement the biodiversity convention that Thailand is expected to ratify sometime during 1998. These laws and regulations will address the fair balance between access to local genetic resources and technology transfer from more advanced nations. The evolution of the technology transfer system in Thailand is therefore still ongoing, with possibly a bright outlook – if the Government doesn't change too often.

References and Further Reading

BOT (1987) *Technology Fee Payment for 1987 (Classified by ISIC)*. Bank of Thailand.

Davidson, F.J. (1983) *Internal Business and Global Technology*. D.C Heath & Co. Lexington, Massachusetts, USA.

ESCAP (1992a) *Technology Transfer: Basic Concepts. Technology Transfer, an ESCAP Training Manual*. Booklet 1. United Nations Economic and Social Commission for Asia and the Pacific, (ST/ESCAP/862) Reprinted from an earlier version by Ministry of Science Technology and Environment, 1992, 50 pp.

ESCAP (1992b) *Intellectual Property and the Transfer of Technology. Technology Transfer, an ESCAP Training Manual*. Booklet 2. United Nations Economic and Social Commission for Asia and the Pacific, (ST/ESCAP/862) Reprinted from an earlier version by Ministry of Science Technology and Environment, 1992, 52 pp.

Kritayakirana (1986) Technological capability. Paper presented at the APCTT Meeting in Bangkok, Thailand, Asia Pacific Center for Technology Transfer, Bangkok, Thailand.

MOSTE (1985) *The Status of Technology Transfer in Thailand.* Ministry of Science, Technology and Environment, Thailand.

MOSTE (c. 1986) *Policy and Plan for Technology Transfer from Abroad.* Internal document, Ministry of Science, Technology and Environment, Thailand, 8 pp.

MOSTE (1994) *Longan Product Technology.* Ministry of Science, Technology and Environment, Thailand, 13 pp.

MOSTE (1995) *Technology Promotion and Transfer Newsletter.* Ministry of Science Technology and Environment, Thailand, 12 issues.

MOSTE (1996a) *Technology Promotion and Transfer Newsletter.* Ministry of Science Technology and Environment, Thailand, various issues.

MOSTE (1996b) *Technology.* Ministry of Science, Technology and Environment, 17(2).

MOSTE (1996c) *Thai Technology Offers.* Ministry of Science, Technology and Environment, Thailand, 105 pp.

MOSTE (1996d) *Technology Listing.* Ministry of Science, Technology and Environment, Thailand, 101 pp.

MOSTE (1996e) *National Science and Technology Development Plan (1997–2006).* Ministry of Science, Technology and Environment, 21 Century Printing, Bangkok, 126 pp.

Panupong, C. (1986) MNC's and the Role of Thai Government. In: *Conference on the Role of Multinational Corporations in Thailand,* organized by Thammasat University in cooperation with Faculties of Commerce and Accountancy, Economics, Political Science, Law and Human Resource Institute, 7–9 July, 1986.

Ramanathan, K, (1996) Introduction to International Technology Transfer. In: *Training Program on Technology Acquisition Through Licensing,* organized by the Office of Promotion and Technology Transfer, Ministry of Science Technology and Environment, Bangkok, 3–4 September, 1996.

Santikarn, M. (1981) *Technology Transfer: A Case Study.* Singapore University Press, Singapore, 270 pp.

Santikarn, M. (1987) *Evolution of Ideas on Technology Transfer to Less Developing Countries.* (Privileged communication) 26 pp.

Tanasugarn, L., Yoonaidharma, S. *et al.* (1996) *Legal Development of Electronic Commerce in Thailand.* National Information Technology Committee, NITC (Thailand). Bangkok, Thailand.

TDRI (1990) *Science and Technology Manpower Status of Thailand. Final Report* (Project supported by funds from USAID).

TDRI (1992) *Thailand Technological Capability in Industry Project Final Report.* Thailand Development Research Institute, Bangkok, Thailand (Project supported by funds from CIDA, Canadian International Development Agency).

UNCTAD (1975) *Transfer of Technology: Technological Dependence, Its Nature, Consequences and Policy Implications.* Report by the UNCTAD Secretariat (TI3/190) 31 December, 1975.

Trade in Conventional and Biotechnology Agricultural Products

Quentin B. Kubicek

United States Department of Agriculture (USDA), Room 1128–5, Washington, DC 20250, USA

The US consumer takes it for granted that the food they purchase is safe, clean, wholesome and relatively inexpensive. Several federal and state agencies have negotiated the entry and export of agricultural products, ensuring that these products are safe, not only to humans but to agriculture as well. This type of negotiation is ongoing and the Animal and Plant Health Inspection Service (APHIS) is primarily responsible for ensuring these imports do not present a risk to US agriculture. It is the role of APHIS that is discussed here, especially within the context of the treatment of biotechnology-based products.

Consumers often ask how they can be sure of the safety of agricultural imports. The US consumer is accustomed to having a variety of food choices and to have these throughout the year. Many take it for granted that the food they purchase is safe, clean, wholesome and relatively inexpensive. The Environmental Protection Agency (EPA), Food and Drug Administration (FDA) and the United States Department of Agriculture (USDA) are continuously working behind the scenes to assure the supply and its safety. It is not 100% secure, and sometimes errors are made. Even with the magnitude of food products that are imported into the US, it is rare to hear of people becoming ill. However, when it does happen it receives considerable publicity.

Several federal and state agencies have negotiated the entry and export of agricultural products. Bananas are imported into the US only after plant health officials in the US and the exporting country have negotiated a system which assures a safe, quality banana is supplied. Safe, not only to humans, but to agriculture as well. US exports must meet similar health standards prior to an

The views expressed are those of the author and not necessarily those of the USDA.

CAB INTERNATIONAL 1998. *Agricultural Biotechnology in International Development*
(eds C.L. Ives and B.M. Bedford)

importing country accepting them. This type of negotiation is ongoing and the Animal and Plant Health Inspection Service (APHIS) is primarily responsible for assuring these imports do not present a risk to US agriculture. While it is the responsibility of APHIS to develop a system which minimizes risk to US agriculture, APHIS does not judge the market potential of these products.

Few know that APHIS has offices in Argentina, Chile, Mexico, Peru and other countries. These offices, for the most part supported by the exporters, are devoted to the inspection and clearance of agricultural products intended for consumption in the US. We expect and negotiate equal opportunities for US exports. From time to time this expectation is met with reluctance by government authorities of the importing country. It is the business of business. APHIS seeks to open, expand or retain markets for US products. Clearly, behind this market drive is a US industry seeking to enter, grow or remain in a market sector. Before APHIS embarks on such negotiations, we trust that a market strategy has been completed which concludes that the market or market share sought can be achieved. APHIS does not have sufficient resources to embark on wishful enterprises.

APHIS will not permit the entry of bananas from Costa Rica until we are assured that they do not harbour pests injurious to American agriculture. Plant health counterparts overseas sometimes consider US quarantine requirements too stringent. US businesses believe that the plant quarantine requirements of our trading partners are too stringent and that APHIS ought to negotiate less stringent requirements. Some in the US say our quarantine requirements are too loose and ought to be tightened. There may be business strategies behind these considerations; however, APHIS's system, requirements and manner of negotiating aims to be open and transparent.

APHIS does not erect barriers to protect US domestic industries. Once a commodity is allowed into the US, a trade pattern develops and movement of commodities begins and does not stop unless that which has been mutually negotiated changes significantly. For example, if samples of the imported commodity are found to harbour injurious pests, then shipments are halted. The shipments can begin again when APHIS is assured that the system to prevent the pest(s) from accompanying the commodity is reliable. Chilean grapes, Costa Rican bananas and Mexican horticultural crops shipped to the US have had existing trade patterns for years. Agricultural products are not sent to US shores without proper approval. A lot of homework has been done before Chilean grapes, Costa Rican bananas or Mexican horticultural crops enter the US. The system works.

Developing a new trade pattern for a new commodity is difficult and lengthy but it can be done. The USDA has animal or plant quarantine concerns; EPA has pesticide residue or environmental concerns; FDA has health concerns; and brokers, exporters and importers have their concerns. But a trade pattern can be developed as evidenced at the supermarket.

APHIS concerns are risk-associated and science-based. Some say politics are involved but, in my view, this is part of business. Argentine citrus exported to the US may displace a US citrus producer who may go bankrupt and their business may collapse. Many believe the best antidote is to protect domestic markets using quarantine concerns and delay the eventuality. It is asked why the US Government doesn't protect its industries from such imports. The US Government believes that in order to trade you have to allow imports on a quid pro quo basis. Competition will cause some individuals to lose, but the US as a whole will win. Market forces will dictate the outcome.

Trade patterns exist for virtually all crops. Some trade patterns cease and renew, depending on their profitability. Markets for good products open and remain open; markets for bad products close. The forces of the market (business) make the system move. For example, trade patterns for maize are well established and are one variable in favour of the commercialization of transgenic maize. These trade patterns are too large and too complex to change. The same can be said for soybeans, cotton and tomatoes.

Biotechnology-based products will follow existing trade patterns. It has been the US Government's position for many years that the products of biotechnology have the same kind of risks as traditionally bred agricultural products. Today, there are commercially available potatoes and squashes that are resistant to certain viruses and this resistance is based on the use of viral coat proteins. This is just one example of the US Government viewing biotechnology-based products under the same risk-based view as similar and traditionally bred products.

The regulatory oversight scheme for biotechnology-based products has supported the US regulatory position. Today, regulatory oversight reflects the collective experience. It is not fair to ask for more or less oversight. It is fair to ask and expect biotechnology regulators to learn from the collective experience. The USDA is doing exactly that in the US as APHIS learns and adapts. Critics wrongly say that APHIS is easing up. APHIS is simply changing based on experience.

A measure of the adaptability of APHIS is the number of transformation events being deregulated. The number of plant species being transformed is growing and the number of field tests is increasing.

Transforming plants and field testing them is becoming conventional, and deregulation is expected to continue.

In the US, public acceptance of biotechnology-based products is increasing. As more and more biotechnology-based products enter the marketplace, they will enter well-established trade patterns and leave our ports to enter world commerce. It is extremely difficult to establish reliable trade statistics, since no distinction is established in trade between biotechnology and conventional products in a given sector.

Without existing trade patterns the success of biotechnology-based products is doubtful. For example, transgenic cassava or sweetpotatoes produced in an African country will not find an existing trade route to the US. Research efforts will not help transgenic cassava or sweetpotatoes to reach a market because these transgenic crops may still harbour pests injurious to US agriculture and APHIS will not allow their importation into the US. A second, unrelated question, would be whether these transgenic cassava or sweetpotatoes have a market in the US. Just because a plant has been successfully transformed with an important gene does not indicate its acceptability. All efforts, time and research funds may yield no commercial success because, in today's world, research without commercial implications must have a marketing component built-in early in the process in order for the finished product to reach its clientele.

Laboratory breakthroughs do not necessarily lead to marketing breakthroughs. The fruits of laboratory success will not reach the market if scientific endeavors are not linked to a marketing strategy. Biotechnology is only a tool. The Cassava Biotechnology Network, for example, may lose significance if alternative crops are cheaper, subsidized or simply better competitors than cassava in the market. Should CBN then continue to be funded? The search for academic answers is noble but, unfortunately, with diminishing resources these quests are becoming rare. The market is a powerful force and we need to come to terms with it – the sooner the better.

Biotechnology-based products will have the same quarantine concerns and tariffs as their non-engineered cousins; licences, quotas, tariffs and other measures will be applied equally. The USDA is opposed to import requirements exclusive to biotechnology, seeing no justifiable scientific reason to enact such requirements. Although commercialization of biotechnology-based products is only just beginning in the agrifood sector, biotechnology-based products seem to encounter nontariff obstacles that were not common for conventionally bred products, particularly relating to public concerns. There is no tariff barrier specific to biotechnology, since the tariff treatment of biotechnology-based products is similar to conventional products.

Regulators must recognize that they are simply the first step in a chain. Once oversight is completed, market forces will do the rest. This is currently functioning for conventional products and should become the conventional way for the products of biotechnology. Irrationality ought not to increase or decrease the technological gap or trade gap. The application of biotechnology has created and will continue to create differences in industries and products.

Can Developing Countries Turn Biotech into Business? Moving Research Results into Products

Wild Biodiversity: the Last Frontier?

Nicolás Mateo

Biodiversity Prospecting Division, National Biodiversity Institute (INBio), Apdo. Postal 22-3100, Santo Domingo, Heredia, Costa Rica

Costa Rica has an exemplary record in biodiversity conservation and development which is the result of strong political support, growing awareness of society of the value of biodiversity, and leadership and vision provided by key individuals and institutions. This achievement has not been painless as indicated by the fact that the country lost one-third of its forest cover during the first 450 years after colonization and another one-third in the last 50 years! Agricultural systems based on high inputs (particularly export crops) and inadequate utilization of marginal lands (particularly subsistence farming) have led in many instances to forest destruction and degradation, soil losses and pollution. It is worth noting that adding value to primary products has also been lacking throughout most of our history.

Biodiversity conservation and development must definitely benefit from past lessons and consider alternative mechanisms and strategies. The National Biodiversity Institute (INBio) was created in 1989, on the recommendation of a national planning commission established by an executive presidential decree, as a non-profit, non-governmental organization for the public good. INBio's mission and objectives, to generate knowledge of biodiversity and develop sustainable uses, complement intensive conservation efforts led by government, private groups and individuals. These efforts have resulted in the protection of about 25% of the national territory.

INBio's operational goals include a national biodiversity inventory to describe and locate the estimated half a million species present in the country; information management systems to handle and analyse high volumes of data; information dissemination mechanisms to share and promote knowledge of biodiversity and a bioprospecting effort to search for sustainable uses.

© CAB INTERNATIONAL 1998. *Agricultural Biotechnology in International Development*
(eds C.L. Ives and B.M. Bedford)

Bioprospecting has been a pioneer experience that preceded the tenets established by the Biodiversity Convention. INBio has established research agreements with private national and international industry based on criteria of access, equity, transfer of technology and training of national scientists. These agreements provide a direct contribution to conservation, over US$2 million to date, and contain royalty obligations in cases where new products may reach the market. Important fields of cooperation with the academic and private sectors include biotechnology, biopesticides, pharmaceuticals, fragrances, ornamentals and tourism. The success of conservation efforts will largely depend on how society perceives the direct and indirect benefits derived from understanding and using biodiversity.

Introduction

The management and expansion of the agricultural frontier have brought mixed blessings to Costa Rica. On the positive side it has promoted democracy, national values and political stability, and it has been the key ingredient of the economic development model during the last century and a half. On the negative side, agricultural expansion has resulted in poor natural resource management in most of the country, a very low level of value-added products and a dangerous dependence on a small number of crops. Figure 24.1 illustrates the paradox of current and potential land use systems in the country.

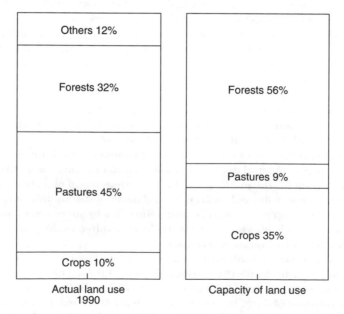

Fig. 24.1. Land use in Costa Rica. (Source: Lücke, 1993, Hartshorn *et al.*, 1983; cited by Proyecto Estado de la Nación, 1995.)

The agricultural sector, while still contributing about 17% to the National Gross Product (NGP), is currently undergoing a severe identity crisis caused by the shifts and pressures of globalization and fluctuating but ever-falling export prices. A few successful exceptions are niche export markets for high-value vegetables, fruits and ornamentals which have brought the country competitive advantages.

Research is a true reflection of the overall agricultural sector picture. It has generated or imported a number of useful technologies, contributed to national food security and developed successful research systems on a few selected crops such as coffee. At the same time, however, Costa Rica's national research strategy, particularly in the early days, adopted and adapted crop and animal production systems not suited to the highly variable tropical environments of Costa Rica. Agricultural research has attempted to maximize production at all costs using a model based on high levels of inputs that cause pollution and contamination of land, water and animal life.

It is essential that we learn from agricultural history in order to conserve and use wild biodiversity intelligently, which is certainly the last research and development (R&D) frontier. There are fundamental criteria to achieve this goal, most of them clearly spelled out in the tenets of the Convention on Biological Diversity. These include access, equity (including research income and royalties allocated for conservation), transfer of technologies and a knowledge base ('know-how'), and advanced training of national scientists. Other criteria are value-added strategies, negotiation skills, understanding of the market and development of strategic alliances with universities, research centres and industry.

On the basis of the existing parallels in the agricultural sector that emphasize the need to add as much value as possible to biodiversity in genetic resources-rich countries, this chapter will describe and discuss experiences and lessons from the National Biodiversity Institute (INBio) in Costa Rica.

The National Biodiversity Institute (INBio)

Approximately 25% of Costa Rica's territory consists of wildlands conserved for their biodiversity and contains approximately half a million species of animals, plants and microorganisms. These organisms are distributed from the nearly desert-dry forest habitat of the northwest to the very wet rainforest habitats of the Costa Rican lowlands, and the >3000-m tall mountain ranges. This biodiversity – representing 4–5% of that of the terrestrial world – is a major renewable resource and a potentially powerful engine for Costa Rica's intellectual and economic development.

In 1988, a National Planning Commission, established by Executive Presidential Decree, recommended that the Instituto Nacional de Biodiversidad-INBio (National Biodiversity Institute) be created as a non-profit, non-governmental organization for the public good. INBio was legally registered in 1989, and is governed by an Assembly of Founders and a Board of Directors.

This legal structure has enabled INBio to satisfy the need for flexibility in handling the rapidly expanding field of biodiversity management.

INBio's activities involve a close integration with both public and private institutions (Costa Rican and international) and are conducted under the assumption that society will conserve a major portion of its wild biodiversity only if protected areas can generate ample intellectual, spiritual and economic benefits. INBio, based on a legally established partnership of cooperative support with the Ministry of the Environment and Energy (MINAE), carries out the following processes in order to fulfil its mission of knowing, conserving, and using Costa Rica's biodiversity in a sustainable fashion:

- biodiversity inventory, with emphasis on Costa Rica's national protected areas *(National Biodiversity Inventory Division)*;
- transfer and dissemination of biodiversity knowledge *(Biodiversity Information Dissemination Division)*;
- organization and administration of biodiversity information *(Biodiversity Information Management Division)*;
- the search for sustainable uses of biodiversity by any and all sectors of society, and the promotion of these uses *(Biodiversity Prospecting Division)*.

The National Biodiversity Inventory

INBio's National Biodiversity Inventory builds on a long history of specialized national and international taxonomic research of Costa Rican fauna and flora, and initiated its activities in 1996, focusing on a limited number of taxa (plants, insects, molluscs and fungi). The field work for the inventory is being conducted by a small army of lay personnel trained in the vocation of 'parataxonomist'. The parataxonomist is not merely a collector, but is also the initial cataloguer of specimens and a direct link to the communities in and around Costa Rican wildlands. Cumulative information on parataxonomist courses is shown in Fig. 24.2.

Parataxonomists bring their collections to INBio on a monthly basis at which time technicians label, process and prepare the material for taxonomic identification by curators. Parataxonomists and technicians receive feedback, planning and guidance from INBio's curators who work within a larger net-work of national and international taxonomy experts. The goal is to generate properly identified reference collections and field guides, and provide electronic identification services that add knowledge to the organisms' natural history and document their distribution thoughout the national territory.

The results are impressive; as of 1996, the Inventory Division has 31 Biodiversity Offices in various conservation areas where 45 parataxonomists are stationed. The process includes 18 curators, 12 technicians, four labellers and a varying number of foreign visiting taxonomists (53 up until 1995). In addition, 373,000 entomological specimens were collected in 1995, increasing

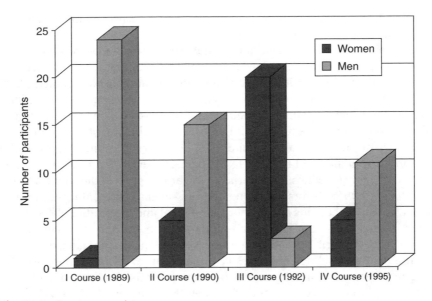

Fig. 24.2. Parataxonomist courses.

the reference collection to over 2.5 million insects. Of these, 2.1 million have been labelled, and 384,000 identified to the species level. Bryophyte specimens numbering more than 1000 were also added to the collection, while the taxonomy of plant species, estimated at 12,000, is almost completed, even though 43 new species of plants were described and 26 others were new records for Costa Rica. The malacology collection now includes 45,000 specimens of which 6600 have been classified to the family level, 16,400 to the genus level and 21,800 to the species level. At least eight are new species for the world and approximately 20 are new records for the country. Finally, a mycology department was opened in 1995, in order to inventory macrofungi as the basis for exploring potentially useful bioactive substances.

Biodiversity Information Dissemination Division

In order to rapidly spread biodiversity literacy and generate appreciation of what biodiversity information can offer, INBio's Biodiversity Information Dissemination Division is developing methods to actively distribute biodiversity information to all levels of society. This distribution comes about through offering natural history and taxonomic information to schools and universities, disseminating information about the commercial possibilities of conserved wildlands, working with legislators, participating in policy-making commissions and symposia, training conservation area staff, producing hard copy field guides and other types of biodiversity literature, holding national and international workshops, and much more. This division is promoting the development of a society whose ethical and moral values are rooted in respect for nature and the wise management of natural resources.

As part of this, in 1995, an educational project, the Biodiversity Education Program (PROEBI), was developed for two local primary schools to provide information on biodiversity conservation and sustainable development. Last year, 2440 primary school students visited INBio to learn about the value of biodiversity and current plans and strategies for its conservation and use. INBio's educational activities have caught the interest of the Ministry of Education which wishes to incorporate elements of biodiversity management in the national curricula. Additionally, INBio participated in 11 fairs and similar events explicitly for educational purposes in various parts of the country.

During 1995, visitors, including 40 mass-media representatives, numerous scientists, politicians and authorities of local and international organizations, joined INBio staff for discussions and an exchange of views and experiences. In April 1995, representatives from Cameroon, Madagascar and Ghana participated in a biodiversity prospecting workshop sponsored by several governments and international organizations. Similarly, a workshop supported by the United Nations Environment Programme (UNEP) was held for participants from 12 English-speaking Caribbean Islands plus Suriname, Guyana and Belize.

Cumulative information on documents published about and by INBio is found in Figs. 24.3 and 24.4, and yearly distributions of articles published about INBio are shown in Fig. 24.5.

Biodiversity Information Management Division

INBio has been described, quite rightly, as an information management organization. The amount of biodiversity information (specimen data, literature and field data) is rapidly growing, and when coupled with relevant support information such as topographic maps, soil maps, climate data, land use and much more, the data package becomes extremely complex. This data requires

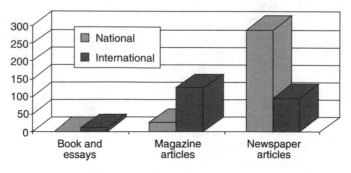

Fig. 24.3. Documents published about INBio.

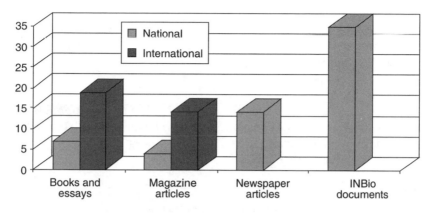

Fig. 24.4. Articles published by INBio Personnel 1995.

a capacity of analysis, management, presentation, distribution and integration not yet achieved by any set of biodiversity users in the world. INBio is bringing cutting edge technology in database management and development to bear on this challenge. This division is currently active in six main areas.

The Biodiversity Information Management System (BIMS)
The BIMS already includes all information generated by the Arthropod Department; botany, malacology and mycology databases are soon to be integrated.

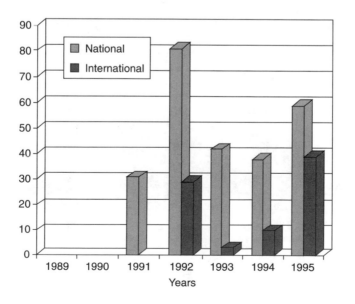

Fig. 24.5. Publications in the media.

Geographic Information Systems (GIS)
A series of aerial photographs of the Osa and Amistad Conservation Areas have
been accumulated as well as photogrametric maps with valuable political-
administrative information.

Multimedia
The Electronic Atlas for Agenda 21 (ELADA 21) is a multinational project
sponsored by International Development Research Centre (IDRC) (Canada) to
include information on environmental issues that permit social and economic
interpretation, among other analyses.

Internet
Electronic mail is used in INBio on a daily basis and more than 9000 screens
with general and specific information have been produced for INBio's World
Wide Web site (http://www.inbio.ac.cr). In partnership with the Missouri
Botanical Garden, an electronic version of the Manual of Costa Rican Flora can
be viewed on the site. Daily access to INBio via the www is presented in Fig. 24.6.

Biodiversity prospecting database
The increase in computer capacity and the acquisition of a Sun computer,
coupled with the integration of the Biodiversity Information Management
System (BIMS) and geographic information system (GIS), allow better handling
and more reliable information.

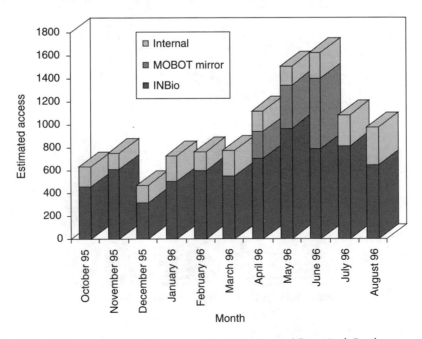

Fig. 24.6. Daily access to INBio www. MOBOT, Missouri Botanical Gardens.

Biodata
A new project for assistance in building Costa Rica's biodiversity data management capacity and networking was initiated in collaboration with UNEP.

The Biodiversity Prospecting Division

The search for new chemicals and genes is essential to INBio's biodiversity conservtion efforts. At present, these efforts are focused on the search for chemicals produced by plants, insects and microorganisms that may be of use to the pharmaceutical, medical and agricultural industries. In this context, the role of the 'bioprospector' is essential. Working closely with the parataxonomists, the bioprospectors follow biological leads, contribute to the natural history of potentially useful organisms collected and add key information to the databases. The bioprospectors collect 'prospectable' specimens (sampling that will not destroy nor will it promote genetic erosion) and make sure that non-damaging resupply is possible.

Prospecting and research processes are carried out in collaboration with local and international research centres, universities and the private sector. The set of criteria used by INBio to define its research agreements in 1991, was ahead of its time and included key elements (access, equity, transfer of technology and training) which were later agreed upon at The United Nations Conference of Environment and Development (UNCED) in Rio de Janeiro in 1992. Agreements stipulate that 10% of research budgets and 50% of any future royalties be awarded to the Ministry of the Environment and Energy (MINAE) for reinvestment in conservation. The remainder of the research budget supports in-country scientific and processing infrastructure and value-added activities, also oriented to conservation and the sustainable use of biodiversity.

INBio's strategy has focused on the development of a diversified portfolio of bioprospecting research agreements that foster innovation, learning and local capacity building. Some of the ongoing agreements are briefly discussed below.

Academic research agreements with Costa Rican universities
These agreements with Universidad de Costa Rica, Universidad Nacional and foreign universities (Strathclyde, Dusseldorf, Lausanne, etc.) vary considerably in scope but all of them are problem-oriented and search for new knowledge and products. The agreement with the University of Costa Rica, for example, allows for collaborative research on malaria as well as extremophilic organisms (archaeobacteria) living in hot volcanic springs.

Chemical prospecting in a Costa Rican Conservation Area
One of five International Cooperative Biodiversity Groups (ICBGs) in the world, this project is carried out in the Guanacaste Conservation Area in collaboration with the University of Costa Rica, Cornell University and The Bristol Myers

Squibb Company. The aim is to obtain potentially useful substances from tropical insects and upgrade human resources capabilities in the fields of ecology, taxonomy and ecochemistry.

INBio–Merck agreement
Considered a landmark, the INBio–Merck project has fully integrated issues of access, equity, technology transfer and capacity building into one agreement. The project searches for new pharmaceutical and agricultural products from plants, insects and environmental samples. The first agreement was initiated in 1992, renewed in 1994 and again in 1996. Promising results on the biological activity and characterization of chemical compounds have been obtained.

Antimicrobial and antiviral activity from natural compounds
This is a new agreement signed with the phytopharmaceutical company INDENA from Milan, Italy. It provides the opportunity to conduct considerable value-added research, including new batteries of microorganisms and bio-assays, for the first time in Costa Rica.

New fragrances and essences for cosmetic and household purposes
This is a small and innovative agreement with Givaudan-Roure Fragrances of New Jersey, USA, which aims to identify and collect interesting odours from the airspace near fragrant forest organisms.

2R, 5R-dihydroxymethyl-3R, 4R-dihydroxypyrrolidine (DMDP)
A nematicide obtained from *Lonchocharpus* sp. is currently being studied for domestication, extraction and field evaluation under an agreement with the British Technology Group and Ecos La Pacifica, in collaboration with the Costa Rican Banana Corporation, Kew Botanical Gardens and the Guanacaste Conservation Area. This is a promising, environmentally-friendly, cost-effective R&D inititative given the high environmental and financial costs associated with the application of synthetic nematicides to tropical crops.

Potential industrial use of extremophilic organisms
In collaboration with Recombinant Biocatalysis Co. of La Jolla, California, USA, INBio is studying microorganisms that thrive in extreme pH and/or temperatures in the search to clone DNA sequences that express enzymes for use in bioremediation and industrial processes. This collaborative research is carried out in conjunction with the University of Costa Rica's Center for Cell and Molecular Biology.

Investment agreement with the Government of Canada
The Governments of Canada and Costa Rica reached a debt-for-nature swap agreement which provides support for the 'Biodiversity and Socio-Economic Development Project', an initiative to sustainably use the country's biodiversity resources and contribute to INBio's consolidation. The project supports the

development of intellectual and industrial uses of biodiversity (Information Dissemination and Bioprospecting Divisions) and INBio's endowment fund during the period 1996–2000, for a total of Can$5.67 million.

Quantifying of direct benefits from bioprospecting

The issue of benefits accrued from bioprospecting is difficult given the inherent complexities of assigning value to the accumulated and increased knowledge of our own biodiversity, to the transfer of know-how and technology that has occurred, and to enhanced human resources capabilities. In Table 24.1, a simplistic approach is presented to quantify the direct financial contributions to other Divisions in INBio, the Guanacaste Conservation Area, the Ministry of the Environment and national universities. It is important to keep in mind that over US$2 since 1993 is significant for a small country like Costa Rica. Interestingly enough, MINAE has used its share to specifically support the management and upkeep of Costa Rica's Cocos Island, a unique and diversity-rich marine sanctuary and a clear-cut example of direct bioprospecting benefits flowing to conservation.

INBio's successes

The overall goals of INBio have been stated as saving, knowing and using bio-diversity for the benefit of society. There are many reasons why INBio has been not only a viable organization, but also a world leader in the context of the Biodiversity Convention. The following aspects are the most relevant ones.

The right circumstances at the right time
The political, scientific and socio-economic environment was ripe in 1989 for creating a flexible association closely linked to the government to assume the national inventory and other biodiversity management activities.

Political support
INBio, given the high priority assigned by all sectors of society to biodiversity management, has been fully supported by the various political factions and different administrations.

Leadership and vision
INBio's founders had a very clear vision and mission in mind and have been able to inspire and maintain leadership.

A commitment to innovation
There were no models and no significant pilot projects to learn from, therefore new concepts such as parataxonomists, databases and datasystems, prospecting

Table 24.1. Estimated direct contributions and payments made by the Biodiversity Prospecting Division[1] (US$).

Contributions and payments	1991–93	1994	1995	1996	January–March 1997	Total
Ministry of the Environment and Energy (MINAE)	110,040	43,400	66,670	51,092	23,531	294,733
Conservation areas	86,102	203,135	153,555	192,035	30,394	665,221
Public universities of Costa Rica	460,409	126,006	46,962	31,265	7,522	672,164
Other groups at INBio	228,161	92,830	118,292	172,591	19,834	631,708
Total	884,712	465,371	385,479	446,983	81,281	2,263,826

[1] The amount of resources going to conservation and research programmes of the national conservation areas and INBio could increase dramatically in the form of royalties, if hits (molecules with positive response in a specific bioassay) already identified within the various agreements are further developed and reach the market.

agreements, bioliteracy campaigns and others ideas have had to be continuously developed.

A diversified portfolio

INBio has worked and experimented with various taxanomic groups, research agreements and R&D modalities to fulfil its mission. It is a learning institution *par excellence.*

Strategic alliances

INBio recognized its own limitations and the need to avoid costly duplications from the beginning. A key strategy has been forging alliances locally and internationally with the government, academic, research and private sectors in order to maximize resources and catalyse processes.

INBio in numbers

Occasionally, INBio has been cited as an organization deriving most of its income from bioprospecting or even from a single research agreement with a private company. To the contrary, as can be seen in Figs. 24.7 and 24.8, INBio has operated on a funding system similar to many conservation non-governmental organizations (NGOs).

INBio's income is highly diversified and the amount generated from agreements with the private sector represents only 16% of the total. Similarly, in terms of expenditures, the largest portions correspond to the national inventory (34%) and to bioprospecting (22%). INBio receives considerable support (except monetary) from the Government; therefore INBio's Board of Trustees is actively developing an endowment fund to buffer budget fluctuations.

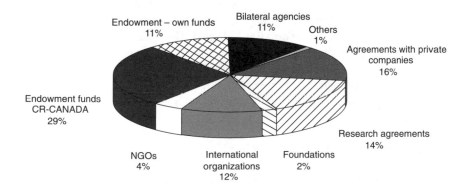

Fig. 24.7. National Biodiversity Institute (INBio) income 1996. CR–CANADA, an agreement between Costa Rica and Canada to support biodiversity conservation and development.

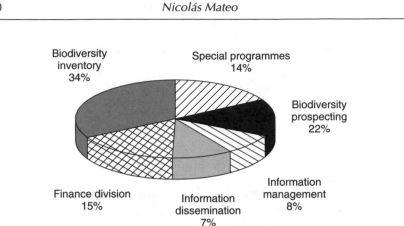

Fig. 24.8. National Biodiversity Institute (INBio) operational costs per division 1996.

Other initiatives and remaining challenges for Costa Rican biodiversity management

The rapid loss of biodiversity worldwide that we see today is a consequence of the destruction of tropical forests and water masses. As a result, biodiversity conservation is tightly linked to the maintenance of forestry and reforestation programmes. Although Costa Rican forests are well conserved as government protected conservation areas, only 180,000 ha of privately owned forests remain. Financial incentives must be provided to private landowners to stimulate reforestation. Unchecked logging often leads to monocultures as well as decreased biodiversity and other negative ecological impacts. Further incentives are required to cultivate mixed stands on private lands. It is in INBio's interest to couple these reforestation programmes with biodiversity conservation and regeneration in order to create stable ecosystems with multiple uses. Table 24.2 summarizes some of these multiple uses and their values, based on the Costa Rican Forest Sector Review (World Bank, 1994). These data provide rough estimates and are not applicable to all types of forests. It is important to note that carbon sequestration is only feasible in deforested areas, ecotourists will prefer certain regions, hydroelectric power can only be generated in regions of high altitude and logging requires that forests are easily accessible.

Carbon sequestration
The first indications of global warming led to hectic activities worldwide to reduce the emission of carbon dioxide (CO_2) and provide additional CO_2 sinks. The dense biomass in tropical forests has the capacity to bind about 150 t CO_2 ha^{-1}. Worldwide, companies with high CO_2 emissions are therefore considering funding reforestation programmes in the tropics for carbon offset. The international market for carbon sequestration is just evolving and the final price for each tonne of fixed CO_2 will ultimately depend on supply and demand and the magnitude of the economic damage that can be avoided by this measure. The

Table 24.2. Potential value of tropical rain forest in US$ (World Bank, 1994).

Use	Value ha^{-1}
Carbon sequestration	845
Sustainable logging	620
Existence and options	295
Ecotourism	209
Hydroelectric power	207
Urban and rural water supply	47
Prospecting	15–22*
Total	2241.5

* From Mendelsohn and Balick, 1995.

value given in Table 24.2 was calculated on the basis of the economic damage created by doubling CO_2 in the atmosphere and is estimated at US$5.63 t^{-1}.

Sustainable logging
Production of wood, paper, paper pulp and firewood is still the most economically important use of forests. Sustainable logging systems must avoid monocultures, and local tree species have to be cultivated to regenerate stable and diverse ecosystems.

Existence and options
These estimations are generated by national and international funds (e.g. World Wildlife Fund for Nature (WWF), International Union for the Conservation of Nature (IUCN), etc.) to finance the conservation of protected areas. They simply reflect the price the world is willing to pay to maintain tropical forests and their biodiversity.

Ecotourism
This value was calculated on the basis of an analysis of tourist expenditures in the Monteverde Natural Reserve.

Hydroelectric power and urban and rural water supply
These values depend on the function of the forest as a water reservoir. Further values could be added for other functions, e.g. the avoidance of flooding, which causes significant damage to the Costa Rican economy each year.

Prospecting for new pharmaceuticals
This value of prospecting for new pharmaceuticals is based on the following assumptions made by Mendelsohn and Balick (1995): with 5.8 % of the world's tropical forest species of higher plants, Costa Rica should also have 5.8 % of the

undiscovered drugs from tropical plants. According to the authors, each drug is worth an average of between US$94 million and US$449 million to society as a whole after subtraction of development, production, marketing and distribution costs. If Costa Rica were to prospect its botanical diversity with one company only, this search would probably yield two or three new drugs. Thus the value of forests ha^{-1} will be $192 to $288 million per 1.3 million ha of current forest in Costa Rica. Prospecting for plant diversity with more than one company will increase this value. Evaluation of further organisms like insects, molluscs, fungi and bacteria will further increase the pharmaceutical value to a degree where it becomes compatible with other uses.

The overall value of 1.3 million ha of Costa Rican tropical forests on the basis of the assumptions and parameters in Table 24.2 is estimated at US$182 million year^{-1}. In comparison, the annual meat export of Costa Rica in 1994, yielded US$32 million (Anon., 1995). To achieve this export volume about 1.2 million ha of tropical forest were turned into pasture land from 1943 to 1987!

Benefits from conservation areas can also be projected to adjacent agricultural lands (buffer zones). Ongoing research in northern Costa Rica attempts to evaluate and measure positive impact of conservation areas on orange plantations in regard to biocontrol and water accessibility. Results are not yet available.

Further values can be expected from non-wood products like nuts, fruits, ornamental plants, fibres, oils, phytopharmaceuticals, and meat and leather from wild animals, which could be produced more economically in a forest than by domestication. Some of these products are already on the market, like the medicinal plant *Smilax* sp. However, these products are collected in a non-sustainable way and have already led to species extinction in certain regions. More innovation and research is required to establish sustainable collection and cultivation techniques in the forest and to develop new products for the international market.

A Final Word

The rapid globalization of economies and the sciences, and the implications of the Convention on Biological Diversity, will bring profound changes in the way agricultural and wild biodiversity R&D will take place in the next few years. Local conditions and strategies will continue to exert a great influence; however, a global scene where bilateral and multilateral agreements and strategic alliances take place will be influential factors in biodiversity conservation and utilization.

The experiences and results obtained by INBio in only a few years are not meant to be a model (conditions, priorities and comparative advantages vary considerably from country to country). Nevertheless, INBio is a very useful pilot project from which several lessons can be easily drawn:

- society will be interested in conservation only if it perceives clear spiritual, intellectual and economic valucs derived from biodiversity;
- the operational flexibility of a private organization with full government backing permits rapid advancements in biodiversity management;
- value-added, negotiation skills, national capacity building and innovation are key ingredients for success.

Mesoamerica has been the cradle of many agricultural crops, some of them of world importance like maize and common beans. These crops were domesticated from wild relatives found in the fast disappearing regional forest. New genes, new crops, new pharmaceutical and agricultural products constitute the last frontier of wild biodiversity. It is our responsibility to guarantee its intelligent use for the benefit of the next generations.

Acknowledgements

The author wishes to express his appreciation to N. Martin, W. Nader, A. Lovejoy, R. Gámez, A. Sittenfeld and H. Ramírez, for information, corrections and editing provided during the preparation of this chapter.

References and Further Reading

Anon. (1995) *Costa Rican Export Directory*. CENPRO, 104 pp.
Instituto Nacional de Biodiversidad (1995) *INBio Memoria Anual*. p. 79.
Mcndelsohn, P. and Balick, M.J. (1995) The value of undiscovered pharmaceuticals in tropical forests. *Economic Botany* 49(2), 223–228.
Proyecto Estado de la Nación (1995) *Estado de la Nación en desarrollo humano sostenible*. Proyecto Estado de la Nación, San José, Costa Rica, 125 pp.
World Bank (1992) Informe del Banco Mundial. No. 11516 *Costa Rica Forestry Sector Review* December 31.
World Bank (1994) *Costa Rican Forestry Sector Review*. World Bank, Washington DC.

Developing an Agricultural Biotechnology Business: Perspective from the Front Lines

Pamela G. Marrone

AgraQuest Incorporated, 1105 Kennedy Place, Davis, CA 95616,USA

AgraQuest Inc. discovers, develops and markets environmentally friendly and effective natural products for farm, home and public health pest management. AgraQuest Inc. recently raised its first major round of financing (US$3.2 million) from venture capital sources after first raising US$500,000 in seed money to start the company. AgraQuest started in 1995, when there were few commercial successes for agricultural biotechnology products, no companies were profitable and venture capitalists preferred to invest elsewhere. This was also before the current astronomical valuations of agricultural biotechnology businesses. How was AgraQuest able to convince investors to invest in the company? The important message for investors is to convince them that you are creating a real business. This means products, not just interesting technology and research. Do you know the market? What is the competition? How much will your product cost? How will it be produced? University or other researchers who want to spin off technology into new companies may want to first 'incubate' their technology to test the feasibility for commercialization to advance from research into development.

Introduction

From 1992 to 1995, capital for starting agricultural biotechnology (agbiotech) companies was at an all time low. Few agbiotech products were on the market and none of the companies that started in the 1980s were profitable. There were no agbiotech versions of Genentech or Amgen. This was also before any of the recent acquisitions of small agbiotech and seed companies (e.g. Calgene, Agracetus, Holdens, Mycogen and Plant Genetic Systems (PGS)) by large companies (e.g. Monsanto, AgrEvo and DowElanco). Agbiotech, with its long-term product horizon and capital intensity, was not as desirable for venture capital investors as internet, telecommunications and healthcare biotechnology companies. In addition, there are few venture capitalists with an in-depth knowledge of agricultural businesses.

AgraQuest, which started in 1995, raised venture capital money by bringing the company to a stage of perceived lower risk to investors and by focusing on investors with a knowledge and interest in agriculture. This chapter outlines the history of AgraQuest and points out different sources of capital for would-be agbiotech entrepreneurs.

Mission and History of AgraQuest

AgraQuest discovers, develops and markets environmentally friendly, safe and effective natural products for farm, home and public health pest management. AgraQuest products will be priced competitively with newer chemicals, while generating good gross margins. They will provide better and more consistent efficacy and will be easier to use than existing biopesticides. Our products will be perceived as a good value for the price.

The history of the company's start-up is outlined below:

June 1994	Conceived of AgraQuest
June–October 1994	Business plan written at nights and weekends
October 1994	Assembled management team with 70 years of combined experience (President & CEO, Chief Financial Officer and Vice-President of Product Development)
January 1995	Incorporated company in the state of Delaware
April 1995	Rented laboratory and office space
April–June 1995	Raised seed capital (US$500,000)
June–December 1995	Hired three key scientists
September 1996	Added Vice-President of Business Development and Marketing to management team
December 1996	Raised US$3.2 million first round financing

The financing completed at the end of 1996 was oversubscribed, and the company sold all stock (Series B preferred) that was authorized by the board of directors (45% ownership of the company was sold). Investors included Rockefeller & Company's Odyssey Fund (a representative joined AgraQuest's board of directors), partners in a US$3 billion investment firm, one of the largest cotton ginning and trading companies in Europe, three large growers of grapes, apples and pears (one joined the board), and several distinguished agricultural and food industry executives. This financing was typical of the recent trend to have 'angels' (high net worth individuals and successful entrepreneurs) in the deal with sophisticated venture capitalists.

Problems for Agbiotech Funding

Although agbiotech is considered more successful now than it was 2–5 years ago, it is still a difficult task raising money for agricultural ventures. First and most important, there are few investors with expertise in agriculture and agbiotech. Venture capitalists invest in industries they know, and in those that have a history of successful investing. Second, although investors who exited early from their investments in agbiotech start-up companies in the 1980s made good returns, there were no blockbuster successes such as agricultural versions of Genentech or Amgen. Third, agbiotech must compete with telecommunications, software, internet and healthcare biotechnology for venture capital. Recent return on investment from software, internet and telecommunications companies has been >50% year^{-1}. Fourth, there are few new agbiotech companies started which continue to generate investor interest (no critical mass). Fifth, investors must see an exit strategy (how will they get their money out of the company). Until recently, the exit strategy for agbiotech companies was unclear. Now that large agri-chemical companies have bought most of the small agbiotech and seed companies (e.g. DowElanco/Mycogen; Monsanto/Calgene; Monsanto/Holdens; Monsanto/Agracetus; Empresas de la Moderna/DNA Plant Technologies (ELM/DNAP); AgrEvo/PGS) and significantly raised company valuations, the exit strategy 'sell to a large company for a good price' is more obvious.

AgraQuest was one of few agbiotech start-ups in 1995. AgraQuest raised money during the worst venture capital funding climate in agbiotech history. What are the factors in this success? One success factor was the focus on investors with interest and experience in agriculture and agbiotech. Also, the focus was primarily on products, not research. By licensing product candidates from universities and small companies, AgraQuest quickly developed a product portfolio that would have taken four years to develop from in-house efforts. This significantly reduced the company's perceived risk to investors. A company's patent portfolio is very important to investors, because it is the asset that can keep competitors at bay. AgraQuest quickly developed a proprietary portfolio by filing seven patents and licensing two others.

Most important in the search for funding, AgraQuest practised and learned from mistakes made when presenting to investors. By listening to feedback, the message was continually honed and the business plan refined weekly. The business plan today is much more analytical and quantitative than the first version, which was perceived as a research and development (R&D) company, not a business. Investors demanded more numbers and data to support statements and assumptions in the plan. Supporting data was provided to the business credo:

> AgraQuest's products will be priced competitively with newer chemicals, while generating good gross margins. They will provide better and more consistent efficacy and will be easier to use than existing biopesticides. Our products will be perceived as a good value for the price.

How this was going to be achieved had to be demonstrated.

Another key success factor is board members with good contacts. One board member provided us an entrée to successful entrepreneurs and retired business executives from food, agribusiness and high technology companies. These investors brought in their colleagues and friends, which included sophisticated partners in a major West Coast investment firm.

Sources of Capital

Investor networks

AgraQuest obtained its lead investor, Rockefeller & Company's Odyssey Fund, from the Investors' Circle. The Investors' Circle (West Chicago, Illinois, USA) is a group of key accredited investors who promote the common goal of developing the socially responsible investment industry, without sacrificing economic returns. Each year, they hold two venture fairs where eight to ten companies present to 100–200 investors. There are other investor networks that successfully raise capital for start-ups by matching entrepreneurs with interested investors, such as 'The Capital Network' (Austin, Texas, USA) and 'International Capital Resources' (ICR, San Francisco, California, USA). ICR specializes in matching 'angels' with entrepreneurs.

'Friends and family'

To start new businesses, more money is used from 'friends and family' than from any other source. For capital intensive businesses such as agbiotech, 'friends and family' money may get a business started, but a larger source of money such as institutional venture capital is usually required to sustain the business. AgraQuest raised US$500,000 in 'friends and family' money, which allowed the company to establish a healthy discovery and product portfolio and reduce the risk to larger investors. In the US, it is easy to accept money from 'friends and family' in a private stock offering via rule 504. This rule specifies that for stock offerings under US$1 million, the company can accept up to 20–30 'unqualified investors' (the exact number varies from state to state). 'Unqualified' investors are individuals you know ('friends and family') who make less than US$200,000 year^{-1} or have less than US$1 million in assets. These 'friends and family investors' should be passive investors who do not want to meddle in the company's day-to-day business, since they would risk tainting the company for more sophisticated institutional investors.

'Angels'

'Angels' are defined as high net-worth individuals with income of more than US$200,000 year^{-1} or assets more than US$1 million year^{-1}. Although most people have not heard of the 'angel' market, more 'angel' money is invested in companies each year than traditional venture capital companies. 'Angels' invest US$10–20 billion in companies compared with US$7–10 billion from the venture capital market (Mehta, 1996). Although rare in the past, more deals now combine 'angel' money with venture capital money in a private placement stock offering (Mehta, 1997).

'Angels' are not easy to find, however, they value their privacy and are not listed in directories like venture capital companies (e.g. *Pratt's Guide to Venture Capital Sources by Testa Hurwitz & Thibeault, Attorneys At Law* (Yong and Weissberg, 1997)). The best sources for 'angels' are people you know, and in particular, board members. Careful selection of board members with a vast network of contacts is critical to any start-up agbiotech company. 'Angels' may be passive or active investors. Some may call and visit frequently, appoint a board member (if the size of the investment merits it), and monitor progress. Others may invest and are never to be heard from again. Likewise, AgraQuest's 'angel' investors vary in their involvement with the company. A large grower (of diversified speciality crops)/investor has a seat on the board of directors; others have no hands-on involvement. 'Angels' invest in people they like and trust, and in businesses they know and like. They can make decisions quickly and can complete a deal much faster than a corporate partner or venture capitalist. A company's return on investment (ROI) is important to an 'angel' investor (typically 20–30% year^{-1}), but they are not as demanding as venture capitalists (30–50% year^{-1}).

Corporate partners

Large agribusinesses, such as agrichemical, food and seed companies, can be excellent sources of funding for research and product development projects. In the pharmaceutical biotechnology arena, large pharmaceutical companies have invested billions of dollars into small early-stage biotechnology companies. In agricultural biotechnology, large companies have been actively investing in and acquiring established agbiotech companies (e.g. Monsanto/Calgene, Monsanto/Agracetus; DowElanco/Mycogen; ELM/DNAP). However, unlike pharmaceutical biotechnology, there have been fewer instances of large agribusinesses funding early-stage agbiotech companies, because agrichemical companies typically outsource less of their discovery than pharmaceutical companies and have less discretionary money for outside deals.

The advantage of having a large company as a corporate partner is that the partnership validates and gives credibility to the smaller agbiotech partner. Unfortunately, negotiating with a larger company can take more than a year. Thus, it is likely that corporate funds will be used for expanding rather than launching a company. There is a higher chance for success if access into the large company is made at high levels (Director, Vice-President and above) in the organization. Lower-level managers and scientists can feel threatened if funds that may have gone to their projects are allocated to an outside partner and they may fight against funding a new project in an outside company (the 'Not Invented Here' syndrome).

Small, early-stage companies may be so eager to get funding that they sell or licence rights to the company's core technology or to the major market for the company's product. This can taint future arrangements with other corporate partners and could be fatal to finding additional funding from venture capitalists. With corporate partners, it is best to start with a deal of narrow scope and then expand the collaboration if the two companies work together successfully.

Customers

Customers can be one of the best sources of funding for new companies. They understand the market and can help in the early stages of the product development cycle. Seeing the potential of AgraQuest's biopesticides to their business, several large growers of diversified speciality crops (AgraQuest's potential customers) invested in AgraQuest. The growers provide AgraQuest with farm sites for field development of the biopesticide products, which gives growers the opportunity to examine and understand the products ahead of their competitors.

Government

Several agbiotech companies have received start-up assistance from local, regional, state or federal economic development funds. Typically funding ranges from US$50,000 to US$400,000. Some states in the US that have venture funds for starting new companies are Oregon, New York, Virginia and Wisconsin. The provincial governments in Canada have funded the start-up of several successful companies. Saskatchewan has become a world-renowned centre for agbiotech. Many of these early stage Saskatchewan agbiotech companies received provincial goverment seed money to prove feasibility and were then able to attract larger venture capital investments.

Several other countries besides Canada have venture funds for investing in new companies that commit to starting and building in that particular country. Israel has funding specifically for agbiotech and biopesticide projects. Denmark has a biotechnology venture fund that has invested in agbiotech companies. Also active in agbiotech are funds from Singapore (e.g. Singapore Bio-Innovations) and Taiwan.

In the US, several government agencies provide small grants for innovative research that eventually leads to a commercial product. These Small Business Innovation Research (SBIR) grants, funded by the United States Department of Agriculture (USDA), Department of Defense, Environmental Protection Agency and National Science Foundation, are usually US$55,000 in phase I (conduct research with commercial potential) and US$250,000 in phase II (demonstrate commercial feasibility). Companies can get trapped into writing SBIR grants on subjects that are ancillary to the company's main focus simply to get money into the company. It is best to apply for SBIR funds for projects already in progress at the company. Successful applicants for phase I grants are companies that already have data demonstrating initial feasibility of a particular technology. For example, AgraQuest was awarded a phase I SBIR grant for R&D of two biofungicides, representing two major ongoing company projects. In the phase I application, we presented greenhouse data which demonstrated the initial potential of the biofungicides.

Venture capital

Institutional venture capital is very difficult to obtain for agbiotech companies and start-up companies in general. Venture capitalists (VCs) fund only 1% of the business plans they see. They fund industries they know and in which they have made money. Very few VCs have agbiotech expertise; therefore, selling an agbiotech company is a harder task than selling a pharmaceutical biotechnology, internet, software or telecommunications company.

VCs typically require 30–50% year^{-1} ROI. Entrepreneurs wishing to start an agbiotech company should make sure that the business plan clearly outlines how the VC is going to make at least a 30% year^{-1} ROI. VCs also demand a clear 'exit strategy'. The exit strategy is how the VCs are going to liquidate their investment in the agbiotech company. Do you intend to take the company public (carry out an 'initial public offering' or sell the company)? The lead VC (the VC that negotiates the price of your stock in the private placement stock offering and coordinates with other VCs in the deal) typically requests at least one seat on the company's board of directors. Besides monitoring the investment, the VC on the board offers skills to supplement and complement the company's management team in areas such as finance, strategic planning and business development.

A record amount of money has flowed into venture capital funds during the past 5 years. As a result, the amount of money invested per company has increased, with fewer VCs investing in smaller, seed-stage deals. Therefore, a good strategy for an agbiotech company that is unable to find a VC for the start-up (seed) stage is to raise money from angels or 'friends and family' to make the company appear more attractive and lower risk to VCs.

Although VCs have more money than ever before, they are more selective in the companies they fund. This means that business plans need to be very analytical, with defensible facts and figures about market size, market penetration, product pricing compared with competitors and margins. This is especially important when describing the company to investors that may not have a deep knowledge of agbiotech and agribusiness.

References

Mehta, S.N. (1996) Angel investors to get on-line service. *The Wall Street Journal* October 28.

Mehta, S.N. (1997) Top-notch start-ups get picky about their partners. *The Wall Street Journal* May 7.

Yong Lim and Weissberg, E. (eds) *Pratt's Guide to Venture Capital Sources by Testa Hurwitz and Thibeault, Attorneys at Law*. Venture Economics, Boston, MA.

Index

2020 Vision (IFPRI) 180

AARD *see* Agency for Agricultural Research
    and Development
Abaca 42
Abiotic stress tolerance 115
ABSP *see* Agricultural Biotechnology for
    Sustainable Productivity
Academic research agreements, in Costa Rica
    325
ACIAR *see* Australian Center for International
    Agricultural Research
ACR *see* Agribiotecnología de Costa Rica
AFLP *see* Amplified fragment length
    polymorphisms
Africa 6, 10, 25, 50, 53, 90, 162, 181, 237,
    243, 286
    see also East Africa; North Africa; South
        Africa; Southern Africa;
        Sub-Saharan Africa; West Africa
African cassava mosaic virus 237
African Regional Conference for International
    Co-operation on Safety in
    Biotechnology, 1993 216
Agency for Agricultural Research and
    Development (AARD) 35, 37, 43
Agency for Technology Assessment, Indonesia
    (ATA) 38
AGERI *see* Agricultural Genetic Engineering
    Research Institute
Agracetus Company 277, 335, 337, 339
AgraQuest Inc. 335–342
Agrarian Reform Law, Peru 250

Agreement on Trade-related Aspects of
    Intellectual Property Rights (TRIPS)
    277, 278, 280
AgrEvo Company 336
Agribiotecnología de Costa Rica (ACR) 5, 7,
    11, 125, 127, 183
Agribusiness systems 44
Agricultural biotechnology research *see*
    Biotechnology research and
    development
Agricultural Biotechnology for Sustainable
    Productivity (ABSP) project 1–14,
    23, 35, 36, 38, 39, 40, 41, 58, 89,
    92, 94, 95, 125, 128, 131, 182, 183
Agricultural Biotechnology Policy Seminars
    174, 175, 180
Agricultural development, in Egypt 17–26
Agricultural Genetic Engineering Research
    Institute (AGERI), Egypt 5, 6, 9, 13,
    17, 19–26
    capacity building 19–21
    goals of 25–26
    human resources development 21–23
    research and scientific collaboration
        23–25
Agricultural products, trade in 311–314
Agricultural Research Center (ARC), Giza,
    Egypt 19–20, 25, 177
Agricultural Research Management 35
Agricultural Research System 8
Agricultural Technology Utilization and
    Transfer (ATUT) project 23
Agriculture Research Management Project 1
    (ARMP 1), Indonesia 37–38, 39

343

Agrobacterium tumefaciens 6, 73, 79, 83, 92, 93, 111, 116
Agroinfection system 79, 84
Alexya stellata 42
Algeria 133, 134, 137, 138, 139
Alternaria sp. 24
Aluminium toxicity 38
America *see* Central America; Latin America; North America; South America
Amplified fragment length polymorphism (AFLP) 68, 81, 165–166
Amylase 141
Angelica acutiloba 42
'Angels' 338–339
Animal and Plant Health Inspection Service (APHIS), USDA 9, 20, 208, 311–313
Animal husbandry 2
Anther culture, of rice 98–99
Antimicrobial activity 326
Antiviral activity 326
Aphids, green peach 94
APHIS *see* Animal and Plant Health Inspection Service
Apple 140
Apricot 140
Arabidopsis thaliana 106
ARBN *see* Asian Rice Biotechnology Network
ARC *see* Agricultural Research Center
Argentina 80, 81, 247, 251, 312
Arizona, University of 10, 23
ARMP 1 *see* Agriculture Research Management Project 1, Indonesia
Artificial insemination 293
ASB *see* Ostrinia furnacalis
Asgrow Seed Company 9, 149–151, 183
Asia 1, 97, 112, 134, 181, 201–212, 286
see also Southeast Asia; West Asia
Asian Development Bank (ADB) 38
Asian Rice Biotechnology Network (ARBN) 35, 38, 39, 40
Asian stem borer *see* Ostrinia furnacalis
Asparagus 29, 141, 249
Aspergillus niger 76
Australia 127
Australian Center for International Agricultural Research (ACIAR) 35, 36, 38, 39, 40, 41
Avocado 140

BAC *see* Bacterial artificial chromosome
Bacillus thuringiensis (Bt) 4, 5, 6, 24, 25, 32, 33, 40, 63, 64, 65–66, 67, 70, 73, 75, 77, 112, 183, 185, 282, 294
Backcrossing 107, 108, 111
Bacterial artificial chromosome (BAC) libraries 98, 106
Bacterial blight (BB) 100, 102–103, 105–106, 107, 108, 109, 110, 113, 237

Bacterial leaf blight (BLB) 36, 38, 40, 41
Bambasa longispiculata (bamboo) 140
Banana 4, 5, 7, 29, 54, 125–131, 178, 183, 311, 312
biotechnology of 126–128
Gran Nain 128, 129, 130
Gros Michel 128, 129
plant, transgenic 128–130
production, economic impact of 126
Valery 128, 129
world production 125
Banana bunchy top virus (BBTV) 130
Banana streak virus (BSV) 130
Banana weevil borer (*Cosmopolites sordidus*) 130
Bandung Technology Institute (ITB), Indonesia 38
Bangladesh 108
Barley 89, 106, 114, 134, 141
Bayoud disease 133, 134, 136–142
BBTV *see* Banana bunchy top virus
Beachy, Dr Roger 89, 90, 91, 149
Bean 29, 114, 178, 333
see also Faba bean; Soybean
Bean Cowpea Collaborative Research Support Programme 10
Bean yellow mosaic virus (BYMV) 24
Belgium 77
Belize 322
Beneficiaries 173–199
BIMS *see* Biodiversity Information Management System
BIO *see* Biotechnology Industry Organization
BioComputing and Networks Unit, AGERI 20, 21
Biodiversity 2, 133, 136, 241, 251, 274, 325
wild 317–333
Biodiversity Convention 318, 327
Biodiversity Education Program (PROEBI), Costa Rica 322
Biodiversity Information Dissemination Division, INBio 320, 321–322
Biodiversity Information Management Division, INBio 322–325
Biodiversity Information Management System (BIMS), INBio 320, 323
Biodiversity prospecting database, INBio 324
Biodiversity Prospecting Division, INBio 320, 325–327
BioLink newsletter 8
Biolistic™ gun 66, 70, 79, 97, 116
Biological nitrogen fixation 40, 54
Bioprospecting 317, 325, 327
Bioreactor technology 4, 125, 128, 183
Biosafety 1, 3, 8, 9–10, 12–13, 14, 20, 29, 31, 33, 36, 41, 94, 186, 208, 213, 214, 219–225, 286, 287, 294
Biosafety Committee, USAID 13
Biosafety regulations 28, 42–43, 49, 56, 213–228

costs 223–225
  development of 218
  harmonization of 217–218, 220
  implementation of 218–219
BioServe database 176, 178
Biotechnology Advisory Commission (BAC),
  Stockholm 59
Biotechnology capacity building *see* Capacity
  building
Biotechnology Career Fellowship 204, 205,
  210
Biotechnology impediments *see* Impediments
  to biotechnology
Biotechnology Industry Organization (BIO) 8,
  11, 153
Biotechnology partnerships *see* Partnerships
Biotechnology research and development
  (R&D) 27–48, 49, 54, 57, 91–95,
  173–199, 229, 257, 285–296
Biotechnology Research Division (BRD), CRIFC
  37, 39
Biotechnology Research Institute (BRI),
  Zimbabwe 285–286, 287, 290,
  291–296
  project proposals 295–296
  research activities 294
Biotechnology, role of, in sub-Saharan Africa
  53–57
Black Sigatoka disease (*Mycosphaerella fijiensis*)
  126, 128, 129, 130
Blast *see* Rice blast
BLB *see* Bacterial leaf blight
Blight *see* Bacterial blight; Late blight disease
Bogor Agricultural Institute (IPB) 38
Bogor Research Institute for Food Crops 38
Bolivia 82
Botrytis fabae 24
BPH *see* Brown plant hopper
Bradyrhizobium japonicum 38, 40
Brazil 11, 233, 237, 239, 240, 241, 247, 251
BRD *see* Biotechnology Research Division,
  CRIFC
BRI *see* Biotechnology Research Institute
Bristol Myers Squibb Company 325–326
British Technology Group 326
Brown plant hopper (BPH) 38, 44, 101, 103,
  107, 108, 109, 110, 113
Bryophytes 321
BSV *see* Banana streak virus
Bt see Bacillus thuringinesis
Burkina Faso 51
Burundi 54
Business development, in agricultural biotech-
  nology 335–342
BYMV *see* Bean yellow mosaic virus

Calgene 30, 149, 335, 337, 339
Callus regeneration 139
Cameroon 54, 322

Campesinos, Peru 250
Canada 326–327, 340
Cananga 42
Cantaloupe 147, 148, 149, 154, 156, 158
Capacity building 3, 35, 42–43, 201–212,
  218, 220, 222–224, 247–252,
  285–296, 326
Capriprox-RVF 55
Carbon sequestration 330–331
Cardamom 42
Caribbean 126, 127, 129
Cashew (*Anacardium occidentale*) 42, 45, 46
Cassava 54, 178, 229–245, 295, 313
  see also African cassava mosaic virus
Cassava Biotechnology Network (CBN) 188,
  229, 230, 231–244, 314
  membership 237–238
  scientific meetings, international
    233–236
  small grants programme 236–237
CBD *see* Convention on Biological Diversity
CBN *see* Cassava Biotechnology Network
CC *see* Country coordinator
cDNA sequencing 161
Cell death, programmed 75, 76
Center for National Biotechnology Network,
  Indonesia 37
Central America 125, 127, 129
Central Research Institute for Food Crops
  (CRIFC), Indonesia 5, 7, 8, 35, 37,
  63, 69, 70, 93, 183
Centres of strength 203, 210–211
Centro de Investigaciones en Ciencas
  Veterinarias, Argentina 81
Centro Internacional de Agricultura Tropical
  (CIAT) 188, 231, 233, 237, 238
CG, CGIAR *see* Consultative Group on
  International Agricultural Research
Chickpea 178
Chile 312
China 54, 90, 99, 108, 204, 206, 207, 209,
  210, 211, 212, 233, 236, 239, 240,
  251
China National Center for Biotechnology
  Development 206
Chitinase 24, 114, 115, 141
Chrysanthemum cenerarefolium 42
CIAT *see* Centro Internacional de Agricultura
  Tropical
CIMMYT *see* International Center for Maize
  and Wheat Improvement
Cinnamon 42
CIP *see* International Potato Center
Citrus vein phloem degeneration (CVPD) 40
Clove 42, 45
CMS *see* Cytoplasmic male sterility
CMV *see* Cucumber mosaic virus
Coat protein technology 4, 6, 44–45, 75, 76,
  89, 90, 91, 112, 114, 147–160,
  183, 185, 313

Cocoa 42, 46, 54, 178
Coconut 45, 46, 163
Cocos Island, Costa Rica 327
Coffee 5, 7, 42, 46, 178, 183
Cohen, Dr Joel 91
Collaboration 3, 4, 6–8, 10, 11, 13, 25, 63,
    174–176, 180, 182, 187, 189, 190,
    205, 209, 221, 236, 237, 238,
    243–244, 325
  international 35, 41, 175–176
  see also Initiatives, international
Colombia 126, 127, 185, 233, 239, 241,
    242
Colorado potato beetle 73
Commercialization 9, 28, 30, 33–34, 54, 64,
    69, 70, 84, 147, 151–159, 161,
    216, 217, 313
Communication, role of 179, 203, 206–208,
    225
Competition, unfair, in Peru 250
Complementarity 175, 187–190
Consultative Group on International
    Agricultural Research (CG, CGIAR)
    8, 229, 230–231, 237
Convention on Biological Diversity (CBD)
    214, 216, 218, 220, 273, 274, 319,
    332
Corn borer 4, 5, 6
Cornell University, Ithaca, New York 5, 6, 23,
    41, 82, 100, 149, 325
Cosmetics 326
Cosmopolites sordidus see Banana weevil borer
Costa Rica 5, 7, 11, 12, 126–129, 312,
    317–333
  conservation areas 324, 326, 327
  economy 319
  land use 318
Costa Rican Banana Corporation 326
Cost–benefit analysis 213–228
Cosuppression 80
Côte d'Ivoire 51
Cotton 24, 27, 29, 55, 163, 249, 313
Country coordinator (CC) 12
Courgette 158, 159
Cowpea 113
CRIFC *see* Central Research Institute for Food
    Crops
Crocus sativus see Saffron
Crop improvement 22, 23
Crop protection 18, 54
Crops *see* Estate crops; Food crops; Industrial
    crops; Medicinal crops
CryV protein 63, 70
CSS *see* Department of Crop and Soil Sciences,
    Michigan State University
Cucumber 147, 148, 156
Cucumber mosaic virus (CMV) 6, 90, 130,
    148, 149, 154, 156, 159, 183
Cucurbits 4, 5, 6, 24
  field trials 151

future developments 156–159
  transgenic variety 156
  virus-resistant 147–160
Curcuma petiolata 42
CVPD *see* Citrus vein phloem degeneration
Cyanogenesis 237
Cytoplasmic hybrids (cybrids) 111
Cytoplasmic male sterility (CMS) 108, 109,
    111

Date palm (*Phoenix dactylifera* L.) 23,
    133–146
  Bou Feggous 137
  cultivars 141
  Deglet Nour 137, 138
  genetic transformation in 141–142
  mass propagation of 139–141
  Mejhoul 136, 138
  products 135
  resistance to Bayoud disease 138–142
  uses of 135
  world production 134
Date production and consumption 135–136
Deforestation 317, 330, 331
Demography *see* Population growth
Denmark 340
Department of Crop and Soil Sciences (CSS),
    Michigan State University 258, 259,
    260, 261–263
Department of Intellectual Property, Thailand
    299, 303
Desertification 135, 136
Developing countries 1, 2, 3, 10, 27–34,
    63–71, 74, 89, 125–126, 131, 173,
    174, 176, 180–186, 187, 190, 201,
    205, 206, 214, 217–219, 221, 222,
    225, 231, 236, 282, 283, 286
DGIS *see* Special Programme on Biotechnology
    of the Netherlands
DGIS/BIOTECH *see* Special Programme on
    Biotechnology and Development
    Cooperation
Directorate of Rice Research, Hyderabad, India
    206
Disease resistance 113–115
Diversification, cytoplasmic 108
DNA Plant Technology Corporation,
    Cinnaminson, New Jersey (DNAP) 5,
    7, 11, 12, 127, 183, 337, 339
DowElanco Company 336
Draa Valley, southern Morocco 136
Drought 19, 40, 41, 45, 46, 54, 89, 104,
    115, 125, 127, 185, 230
Drugs *see* Pharmaceutics
Dupont gun patent 66

East Africa 89, 175, 188
ECB *see Ostrinia nubilalis*

Ecogen, Langhorne, Pennsylvania 12
Economic development, in sub-Saharan Africa
    50–53
Ecos La Pacífica 326
Ecotourism 331
Ecuador 126, 127, 239
Egypt 5, 6, 9, 10, 11, 12, 13, 14, 17–26, 55,
    58, 59
Elaeis guineensis 162, 166
Elaeis oleifera 162, 166
Electronic Data Exchange (EDI) 307
ELISA *see* Enzyme-linked immunosorbent
    assay
Embryo rescue 46, 97, 107
Embryogenesis, somatic 128, 139
Embryogenic callus culture 40
Environment, enabling 58–59
Environmental Protection Agency (EPA), USA
    9, 20, 153, 311, 312, 340–341
Enzyme-linked immunosorbent assay (ELISA)
    5, 23, 67, 68, 69, 93–94, 153, 154,
    155, 183
EPA *see* Environmental Protection Agency
Escalant's modified methodology 128
Escherichia coli 65, 83, 115
Estate crops 37, 38
    in Indonesia 42, 46
Ethiopia 51, 54
Europe 77, 139, 203, 280, 299
European Community (EC) 126, 215, 220
European corn borer *see Ostrinia nubilalis*
European Economic Community (EEC) *see*
    European Community
European Patent Convention 276, 278, 280
    see also Patents
European Union (EU) 274
Extremophilic organisms 325, 326

Faba bean 24, 55
Faba bean necrotic yellow virus (FBNYV) 24
Famine 90
FBNYV *see* Faba bean necrotic yellow virus
FDA *see* Food and Drug Administration
Federal Food, Drug, and Cosmetic Act
    (FFDCA), USA 154
Federal Government, USA 257, 258
Federal Insecticide, Fungicide, and Rodenticide
    Act (FIFRA), USA 154
Fermentation 36
Fertility 115
FHIA *see* Fundación Hondureña de
    Investigación Agrícola
Field efficacy testing 67–68
Financial support 224
Fitotek Unggul, Indonesia 5, 7, 11, 183
Flooding 45, 331
Fluorescence *in situ* hybridization (FISH) 167
Food and Drug Administration (FDA), USA 9,
    28, 31, 149, 156, 311, 312

Food crops 37
    in Indonesia 41, 44–45
Food security 1, 4, 50, 51, 90, 237
Forestry 178, 295, 331
Foundation Peru 247–252
Fragrances 326
France 137, 280
Fundación Hondureña de Investigación
    Agrícola (FHIA) 129
Fundación Perú *see* Foundation Peru
Funding, of agbiotech 337–338
Fungi 17, 19, 44, 75, 76, 83, 177, 242, 321
Fusarium oxysporum f. sp. *albidinis* Malençon
    136
Fusarium oxysporum f. sp. *cubense see* Panama
    disease
Fusarium sp. 24

Gadjah Mada University (UGM), Indonesia 38
Gall midge 101, 103
Ganoderma root infection 42
Garlic 29
Garst Seed Company, Slater, Iowa 5, 9, 11,
    12, 41, 63, 66, 70, 183
Gene flow 34
Gene mapping 100, 115, 161
Gene silencing 80
Gene tagging 100–101
Gene transfer 79–81, 107–108
General Agreement on Tariffs and Trade
    (GATT), Uruguay Round 275, 277,
    280
Genes, pyramiding 80, 84, 101, 105–106,
    116
Genetic engineering 6, 17–26, 27, 41, 46,
    49, 53, 54, 79, 81–83, 127, 148,
    161, 164–165, 236, 240, 273–283,
    285, 286
Genetic fingerprinting 129, 166
Genetic mapping 78, 81
Geographic(al) Information Systems (GIS) 53,
    289, 324
Germany 83
Germplasm 12, 44, 45, 63, 64, 68, 70, 201,
    291
    insect-resistant 69
    maize 5, 6
    potato 4, 78–79
    Solanum 75, 76, 77, 83
    tropical 64
Ghana 322
Ginger (*Zingiber officinale*) 42, 45, 46
GIS *see* Geographic(al) Information Systems
Givaudan-Roure Fragrances, New Jersey 326
Globodera pallida 83
Globodera rostochiensis 83
Glucanase 141
Glucose oxidase (*GO* gene) 75, 76
Gnettum 42

GO gene *see* Glucose oxidase
Gonsalves, Dr Dennis 149, 153
Grafting 140
Grape 140, 249, 312
Grassy stunt virus 107, 109
Groundnut 36, 40, 41, 163, 178, 248, 251
Guatemala 127
Guyana 322
Gynura procumbens 42

Harmonization 214
Health, human 216, 285, 290, 293
Heliothis zea 185
Henna (*Lawsonia inermis*) 133, 134
Herbicides 177
Hobson Interbuana Indonesia Co. 40
Holdens Company 335, 337, 339
Honduras 127
Horsch, Dr Robert 91
Host-plant resistance 4, 78, 84, 111
Host–pathogen interactions 141, 237, 240
Human resources 35, 44, 56–57, 179, 201, 288, 293–294
    in Indonesia 40–41
Humboldt cold water current 251
Hybridization 138
    of rice 97, 107–110, 116
    somatic 110–111
Hybrids, interspecific 107
Hydroelectric power 331

IARC *see* International agriculture research centre
IBS *see* Intermediary Biotechnology Service
ICI Seed Company/Zeneca Seeds *see* Garst Seed Company
IITA *see* International Institute of Tropical Agriculture
Immunodiffusion assay 153
Impediments to biotechnology 57–58
Implementation workshops 8
INBio *see* National Biodiversity Institute
INBio–Merck agreement 326
India 108, 185, 206, 207, 209, 210, 211, 212, 251
Indian Council for Agricultural Research 206
Indigenous technology 298
Indonesia 5, 6, 8–11, 35–48, 63, 64, 68–70, 93, 186, 189
Industrial crops 37, 38, 41, 178
    in Indonesia 41–42, 45–46
Industrial development, in Zimbabwe 285–296
Industry, in USA 256, 257–258
Information, role of 43–44, 49, 56, 179, 206–208, 220–223, 224, 303

Information Systems for Biotechnology Project, Virginia Polytechnic and State University 10
Infrastructure, role of 2, 56–57, 179
INIFAP *see* National Institute for Forestry, Agriculture and Livestock
Initiatives, international, in agricultural biotechnology 192–199
Insect pests 17, 19, 44, 75, 77–78, 81, 185, 249
Insect resistance 112–113
Institute of International Agriculture 11
Instituto Agronomico de Campinas, Brazil 237
Instituto Nacional de Biodiversidad *see* National Biodiversity Institute
Integrated pest management programmes (IPM) 32
Integrated Superior Research (ISR), Indonesia 35, 38, 39
Intellectual Property Institute, Chulalongkorn University, Thailand 302
Intellectual property rights 1, 3, 8–13, 36, 41–43, 56, 166, 255, 273, 274, 280, 282, 287, 291, 293, 297, 307–308
    CD-ROM 279
    implications of 281–283
    international 273–283
    in Peru 250
Inter-University Centers (IUC), Indonesia 38
Intermediary Biotechnology Service (IBS), ISNAR 180, 182, 184, 187
International agriculture research centre (IARC) 8
International biotechnology programmes 173, 174, 176, 177, 181, 184, 187, 190
International Center for Agricultural Research in the Dry Areas (ICARDA), Aleppo, Syria 23, 188
International Center for Maize and Wheat Improvement (CIMMYT), Mexico 29, 30, 31, 294
International Institute of Tropical Agriculture (IITA) 231, 243
International Monetary Fund (IMF) 302
International Potato Center (CIP) 73–78, 81, 183, 188
    Midterm Plan 75
International Program on Rice Biotechnology (IPRB) 201–212
    fellowships 202–205
    field staff 211–212
    research grants, renewable 203, 208–210
    research priorities 208
    training courses 202–205
International Rice Research Institute (IRRI) 35, 38, 41, 99, 100, 109, 188, 208, 211

International Service for the Acquisition of Agri-Biotech Applications (ISAAA) 58, 59, 92, 95, 183
International Service for National Agricultural Research (ISNAR) 173–176, 187, 189
International Union for the Conservation of Nature (IUCN) 331
International Union for the Protection of New Plant Varieties (UPOV) 275, 276
Internet 167, 299, 324
Introgression 30, 32, 33, 63, 64, 68, 81, 82, 107–110
Investors' Circle, West Chicago, Illinois 338
Ipomoea batatas L. *see* Sweetpotato
IPRB *see* International Program on Rice Biotechnology
Iraq 137, 139
Iron toxicity 38
IRRI *see* International Rice Research Institute
Irrigation, in Peru 248, 251
ISAAA *see* International Service for the Acquisition of Agri-Biotech Applications
ISNAR *see* International Service for National Agricultural Research
ISR *see* Integrated Superior Research
Israel 127, 340
Italy 137

Jamaica 10, 12
Jamaica Agricultural Research Programme 10
Japan 35, 99, 109, 202, 299
Japan International Cooperation Agency (JICA) 35, 37
Japan International Research Center for Agricultural Sciences (JIRCAS) 35, 38, 39, 41, 174
Java 38, 43
Jaworksi, Dr Ernest 91
JICA *see* Japan International Cooperation Agency
JIRCAS *see* Japan International Research Center for Agricultural Sciences
John Innes Centre, UK 204, 294

Kaltimex-Jaya, Indonesia 69
Kapok 42
KARI *see* Kenya Agricultural Research Institute
Kenya 5, 9, 11, 12, 51, 54, 55, 58, 59, 89, 90, 94, 95, 182, 185, 239
Kenya Agricultural Research Institute (KARI) 5, 7, 59, 89, 91, 92, 94, 95, 182, 183
Kew Botanical Gardens, UK 326
Korea 99, 210, 211, 212

Laboratory analysis 67
Land degradation 50, 51, 54, 136, 317
Land Grant universities, USA 255, 258
Land reclamation 142
Landraces 29, 30, 32–34
Late blight disease, potato 73, 75–76, 78, 81, 82
Latin America 9, 10, 126, 127, 134, 175, 181, 188, 286
Lawsonia inermis see Henna
Leafminer fly 78
Least developed countries (LDCs) 50
Libya 133, 134, 137, 139
Linkages, private sector 11–12
Logging *see* Forestry
Lonchacharpus sp. 326
Lucerne 134
Lycopersicon sp. 30

Madagascar 322
MAES *see* Michigan Agricultural Experiment Station
Maize (*Zea mays*) 4, 5–6, 27–29, 31, 33, 36, 54, 55, 63, 69, 70, 89, 106, 112–113, 141, 178, 183, 185, 294, 295, 333
  insect-resistant 63–71
  MON 80100 32
  transgenic, in Mexico 30–33
Malawi 233, 239
Malaysia 163, 164, 166, 189, 207, 286
Mali 51
Marker assisted selection (MAS) 101, 105, 167, 185, 204
Marrakech 138
Maryland, University of 23
MAS *see* Marker assisted selection
Mass propagation 139, 140
Material transfer agreement (MTA) 274
Mauritania 133, 134, 137, 139
Max Planck Institute, Cologne, Germany 83
MCIA *see* Michigan Crop Improvement Association
MDA *see* Michigan Department of Agriculture
Medicinal crops 45
Melon 24, 183
  honeydew 158
  see also Cucurbits
Mendelian segregation 67, 149
Mentha spp. 42
Merck 326
Mesocarp oil 165
Methionine 24
Mexico 27–34, 77, 127, 149, 294, 312
Michigan Agricultural Experiment Station (MAES) 259, 260, 261, 264–265
Michigan Crop Improvement Association (MCIA) 259, 270

Michigan Department of Agriculture (MDA),
    USA 9
Michigan State University (MSU) 3, 5, 6, 9,
    23, 35, 38, 40, 41, 42, 58, 65, 92,
    256, 257, 258–272
    Office of Intellectual Property *see* Office of
        Intellectual Property
Micrografting 139, 141
Micropropagation 1, 4, 7, 23, 42, 45, 46, 54,
    93, 125, 127, 131, 178, 181, 183,
    292
Microsatellites 166
Middle East 6, 9, 10, 25, 133, 134
Ministry of Agriculture, Indonesia (MOA) 35,
    37
Missouri Botanical Garden, USA 324
MOA *see* Ministry of Agriculture, Indonesia
Moko (*Pseudomonas solanacearum*) 129,
    130
Molecular breeding 73, 83
Molecular cytogenetic analysis 161
Molecular genomics 161, 164
Molecular mapping 97, 98, 100, 231, 232
Molecular markers 36, 38, 40, 45, 46, 73,
    76, 82, 84, 97–100, 105, 106, 115,
    165, 185, 240, 241, 292
Monoclonal antibodies (MCA) 36
Monsanto Company 5, 7, 11, 32, 40, 65, 77,
    89, 90, 91–95, 182, 183, 336, 337,
    339
Morocco 11, 133–138
Moyer, Dr Jim 89, 91
MSU *see* Michigan State University
MTA *see* Material transfer agreement
Multimedia 324
Mungbean 113
Musa acuminata Colla 126
Musa balbisiana Colla 126
Mycogen Company 335, 337, 339
Mycosphaerella fijiensis see Black Sigatoka
    disease

NAGEL *see* National Agricultural Genetic
    Engineering Laboratory
National Agricultural Genetic Engineering
    Laboratory (NAGEL) 19, 23
National Agricultural Research Centres,
    Zimbabwe 288
National biodiversity inventory 317
National Biodiversity Institute (INBio), Costa
    Rica 317–318, 319–332
National Biodiversity Inventory, INBio
    320–321
National Biological Impact Assessment
    Programme (NBIAP) 9
National Center for Genetic Engineering and
    Biotechnology, Thailand 206
National Committee for Agricultural Biosafety
    (NCAB), Mexico 30

National Development Programme, Indonesia
    35
National Institute for Forestry, Agriculture
    and Livestock (INIFAP), Mexico 29,
    31, 33
National Philippine Rice Research Institute
    (PhilRice), Maligaya 210
National Policy on Biotechnology Research,
    Zimbabwe 287
National Research Council (NRC), Indonesia
    37
National Science Foundation, USA 341
National Technology Transfer Policy, Thailand
    304–308
NBIAP *see* National Biological Impact
    Assessment Programme
NCAB *see* National Committee for Agricultural
    Biosafety
Nematode, burrowing (*Radopholus similis*)
    130
Nematodes 75, 82, 83
Nephthytis afzelii 140
Netherlands, The 74, 176, 294, 295
Nicaragua 127
Nicotiana benthemana 92
Nigeria 51, 54
North Africa 133, 134, 175, 188
North America 134
North American Free Trade Association
    (NAFTA) 31
North Carolina State University 89, 91

Oasis 134, 136, 137
Office of Intellectual Property (OIP) 11, 259,
    260, 264, 267, 268
Ohio State University 208
Oil palm 42, 46, 161–170
    biotechnology of 164–167
    cultivation of 163
    genetic engineering 164–165
    genome, molecular analysis of 165–167
    oil quality 164–165
    and palm oil 161–162
    yield 163
Oilseed rape 115
OIP *see* Office of Intellectual Property
Oleic acid 164, 165
Oleochemical industry 162
Onion (*Allium cepa*) 29, 140
Orange 332
Organogenesis 139, 140
Oryza australiensis 108, 110
Oryza brachyantha 108, 109
Oryza glaberrima 107
Oryza granulata 109
Oryza latifolia 108
Oryza longistaminata 100, 101
Oryza minuta 108
Oryza nivara 107

Oryza officinalis 108
Oryza perennis 108
Oryza rufipogon 108
Oryza sativa 100, 107, 108, 110
Oryza spp. 109
Osmoregulation 24
Osmotins 75
Ostrinia furnacalis (Asian stem borer, ASB) 63, 65, 68, 70, 183
Ostrinia nubilalis (European corn borer, ECB) 63, 65, 66

Pac Seeds 69
PAL (phenylalanine ammonia-lyase) 82
Palm *see* Date palm; Oil palm; Palm oil; Palm, ornamental
Palm oil 42, 161–163
    hybrid *dura* 163
    hybrid *pisifera* 163
    hybrid *tenera* 163
    iodine value (IV) 162, 165
Palm Oil Research Institute of Malaysia (PORIM) 161, 164–167
Palm, ornamental 7, 183
PalmGenes database 167
Palmitic acid 164, 165
Panamá 127
Panama disease (*Fusarium oxysporum* f. sp. *cubense*) 129, 130
Panicum maximum (guinea grass) 140
Panicum purascens (para grass) 140
Papaya ringspot virus (PRV) 148, 149, 152, 154, 156
Papaya ringspot virus, watermelon strain (PRV-W) 4, 6
Parataxonomist 320, 321
Paris Convention, 1979 280
Partnership Superior Research (PSR), Indonesia 35, 36, 38–40, 41
Partnerships 57–58, 95, 182–184, 190, 203, 206, 247
    see also Collaboration
Patchouli 42
Patent Cooperation Treaty (PCT), 1970 280
Patents 10, 28, 41, 65, 207, 276–278, 299, 302
PCR *see* Polymerase chain reaction
Peach 140
Peanut mottle virus 36, 41
Pelargonium spp. 42
Pennisetum purpureum var. *merkerii* (merker grass) 140
Pepper (*Piper nigrum*) 29, 42, 45, 46
Perlak, Dr Fred 91
Perspectives, public/private sector 255–272
Peru 80, 81, 247–252, 312
Pesticides 74, 185
Pests *see* Insect pests
PetoSeed Company 156

Pharmaceutics 167–168, 248, 251, 318, 326, 331–332, 339
Philippines, The 41, 99, 108, 127, 189, 210, 211, 212
Phoenix canariensis L. 137–138
Phoenix dactylifera L. *see* Date palm
Phthorimaea operculella see Potato tuber moth
Phylodendron nachodomi 140
Physico-biology 53
Phytophthora capcici 46
Phytophthora infestans 75, 76, 82, 83
Phytophthora oxysporum 46
Phytophthora spp. 40, 298
Phytosanitary requirements 76
Pimpinella pruacan 42
Pineapple 4, 5, 7, 183
Pioneer Hi-Bred 5, 6, 11, 23, 183
Plant analysis 67–68
Plant breeders' rights (PBR) 43
Plant Genetic Systems (PGS), Belgium 77, 335, 337, 339
Plant pathology 20
Plant variety protection (PVP) 43, 273, 274–276, 278, 279, 280, 282
Plant Variety Protection Act (PVPA), Indonesia 36, 43
Plantain 125, 126, 129, 131, 178
PLRV *see* Potato leaf roll luteovirus
Plum 140
Pod borer 46
Policy 7–13, 56, 180
Pollination 32
Polymerase chain reaction (PCR) 5, 23, 69, 98
Population growth 17, 18, 45, 50, 51–52
PORIM *see* Palm Oil Research Institute of Malaysia
Potato (*Solanum* spp.) 4–5, 24, 25, 27, 29, 36, 54, 55, 73–88, 112, 129, 178, 248, 251
    breeding of 81–83
    improvement of 74–83
    Irish 81
    production of 83–85
    Russet Burbank 77
    see also *Solanum*
Potato leaf roll luteovirus (PLRV) 24, 73, 76, 77, 90
Potato potyvirus X (PVX) 24, 76, 73, 77, 78, 83, 90, 183
Potato potyvirus Y (PVY) 24, 73, 76, 77, 78, 83, 90, 183
Potato tuber moth (*Phthorimaea operculella*, PTM) 4, 5, 36, 73, 75, 77, 78, 80
Poverty 1, 50, 230–231
Prospecting, chemical, in Costa Rica 325–326
Protease 75, 113
Proteinase 4, 40, 113, 141
Protoplasts 40, 107, 112, 116, 129

PRV *see* Papaya ringspot virus
PRV-W *see* Papaya ringspot virus, watermelon
    strain
Pseudomonas solanacearum see Moko; Wilt,
    bacterial
PSR *see* Partnership Superior Research
PTM *see* Potato tuber moth
Public/private sector perspectives 2–5, 8, 11,
    13, 40, 42, 58, 182, 184, 217
Pumpkin 148
PVP *see* Plant variety protection
PVPA *see* Plant Variety Protection Act,
    Indonesia
PVX *see* Potato potyvirus X
PVY *see* Potato potyvirus Y
Pyrethrum 54
Pyricularia oryzae 40, 100, 110

Quantitative trait loci (QTL) 73, 76, 81, 82,
    84, 98, 99, 101, 104–105, 115, 294
Queensland, University of 41

R&D *see* Biotechnology research and
    development
Radiobiology 53
Radopholus similis see Nematode, burrowing
Rami 42
Random amplified polymorphic DNA (RAPD)
    81, 97, 165, 292
Rapeseed 25, 163, 164
Recombinant Biocatalysis Co., La Jolla,
    California 326
Recombinant DNA technology 36, 45, 185,
    285, 286
Red flour beetle 113
Regulations, international 213–228
Repelita, Indonesia 36
Replicase 4, 75, 76, 77
Research *see* Biotechnology research and
    development
Research Institute for Food Crops
    Biotechnology (RIFCB) 36, 38, 39,
    40, 42, 46, 47
Restriction fragment length polymorphism
    (RFLP) 23, 67, 97, 104, 108, 165,
    166, 292
RF *see* Rockefeller Foundation
RFLP *see* Restriction fragment length
    polymorphism
Rhizo-plus 36, 38, 39, 41
Rhizoctonia solani 114
Rhoeo discolor 140
Rice 38, 40, 44, 54, 89, 97–121, 134, 141,
    178, 185, 201–212, 249
    Chinsurah Boro II 111
    Cisidane 40
    CT9993 × IR62266 99
    Dama 110

Fujiminori 111
Hatsuyume 110
indica 97–100, 111, 116
IR64 × Azucena 99
japonica 97–100, 111, 113, 114, 116
Koshihikari 110
Moroberekan 101
Nipponbare 113
world production 97, 98, 115
Rice Biotechnology Quarterly 207
Rice blast 36, 38, 40, 41, 100–102, 104,
    107–110, 113
Rice stem borer 44
Rice stripe virus 112, 114
Rice tungro spherical virus 101, 103, 107,
    109, 114
Rice weevil 113
RiceGenes database 208
RIFCB *see* Research Institute for Food Crops
    Biotechnology
Ripening 177
Risk assessment 219, 221
Rockefeller & Company, Odyssey Fund 336,
    338
Rockefeller Foundation, The (RF) 35, 36,
    38–42, 202–204, 208, 209, 212
Rouwolfia serventina 42
Rubber 42, 46
Rutta angustifolia 42

Sabal sp. 138
Saccharum officinarum see Sugarcane
Saffron (*Crocus sativus*) 133, 134
Sainsbury Laboratory, UK 77, 83
Salinity 17, 19, 104, 115, 137, 142
San Ramón, Peru 80, 81
Saudi Arabia 137
Scientific and Industrial Research and
    Development Center (SIRDC) 285,
    287–291
    constituent institutes 289–291
Scirpophaga incertulas (Walker) *see* Yellow stem
    borer
Scottish Crop Research Institute 81
Scripps Research Institute 5, 23, 41, 89, 91,
    204
Secrecy *see* Trade secrecy
Selaginella 140
Seminis Vegetable Seeds (SVS) 156
Sheath blight 107, 109, 112, 114, 115
Singapore 340
SIRDC *see* Scientific and Industrial Research
    and Development Center
Sitophilus zeamais 112, 113
Small Business Innovation Research (SBIR)
    grants 340–341
Smallholders 182, 184, 185, 231, 232,
    236–238, 244
Smilax sp. 332

Solanum acaule 77
Solanum berthaultii 78, 82
Solanum berthaultii × *tuberosum* 82
Solanum phureja *76, 81*
Solanum sparsipilum 78
Solanum spp. 75, 83
Solanum stoloniferum 77
Solanum tuberosum subsp. *andigena* 76, 77
Solomon Island ivyarum (*Schindapsus aureus*)
    140
Sorghum 54, 106, 178, 294
South Africa 54, 55, 58, 59, 134, 139, 243,
    294
South African Committee for Genetic
    Experimentation 224
South America 76, 125, 127, 237, 244
Southeast Asia 6, 163, 174, 175, 188, 212,
    238, 244
Southern Africa 175, 188
Southern blot analysis 67, 68, 69, 93, 94,
    114
Soya 163
Soybean 27, 38, 40, 163, 313
Special Programme on Biotechnology and
    Development Cooperation, The
    Netherlands (DGIS/BIOTECH) 232,
    233
Special Programme on Biotechnology of the
    Netherlands (DGIS) 59, 185
SPFMV *see* Sweetpotato feathery mottle virus
Sporangial lysis 75
Squash 24, 148, 149, 154–156
    'attached tendril' 156
    Freedom II 147, 152, 156, 159
    Pavo 152
    weediness of 152, 153
    yellow crookneck 147, 149, 158
    see also Cucurbits
Stanford University Law School 10
Stem borer *see* Asian stem borer; Rice stem
    borer; Striped stem borer; Yellow
    stem borer
Steppes 136
Sterility, nuclear male 115
Stockholm Environment Institute (SEI) 59
Strawberry 54, 249
Striped stem borer 112
Sub-Saharan Africa (SSA) 1, 49–60, 74
    economic development in 50–53
    population and land resource 52
Submergence, tolerance of 101, 103, 104,
    105
Subsistence farming 317
Sugarcane (*Saccharum officinarum*) 54, 140,
    178, 249, 294
Sunflower 163
Supermarket fruit 158
Surinam 322
Sustainability 44, 46, 50, 56, 74, 173–199,
    249, 285–296, 317

Sweden 224, 233
Sweetpotato (*Ipomoea batatas* L.) 4, 5, 7, 40,
    54, 55, 89–95, 178, 182, 183, 185,
    295, 313
Sweetpotato feathery mottle virus (SPFMV) 4,
    7, 89, 91, 93–94, 95, 183, 185
Sweetpotato weevil 4, 7
Syria 23

Taiwan 127, 340
Tamarind 42
Tanzania 51, 59, 233, 239
Tapping panel dryness (TPD) 42
Tea 42, 46
Technology 1, 3, 7–13, 68–69, 217,
    274–281, 297, 298, 306–307
Technology Information Access Center (TIAC),
    Thailand 303
Technology transfer 1, 3, 12–13, 22, 35, 49,
    204, 214, 215, 232, 291, 297–309,
    326
Technology Transfer Center, Thailand 299,
    302
Technology transfer coordinator (TTC) 12
Temperature tolerance 19, 125, 127
Teosintes 29, 30, 32, 33
Texas A&M University 204
Thailand 11, 189, 206, 212, 223, 240,
    297–309
Theobroma cacao 40
Tissue culture 2, 22, 23, 24, 37, 41–42, 45,
    46, 49, 53, 54, 110, 134, 142, 166,
    241, 249, 292
TMV *see* Tobacco mosaic virus
Tobacco 55, 76, 110, 112, 113, 114, 115,
    149
Tobacco mosaic virus (TMV) 90, 114
Tomato 4, 5, 6, 27, 29, 31, 55, 76, 81, 112,
    114, 129, 248, 251, 313
    FLAVR SAVR (Calgene) 30
Tomato golden mosaic virus 114
Tomato mosaic virus (ToMV) 90
Tomato yellow leaf curl virus (TYLCV) 4, 6,
    24
ToMV *see* Tomato mosaic virus
Trade in agricultural products 311–314
Trade secrecy 278–279
Tradescantia fluminensis 140
Training 23, 68–69, 201, 202, 203,
    220–223, 292, 321
Transformation, genetic 45, 63, 64–68, 97,
    106, 111–115, 131, 134, 140,
    141–142, 181, 231, 292
Transgenic varieties, release of 27–34
Trichoderma sp. 298
Trichomes, glandular 73
TRIPS *see* Agreement on Trade-related Aspects
    of Intellectual Property Rights
TTC *see* Technology transfer coordinator

Tumer, Dr Nilgun 91
Tungro see Rice tungro spherical virus
Tunisia 133, 134, 137, 138, 139
TYLCV see Tomato yellow leaf curl virus

Uganda 54, 55, 58, 59, 237, 242
UK 77, 236, 280
UNDP see United Nations Development
       Programme
UNEP see United Nations Environment
       Programme
United Nations Conference on Environment
       and Development (UNCED), 1992
       213, 325
United Nations Development Programme
       (UNDP) 23
United Nations Environment Programme
       (UNEP) 10, 218, 294, 322, 325
United States see USA
UPOV see International Union for the
       Protection of New Plant Varieties
Uruguay Round see General Agreement on
       Tariffs and Trade
US Agency for International Development
       (USAID) 1, 3, 8, 11, 13, 23, 38, 41,
       58, 91, 94, 95, 125, 127–128, 243
US Department of Agriculture (USDA) 2, 9,
       20, 23, 32, 92, 151, 153, 208,
       311–314, 340–341
USA 5, 9, 27, 29–30, 31, 32, 41, 58, 63,
       68–69, 77, 99, 126, 138, 139, 149,
       203, 215, 250, 251, 255–272,
       276–280, 299, 312, 313, 340
       universities 256–258
USAID see US Agency for International
       Development
USDA see US Department of Agriculture
Ustilago maydis (Huitlacoche) 29

Vanilla (Vanilla planifolia) 42, 45, 46, 140
Variation, genetic 108–110
Variation, somaclonal 110
Venezuela 127
Venture capital 341
Vietnam 189, 212
Viruses 17, 19, 75, 76–77, 83, 90
       see also African cassava mosaic virus;
       Banana bunchy top virus; Banana
       streak virus; Bean yellow mosaic
       virus; Cucumber mosaic virus; Faba
       bean necrotic yellow virus; Grassy
       stunt virus; Papaya ringspot virus;
       Papaya ringspot virus, watermelon
       strain; Peanut mottle virus; Potato
       leaf roll luteovirus; Potato potyvirus
       X; Potato potyvirus Y; Rice stripe

virus; Rice tungro spherical virus;
       Sweetpotato feathery mottle virus;
       Tobacco mosaic virus; Tomato golden
       mosaic virus; Tomato mosaic virus;
       Tomato yellow leaf curl virus;
       Watermelon mosaic virus;
       Watermelon mosaic virus 2;
       Zucchini yellow mosaic virus

Wambugu, Dr Florence 91, 92
Washington University, St Louis, Missouri 90
Washingtonia robusta 138
Water supply 331
Watermelon 147, 148, 156
Watermelon mosaic virus (WMV) 4, 6
Watermelon mosaic virus 2 (WMV 2)
       148–150, 151, 154, 156–159
West Africa 107, 128, 240
West Asia 175, 188
Western blot analysis 67, 69, 92, 114
Wheat 89, 106, 134, 135
       Bavaria 261–267
'Whiskers' technology 5, 66, 69, 70
Whitebacked plant hopper 103, 109
Whitefly 156
Wild biodiversity 317–333
Wild relatives/species 29–32, 34, 45, 78, 81,
       82, 100, 105, 107–108, 110, 111,
       126, 151, 153, 155, 248, 251, 333
Wilt, bacterial (Pseudomonas solanacearum)
       40, 75, 78
WMV 2 see Watermelon mosaic virus 2
WMV see Watermelon mosaic virus
World Bank, The 10, 11, 50, 208
World Trade Organization (WTO) 281, 282,
       307–308
World Wildlife Fund for Nature (WWF) 331
WTO see World Trade Organization
Wyoming, University of 23

Xanthomonas oryzae pv. oryzae 40, 100

Yam 54, 178
Yellow mealworm 113
Yellow stem borer (Scirpophaga incertulas
       (Walker)) 107, 109, 111–113

Zebrina pendula 140
Zeneca Seeds 63, 64
       see also Garst Seed Company
Zimbabwe 51, 54, 55, 58, 59, 185,
       285–296
Zucchini yellow mosaic virus (ZYMV) 4, 6,
       24, 148–150, 151, 154, 156–159